MONOGRAPHS ON STATISTICS AND APPLIED PROBABILITY

General Editors

D.R. Cox, D.V. Hinkley, N. Keiding, N. Reid, D.B. Rubin and B.W. Silverman

1 Stochastic Population Models in Ecology and Epidemiology *M.S. Bartlett* (1960)

2 Queues *D.R. Cox and W.L. Smith* (1961)

3 Monte Carlo Methods *J.M. Hammersley and D.C. Handscomb* (1964)

4 The Statistical Analysis of Series of Events *D.R. Cox and P.A.W. Lewis* (1966)

5 Population Genetics *W.J. Ewens* (1969)

6 Probability, Statistics and Time *M.S. Bartlett* (1975)

7 Statistical Inference *S.D. Silvey* (1975)

8 The Analysis of Contingency Tables *B.S. Everitt* (1977)

9 Multivariate Analysis in Behavioural Research *A.E. Maxwell* (1977)

10 Stochastic Abundance Models *S. Engen* (1978)

11 Some Basic Theory for Statistical Inference *E.J.G. Pitman* (1979)

12 Point Processes *D.R. Cox and V. Isham* (1980)

13 Identification of Outliers *D.M. Hawkins* (1980)

14 Optimal Design *S.D. Silvey* (1980)

15 Finite Mixture Distributions *B.S. Everitt and D.J. Hand* (1981)

16 Classification *A.D. Gordon* (1981)

17 Distribution-free Statistical Methods *J.S. Mariz* (1995)

18 Residuals and Influence in Regression *R.D. Cook and S. Weisberg* (1982)

19 Applications of Queueing Theory, 2nd edition *G.F. Newell* (1982)

20 Risk Theory, 3rd edition *R.E. Beard, T. Pentikainen and E. Pesonen* (1984)

21 Analysis of Survival Data *D.R. Cox and D. Oakes* (1984)

22 An Introduction to Latent Variable Models *B.S. Everitt* (1984)

23 Bandit Problems *D.A. Berry and B. Fristedt* (1985)

24 Stochastic Modelling and Control *M.H.A. Davis and R. Vinter* (1985)

25 The Statistical Analysis of Compositional Data *J. Aitchison* (1986)

26 Density Estimation for Statistical and Data Analysis *B.W. Silverman*

27 Regression Analysis with Applications *G.B. Wetherill* (1986)

28 Sequential Methods in Statistics, 3rd edition *G.B. Wetherill and K.D. Glazebrook* (1986)

29 Tensor Methods in Statistics *P. McCullagh* (1987)

30 Transformation and Weighting in Regression *R.J. Carroll and D. Ruppert* (1988)

31 Asymptotic Techniques for Use in Statistics *O.E. Barndoff-Nielson and D.R. Cox* (1989)

32 Analysis of Binary Data, 2nd edition *D.R. Cox and E.J. Snell* (1989)

33 Analysis of Infectious Disease Data *N.G. Becker* (1989)

34 Design and Analysis of Cross-Over Trials *B. Jones and M.G. Kenward* (1989)

35 Empirical Bayes Method, 2nd edition *J.S. Maritz and T. Lwin* (1989)

36 Symmetric Multivariate and Related Distributions *K.-T. Fang, S. Kotz and K. W. Ng* (1990)

37 Generalized Linear Models, 2nd edition *P. McCullagh and J.A. Nelder* (1989)

38 Cyclic and Computer Generated Designs, 2nd edition *J.A. John and E.R. Williams* (1995)

39 Analog Estimation Methods in Econometrics *C.F. Manski* (1988)

40 Subset Selection in Regression *A.J. Miller* (1990)

41 Analysis of Repeated Measures *M.J. Crowder and D.J. Hand* (1990)

42 Statistical Reasoning with Imprecise Probabilities *P. Walley* (1991)

43 Generalized Additive Models *T.J. Hastie and R.J. Tibshirani* (1990)

44 Inspection Errors for Attributes in Quality Control *N.L. Johnson, S. Kotz and X. Wu* (1991)

45 The Analysis of Contingency Tables, 2nd edition *B.S. Everitt* (1992)

46 The Analysis of Quantal Response Data *B.J.T. Morgan* (1993)

47 Longitudinal Data with Serial Correlation: A State-Space Approach *R.H. Jones* (1993)

48 Differential Geometry and Statistics *M.K. Murray and J.W. Rice* (1993)

49 Markov Models and Optimization *M.H.A. Davies* (1993)

50 Networks and Chao—Statistical and Probabilistic Aspects *O.E. Barndorff-Nielsen, J.L. Jensen and W.S. Kendall* (1993)

51 Number Theoretic Methods in Statistics *K.-T. Fang and Y. Wang* (1994)

52 Inference and Asymptotics *O.E. Barndorff-Nielsen and D.R. Cox* (1994)

53 Practical Risk Theory for Actuaries *C.D. Daykin, T. Pentikainen and M. Pesonen* (1994)

54 Biplots *J.C. Gower and D.J. Hand* (1996)

55 Predictive Inference: An Introduction *S. Geisser* (1993)

56 Model-Free Curve Estimation *M.E. Tarter and M.D. Lock* (1993)

57 An Introduction to the Bootstrap *B. Efron and R.J. Tibshirani* (1993)

58 Nonparametric Regression and Generalized Linear Models *P.J. Green and B.W. Silverman* (1994)

59 Multidimensional Scaling *T.F. Cox and M.A.A. Cox* (1994)

60 Kernel Smoothing *M.P. Wand and M.C. Jones* (1995)

61 Statistics for Long-Memory Processes *J. Beran* (1995)

62 Nonlinear Models for Repeated Measurement Data *M. Davidian and D.M. Giltinan* (1995)

63 Measurement Error in Non-Linear Models *R.J. Carroll, D. Ruppert and L.A. Stefanski* (1995)

64 Analyzing and Modeling Rank Data *John I. Marden* (1995)

(Full details concerning this series are available from the Publishers.)

Analyzing and Modeling Rank Data

JOHN I. MARDEN

Department of Statistics,
University of Illinois at Urbana-Champaign, USA

CHAPMAN & HALL

London • Glasgow • Weinheim • New York • Tokyo • Melbourne • Madras

Published by
Chapman & Hall, 2–6 Boundary Row, London SE1 8HN, UK

Chapman & Hall, 2–6 Boundary Row, London SE1 8HN, UK

Blackie Academic & Professional, Wester Cleddens Road, Bishopbriggs, Glasgow, UK

Chapman & Hall GmbH, Pappelallee 3, 69469 Weinheim, Germany

Chapman & Hall USA, 115 Fifth Avenue, New York, NY 10003, USA

Chapman & Hall Japan, ITP-Japan, Kyowa Building, 3F, 2-2-1 Hirakawacho, Chiyoda-ku, Tokyo 102, Japan

Chapman & Hall Australia, 102 Dodds Street, South Melbourne, Victoria 3205, Australia

Chapman & Hall India, R. Seshadri, 32 Second Main Road, CIT East, Madras 600 035, India

First edition 1995

ISBN 0412995212

Library of Congress Cataloging-in-Publication Data

Marden, John I.
Analyzing and modeling rank data / John I. Marden.
p. cm.
Includes bibliographical references and index.
ISBN 0-412-99521-2
1. Ranking and selection (Statistics). I. Title.
QA278.75.M37 1995
519.5'4—dc20 94-49214
CIP

Printed on acid-free text paper, manufactured in accordance with ANSI/NISO Z39.48-1992 (Permanence of Paper).

Contents

Preface

Ranking has always been an integral part of statistics, both in nonparametric analysis and in the analysis of judges' rankings of objects. Myriad data-analytic techniques and probability models have been developed over the years. The present volume attempts to summarize and unify much of the statistical work on rank data. The state of the art *circa* 1950 is summarized in the "Symposium on Ranking Methods" conducted by the Royal Statistical Society. (See *Journal of the Royal Statistical Society B*, 1950, Volume 12, pages 153–191.) The papers and discussions address the fundamental the questions: How best to elicit rankings? What goes on in a ranker's mind? How does a statistician model and analyze such highly structured data? Since then, high points include Luce's work on choice (*Individual Choice Behavior*, 1959), Arrow's on combining rankings (*Social Choice and Individual Values*, 1963), and Coombs' on unfolding models (*A Theory of Data*, 1964). Critchlow's book, *Metric Methods for Analyzing Partially Ranked Data* (1985), unifies an important aspect of ranking. The book by Diaconis, *Group Representations in Probability and Statistics* (1988), has an amazing wealth of information and insights, on ranking and many other subjects. Most of the probability models and statistical analyses resulting from these years of activity gathered (with their creators) in Amherst, Massachusetts in 1990. An AMS-IMS-SIAM Research Conference at the University of Massachusetts turned the spotlight on the broad range of ranking activities. Michael Fligner and Joseph Verducci, in addition to arranging the conference, edited the collection of papers presented in the book *Probability Models and Statistical Analyses for Ranking Data* (1993). I think all participants were surprised and delighted at the variety of work in the area.

The literature is rich with methods for analysis and modeling of rank data. People already in the field know the basic models and methods, but the scientific community at large has no single source for this information. The main goal of this book is to collect and present systematically many of the popular and useful methods. I try to reflect the spirit of the 1950 symposium and 1990 conference. Models and analyses should be realistic enough to make sense, but practical, too. The objective is to gain insight into the data and the mechanism that produced them. Beautiful models that do not make sense and cannot be calculated with current technology are not helpful.

The material should be accessible to anyone with a basic course in mathematical statistics, and familiarity with linear algebra and the basic elements of multivariate statistical analysis. The texts by Hogg and Craig (1978) and Bickel and Doksum (1977) are sufficient to understand and apply most of the results, although a few of the proofs need more advanced material such as Lehmann (1983), Rao (1973) and Serfling (1980). Mardia, Kent and Bibby (1979) is an excellent introduction to multivariate analysis. To appreciate the models in the book, one should have familiarity with linear regression and analysis of variance (e.g., Weisberg, 1985) and contingency table models (e.g., Agresti, 1990, or Bishop, Fienberg, and Holland, 1975).

The book is quite dense. To use it as a text in a one-semester graduate-level course, one may wish to cover about half the book. I recommend choosing something like the following material: all of Chapter 1; all of Chapter 2 through Section 2.5, and the beginning of Section 2.6, leaving out Sections 2.6.1 and 2.6.2; Sections 3.1 and 3.2, Section 3.4 through 3.4.1, and Sections 3.5 and 3.6; Chapter 5 through Section 5.11, plus Section 5.14; Sections 6.1 and 6.2; Chapter 7 through Section 7.7, but skipping Sections 7.4.2, 7.4.3, and 7.6; Section 8.1; Sections 9.1, 9.2, and 9.4.

Many people have spurred my interest in ranking, most notably Douglas Critchlow, Persi Diaconis, Michael Fligner, Georgia Lee Thompson, and Joseph Verducci. Robert Bohrer, who recently passed away, had been terrifically inspiring. I especially want to thank Lyinn Chung for her fine work in this area, and Kyungmee Choi, Michael Walker, Colin Mallows, and Louis Roussos for providing fascinating data. The University of Illinois deserves thanks for granting me a sabbatical leave, as does the National Science Foundation for support during my career. The reviewers made many valuable comments, and TeX and DraTeX (Gurari, 1994) helped create the typesetting and figures.

This book is dedicated to my mother, Ann, and Phil.

John Iglehart Marden
Urbana, March 1995

CHAPTER 1

Introduction

1.1 Rank data

Humans seem to be unable to avoid ranking things. Top Five/Ten/Twenty/Hundred lists abound: Best (Worst) Dressed, Best Suburbs, Most Watched TV Shows, Movies, Videos, Best Football Teams, and so on. Rankings have very serious uses as well. Companies need to know what products consumers prefer; social and political leaders need to know what the society values; elections need to be conducted; young people need to be drafted (?); prospective students need to be assessed. Data consisting of rankings appear in Psychology, Animal Science, Educational Testing, Sociology, Economics, and Biology. In fact, for almost any situation where there are data it can be helpful to transform the data into ranks. Nonparametric statistical analysis was initially based mainly on ranks.

This book is aimed at preference rankings. The basic unit of analysis consists of n judges ranking a set of m objects, the set of objects denoted by $\mathcal{O} = \{O_1, O_2, \ldots, O_m\}$. The objects can be almost anything: words, people, sports, soft drinks, ethnic groups, styles of literary criticism, crackers, political goals, occupations. Judges are usually people, but they can be animals or even inanimate sensing devices.

A full *ranking* of the objects assigns a complete ordering to the objects: There is a best object, second best, ..., and finally a worst one. The two most useful representations of a ranking are the *rank vector* and the *order vector*. A rank vector lists the ranks given to the objects, where "1" denotes best and "m" denotes worst. It presumes the objects are listed in a prespecified order. An order vector lists the objects themselves, in order from best to worst. For example, suppose the objects are $m = 4$ popular soft drinks, $\mathcal{O} = \{$*Coke, Pepsi, 7-Up , Sprite*$\}$, and a judge likes 7-Up best, then Sprite, then Pepsi, then Coke. The ordering is (*7-Up, Sprite,*

Table 1.1 *Two representations of the same ranking.*

The Ranking		**The Ordering**	
Coke	**4**	**7-Up**	**1**
Pepsi	**3**	**Sprite**	**2**
7-Up	**1**	**Pepsi**	**3**
Sprite	**2**	**Coke**	**4**

Pepsi, Coke). What is the rank vector? It depends on the listed order of the drinks. If this list is taken as in the first column of Table 1.1, i.e., Coke, Pepsi, 7-Up, Sprite, then the ranking is $(4, 3, 1, 2)$. (We will sometimes write it 4312.) As in the last two columns of Table 1.1, the ordering is the listing of the objects for which the rankings are in the order $1, 2, \ldots, m$. Whichever way is chosen, the basic idea is to match each integer from $1, \ldots, m$ with an object.

Abstractly, the objects themselves will be identified with integers, so that both rankings and orderings will be permutations of the first m integers. Through bitter experience, I know how easy it is to confuse rankings and orderings. To alleviate some of the confusion, rankings will generally be called "y" (or maybe "x" or "z"), and will reside in the space

$$\mathcal{S}_m \equiv \{\text{Permutations of the Ranks } \{1, 2, \ldots, m\}\}, \tag{1.1}$$

whereas orderings will generally be called "w" and reside in the space

$$\mathcal{T}_m \equiv \{\text{Permutations of the Objects } \{1, 2, \ldots, m\}\}. \tag{1.2}$$

For either, there needs to be an *a priori* identification of the objects with the integers.

We will typically be dealing with a sample of n rank vectors:

$$y^{(1)}, \ldots, y^{(n)} \in \mathcal{S}_m. \tag{1.3}$$

What does one do with such rank data? Chapter 2 takes a data-analytic approach. Rank data are multivariate data, the objects representing the variables; hence any multivariate method can be applied to the rank data. Means, standard deviations, histograms, box plots, cluster analyses, multidimensional scaling, factor analysis, etc., all have potential to provide insight into the data. For example, simple statistics such as the number of judges who rank a particular object as first, or the number who prefer one object to another, arise naturally.

Since rank data are so highly structured, methods that respect the peculiarities of rank data can be especially useful. Our fundamental structure is the "permutation polytope" created by plotting the rank vectors in Euclidean space, then connecting the points. Thompson (1993a,b) represents a set of ranking data on the polytope by placing a ball at each ranking, where the radius of the ball is indicative of the frequency of that ranking in the sample. Much of Chapter 2 is involved with trying to visualize (or numericize) these multidimensional plots. Distances defined on $\mathcal{S}_m$ or $\mathcal{T}_m$ allow defining the center and spread of a data set, and finding clusters of judges. In the final section of Chapter 2, we consider functions on $\mathcal{S}_m$ as vectors in $m!$-dimensional space, and try to decompose them by projecting onto interesting subspaces. The popular descriptive statistics fit into this framework. Diaconis (1988, 1989) develops a spectral analysis along these lines, decomposing frequencies into components based on the number of rankings with each object attaining each rank, or each pair of objects attaining each pair of ranks, etc. McCullagh (1993a) introduces inversions, which similarly decompose the data, but into factors based on paired comparisons of objects, triples of objects, etc.

Along with descriptions, a good statistician wants to be careful not to see structure in the data that is basically just random variation. Chapter 3 provides some tools, mainly for testing whether the uniform distribution is tenable. (Under the uniform distribution, all rankings have the same probability.) The statistics in Chapter 2 are turned into formal test statistics. In addition, *concordances* based on distance measures prove to be very useful. Chapter 4 continues by comparing groups of judges.

Modeling begins in Chapter 5. Some models arise from theoretical constructs, some from experimental methods, and others from attempts to find a simple description of the population of rankers. The basic models are summarized, including those named after

Thurstone, Babington Smith, Bradley, Terry, Mallows, Plackett, and Luce. Many of the models are built, either on purpose or by happenstance, on the statistics in Chapters 2 and 3. Chapters 6 through 10 plunge more deeply into the models, finding specializations and generalizations, and methods for assessing fit.

Partial rankings, incomplete rankings, and tied rankings challenge the material up to Chapter 11. Rank data commonly arrive in form different than n complete rankings of m objects. It may be impractical to have every judge assess every object, so that the experiment will present different judges with different subsets of objects to rank. Often, people are asked to list only their favorite q out of m rankings. (Such partial rankings are called "q-permutations" by Silverberg, 1980.) For example, in ranking college football teams, the judges are not expected to list all hundreds or thousands of college teams, but just their top 25, or a similar number. Judges are also wont to not provide full ranks even when asked. For example, when the American Psychological Association voted for President in 1980, the ballots asked for a complete ranking of the candidates, but most people provided only their top one, two, or three candidates. See Diaconis (1988, 1989). When analyzing such data, some models adapt well to some types of partial rankings, but none adapt well to all types, and some to no types.

Rank data have close connections with many other areas of statistics. Nonparametric analysis makes heavy use of ranks. See Gibbons and Chakraborti (1992). Paired-comparison analysis uses data in the form of pairwise comparisons of several objects or treatments, rather than complete rankings. See David (1988). Ordinal analysis deals with variables whose values are given only on an ordered scale. For example, a question on a survey may be "How important is it for the government to fund education?" The answers are chosen among the ratings *very important, somewhat important, slightly important, not important at all.* There is a definite order to the possible answers, but no obvious numerical scale. See Agresti (1984).

The models and exploratory techniques covered in this book are meant to be building blocks. Rankings may be only a part of the data, or may be collected in complicated designs. The hope is that students and researchers will use this book as a starting point for further explorations.

CHAPTER 2

Looking at Data

2.1 Introduction: Permutation polytopes

Rank data have a natural structure that presents challenges and opportunities unavailable in typical multivariate samples. This chapter develops several strategies for uncovering the structure present in one or more samples of rankings. Common univariate and multivariate calculations can be very informative, but the real intrigue lies in the unique features of rank data. We will start with a small example.

Example 2.1. Leisure Time
Hollander and Sethuraman (1978) report on a study by C. Sutton at Florida State University. The study focuses on the leisure time activities of elderly retired people, but rather than trying to determine which activities the participants prefer, it attempts to evaluate the circumstances under which the activities take place. For example, the makeup of the group engaging in an activity may be more important than the activity itself. A snippet of the study asked 27 women aged 70–79 with whom they would prefer to spend their leisure time: M = Male(s), F = Female(s), B = Both. Table 2.1 contains the data, where the women are categorized by race.

Each group basically splits its choices between two orderings. For the black women, the two favorite orderings are (*Both, Male, Female*) and (*Both, Female, Male*). Thus they prefer both genders to be present, followed by either gender. The white women favor (*Female, Both, Male*) and (*Both, Female, Male*), i.e., they are happy as long as some females are present. □

How can we picture such data? First, what does the sample space look like? For regular complete rankings, the sample space is $\mathcal{S}_m$ of (1.1), which is certainly a subset of $\Re^m$, but a very special subset. The points $y \in \mathcal{S}_m$ form the vertices of a *permutation polytope*, as termed by Yemelichev, Kovalev, and Kravtsov (1984).

Table 2.1 *Leisure Time: raw data.*

Ranking	Ordering	No. of Blacks	No. of Whites	Total No.
123	(Male, Female, Both)	1	0	1
132	(Male, Both, Female)	1	0	1
213	(Female, Male, Both)	0	1	1
231	(Female, Both, Male)	0	7	7
312	(Both, Male, Female)	5	0	5
321	(Both, Female, Male)	6	6	12
Total		13	14	27

Definition 2.1
The *permutation polytope* for given m is the convex hull of the points $y \in \mathcal{S}_m \subset \Re^m$. □

[The convex hull is the set of all weighted averages of the points, i.e., the set of all points $\sum_{y \in \mathcal{S}_m} \alpha_y y$ for weights $\alpha_y \geq 0$, $\sum_{y \in \mathcal{S}_m} \alpha_y = 1$.] Ye and McCullagh (1993) and McCullagh (1993b) relate the polytopes to ranking data, and Cohen and Mallows (1980) and Thompson (1993a,b) further explore representing sample frequencies on the polytope using balls of varying diameters. Such graphs are the analogs of histograms for continuous data, providing an overall snapshot of the data. Much of this chapter is dedicated to visualizing these polytopes.

The polytope for $m = 3$ objects is the hexagon in Figure 2.1. The vertices are the points y for $y \in \mathcal{S}_3$, where two points y and $y*$ are connected if they differ only by the interchange of consecutive rankings, i.e., if for some $i \neq j$, $y_i = y_j^*$, $y_j = y_i^*$, $|y_i - y_j| = 1$, and $y_k = y_k^*$ for $k \neq i, j$. Since the values in each vector y are the same, they always have the same sum, so that they all lie on the same $(m-1)$-dimensional hyperplane,

$$\mathcal{H}_m = \{x \in \Re^m | \sum_{i=1}^{m} x_i = \frac{m(m+1)}{2}\}. \tag{2.1}$$

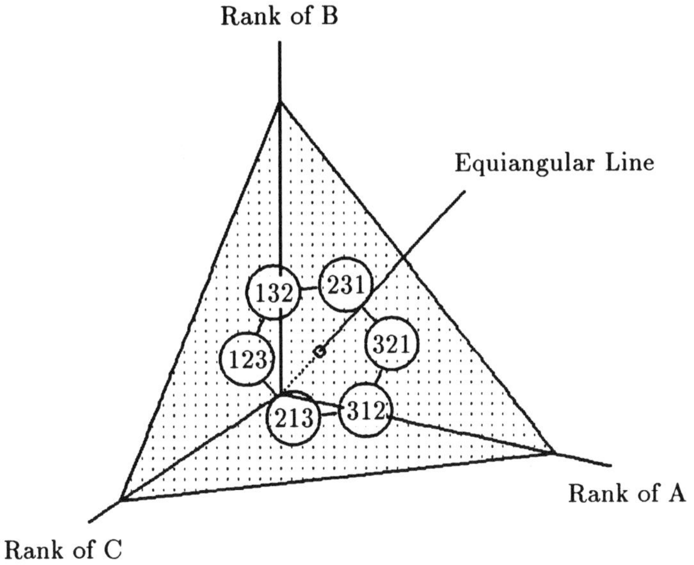

Figure 2.1 *The permutation polytope for three objects in three dimensions.*

In Figure 2.1, the shaded triangle is the part of this plane in the positive orthant. The hyperplane is perpendicular to the equiangular line, intersecting the line at the center point c_m given by

$$c_m = \frac{m+1}{2} 1_m, \quad 1'_m = \overbrace{(1, 1, \ldots, 1)}^{m}. \tag{2.2}$$

In Figure 2.1, this point is $(2, 2, 2)$, indicated by the small circle.

Again since the values in each y are the same, they have the same sum of squares, or equivalently, the same $\sum_{i=1}^{m}(y_i - \frac{m+1}{2})^2$. That is, the points in $\mathcal{S}_m$ are all the same distance from the center, lying on the $(m-1)$-dimensional sphere

$$\mathcal{C}_m = \{x \in \mathcal{H}_m | \ \|x - c_m\|^2 = v_m\}, \quad v_m = \frac{m(m^2-1)}{12}. \tag{2.3}$$

Here, $\|\cdot\|$ is the Euclidean norm, $\|x\|^2 = \sum_{i=1}^{m} x_i^2$. See Exercise 2.1. In Figure 2.1, this sphere is the circle centered at $(2, 2, 2)$ with radius $\sqrt{2}$ lying in the plane $\mathcal{H}_3$.

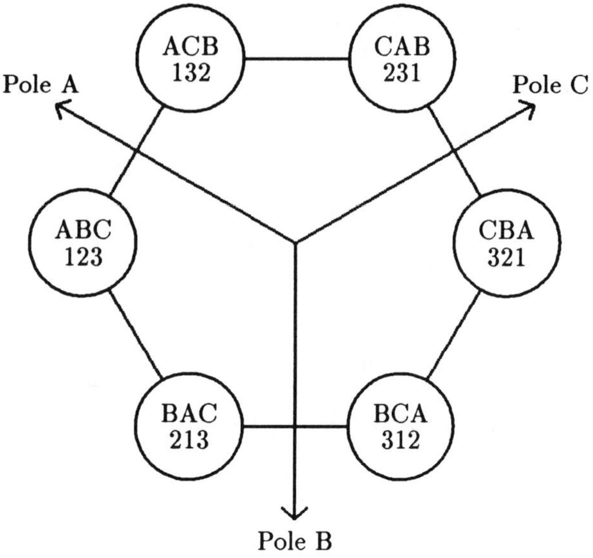

Figure 2.2 ***The permutation polytope for three objects represented in two dimensions, with poles.***

Since $\mathcal{S}_m \subset \mathcal{H}_m$, it can be represented in $(m-1)$ dimensions. Figure 2.2 shows the $m = 3$ polytope in two dimensions. [The horizontal and vertical axes are $(y_1 - y_3)/\sqrt{2}$ and $(2y_2 - y_1 - y_3)/\sqrt{6}$, respectively.] Rankings and orderings represented by each vertex are indicated. One can imagine "poles of attraction" for the objects. That is, the upper left has the pole for object A, so that the two vertices closest to A rank A first, the two farthest away rank A third, and the other two rank A second. The other poles are handled similarly. Formally, we have the following.

Definition 2.2
Given the permutation polytope for m, the *pole of attraction* for Object i is the half line beginning at the center c_m and passing through the point $x^{(i)}$ with 1 as the ith element and $1 + \frac{m}{2}$ for the other elements:

$$z^{(i)}(c) = c_m + c(x^{(i)} - c_m) \text{ for } c > 0. \qquad \square \quad (2.4)$$

The point $x^{(i)}$ is the average of the points $y \in \mathcal{S}_m$ that rank Object i first. The farther a vertex is from the pole for Object

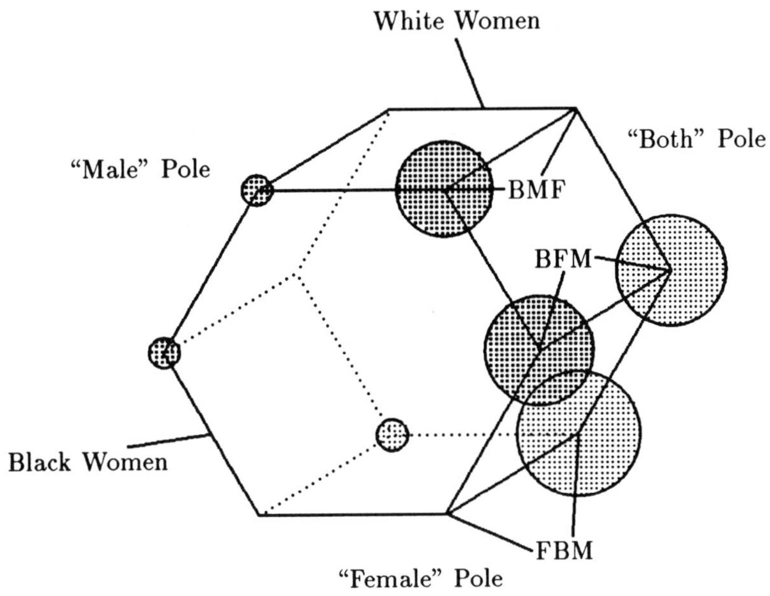

Figure 2.3 ***Leisure Time: comparison of the polytopes.***

i, the worse the rank assigned to Object i. In fact, the squared distance is a linear function of the rank:

$$||y - z^{(i)}(c)||^2 = \eta_{m,c} + cmy_i, \; y \in \mathcal{S}_m, \tag{2.5}$$

where η depends only on m and c. (See Exercise 2.2.) The poles are helpful in orienting oneself, especially in higher dimensions. Note that for $c < 0$, the farther the vertex the better the rank.

Now that we have the sample space, we can find the permutation polytope-o-gram of Thompson (1993a). (The "o-gram" part of the terminology is my designation.) At each vertex, place a ball whose radius is an increasing function of the observed frequency of rankings corresponding to the vertex. She recommends the radius be proportional to $(frequency)^{5/7}$ as per Steven's Law (Cleveland, 1985). This procedure is used by Cohen and Mallows (1980) to illustrate the data in Example 2.1, as in Figure 2.3.

Example 2.2. Leisure Time: Polytope-o-grams
Figure 2.3 exhibits the polytope-o-grams for the two groups of women in Example 2.1. The graph clearly shows what we saw in

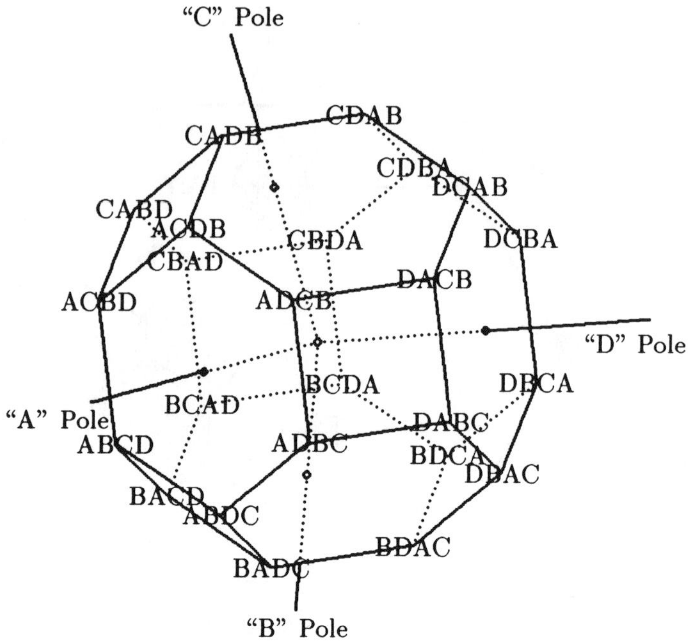

Figure 2.4 ***A generic polytope with $m = 4$ objects represented in three dimensions.***

Example 2.1, i.e., that the White Women are far from the Male pole, and the Black Women are close to the Both pole. □

The larger the m, the more difficult it is to picture the polytope. When $m = 4$, the polytope exists in three dimensions. Professor G. L. Thompson has provided software that creates the plot and interactively rotates the plot for $m = 4$. The animation is amazingly effective at portraying three dimensions.

Example 2.3. Political Goals: Polytope-o-gram

Croon (1989b) presents data from a study on change in mass politics described in Barnes and Kaase (1979). A sample of 2262 Germans ranked the following four political goals based on their importance: O = Order = Maintain order in the nation; S = Say = Give people more say in the decisions of the government; P = Prices = Fight rising prices; F = Freedom = Protect freedom of speech.

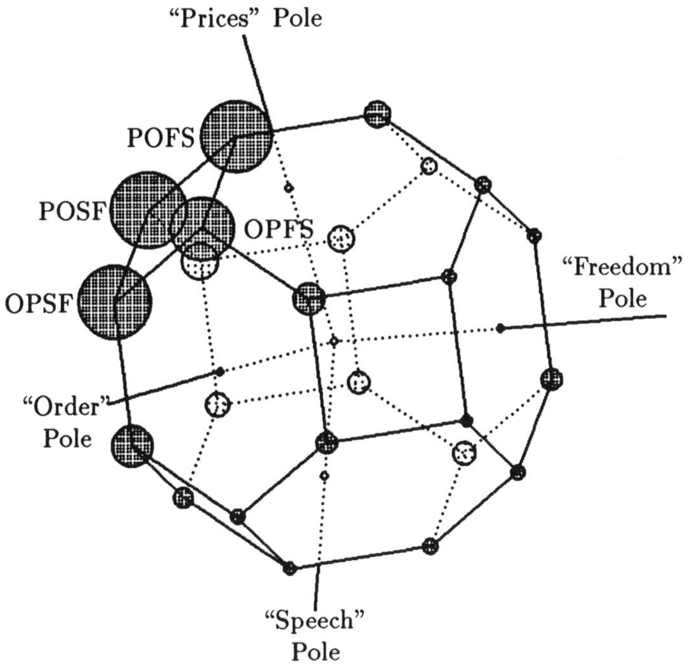

Figure 2.5 ***Political Goals: polytope as in Figure 2.4.***

Barnes and Kaase classify the goals into Materialist (Order and Prices) and Post-materialist (Say and Freedom), the latter being more "Green."

Figure 2.4 shows a generic permutation polytope for $m = 4$. Contemplating the polytope, we see that it is bounded by eight hexagons and six squares. Each pole of attraction passes through the center of the hexagon for which all rankings have the pole's object in first place. Exactly opposite each pole is the hexagon for which the pole's object is always ranked last. Between any two poles is the square for which the two poles' objects are always first and second, in either order. Figure 2.5 adds the frequency balls for the present data. The main feature jumps out at the viewer: the bulk of the mass is in the Materialist square, and much of the rest is just one vertex away. Relatively few judges are near the Post-materialist poles. □

Polytopes are more difficult to visualize in higher dimensions. Low-dimensional projections and some numerical summaries help.

See Sections 2.2, 2.3, and 2.4. Finding the one or more points about which the data cluster, and determining how closely they cluster, is covered in Section 2.5. Section 2.6 takes a different approach. Instead of looking at the space $\mathcal{S}_m$, we look at the space of functions on $\mathcal{S}_m$. Diaconis (1988, 1989) develops a spectral theory for such functions that decomposes the space into an interesting hierarchical set of subspaces based on types of marginals. McCullagh (1993a) proposes a different decomposition based on inversions.

2.2 Projections of polytopes

A common approach to visualizing highly multivariate data is first to reduce the data to a few linear combinations of the original variables. Since the permutation polytopes are so nicely structured, we will use orthonormal linear combinations to prevent warping. We are combining the rankings; hence each linear combination can be represented by an $m \times 1$ vector. With k linear combinations, we have the vectors $g^{(1)}, \ldots, g^{(k)}$, where orthonormality means that the vectors have length 1 and are orthogonal:

$$||g^{(i)}|| = 1, \;\; g^{(i)\prime} g^{(j)} = 0 \text{ for } i \neq j. \tag{2.6}$$

We will restrict to contrasts, i.e., vectors for which the elements sum to zero, since the rank vectors all have elements summing to the same value. Define the $m \times k$ matrix $G = (g^{(1)}, \ldots, g^{(k)})$. Then the data $y^{(1)}, \ldots, y^{(n)}$ as in (1.3) is reduced to the $k \times 1$ vectors

$$\widetilde{y}^{(i)} = G' y^{(i)}, \;\; i = 1, \ldots, n. \tag{2.7}$$

We also bring along the poles, which are projected to $q^{(i)}(c) = G' z^{(i)}(c)$, $c > 0$. The $\widetilde{y}^{(i)}$'s will be contained in the k-dimensional unit ball, the center of which is the projection of the center of the original polytope.

There is an infinite number of linear combinations we could take. Multivariate analysis provides several techniques for choosing them, including those in the realm of projection pursuit (Huber, 1985.) Of course, the best projections may be those suggested by the substantive considerations in the experiment. For exploratory analysis, we will use Gabriel's (1982) biplot suggested by Cohen and Mallows (1980). It is similar to principal components and multidimensional scaling. (See, e.g., Mardia, Kent, and Bibby, 1979.) The idea is to find projections for which the data are far from the center of the polytope. If most of the data are clustered near a single ranking, or if there are groups of judges clustered around disparate rankings, then the linear combinations should pick out these features.

The first vector $g^{(1)}$ is that which maximizes

$$ss(g) \equiv \frac{1}{n} \sum_{i=1}^{n} (g'(y^{(i)} - c_m))^2 \tag{2.8}$$

over g with $||g|| = 1$. The jth, $j = 2, \ldots, k$, vector is that which maximizes $ss(g)$ subject to $||g|| = 1$ plus being orthogonal to the previous vectors, i.e., $g'g^{(i)} = 0$, $i = 1, \ldots, j-1$. (The difference between the process here and principal components is that in the latter, the average of the observed $y^{(i)}$'s is used instead of the center c_m.) We can rewrite $ss(g)$ as

$$ss(g) = g'Wg, \quad W = \frac{1}{n} \sum_{i=1}^{n} (y^{(i)} - c_m)(y^{(i)} - c_m)' \tag{2.9}$$

(Exercise 2.3). Thus as in, e.g., Mardia, Kent, and Bibby (1979, Definition 8.2.1), or Section 12.1.1, $g^{(1)}, \ldots, g^{(k)}$ are the first k eigenvectors of W, and $ss(g^{(1)}), \ldots, ss(g^{(k)})$ are the corresponding eigenvalues. See Exercise 2.4. [Since each $y^{(i)} - c_m$ has elements summing to zero, the eigenvectors will, too. Thus our requirements in (2.6) and below will automatically be satisfied. Also, there will be at most $m - 1$ non-zero eigenvalues.]

Some examples now follow. In each case, we hope the projections provide an informative first picture of the main features of the data.

Example 2.4. Occupations: Projections

Goldberg (1976) surveyed graduates from Technion – Israel Institute of Technology. One question asked them to rank 10 occupations in order of their perceived prestige. There are 143 complete rankings, which are some of the most analyzed rank data to date. See especially Cohen (1982) and Cohen and Mallows (1983). The

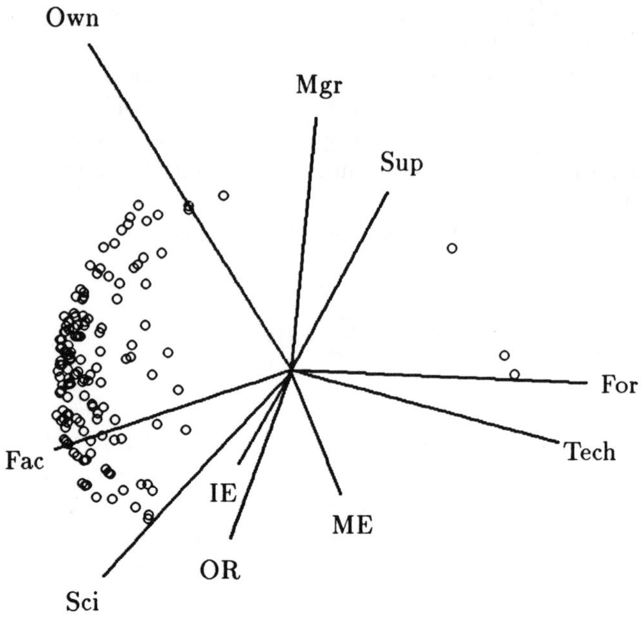

Figure 2.6 ***Occupations: projection.***

occupations are Fac = Faculty Member; Own = Owner of a business; Sci = Applied Scientist; OR = Operations Researcher; IE = Industrial Engineer; Mgr = Manager; ME = Mechanical Engineer; Sup = Supervisor; $Tech$ = Technician; For = Foreman.

The polytope is 9-dimensional, with $10! = 3628800$ vertices; hence reduction is called for. Figure 2.6 plots the $\widehat{y}^{(i)}$ for $k = 2$ linear combinations. The poles are labeled with their objects' names. Note first that most of the data are on the left, but there are three points on the right. Feigin and Cohen (1978) identified these points, and after consulting with Professor Goldberg, decided these three judges had misunderstood the instructions and gave 10's to their top occupations and 1's to their bottom ones. We will follow these authors and replace the rankings with their reverse rankings.

Recalculating the projection gives essentially the same picture, but the three reversed observations are reflected through the center to join the others. The horizontal variable shows that Faculty, Owner, and Scientist are all ranked high; Technician and Foreman are ranked low; and the others are in between. The vertical vari-

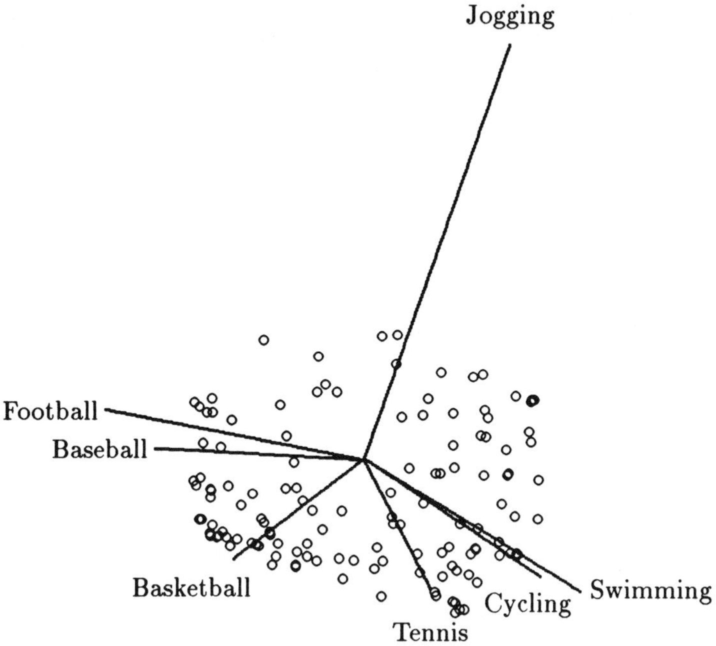

Figure 2.7 ***Sports: projection.***

able shows an interesting Professional/Managerial split: The four occupations above (approximately) zero are heavily involved with managing people, whereas those below zero are much less so. We may even risk declaring that the larger the value, the greater the number of people the occupation manages. Fligner and Verducci (1988) and van Blokland-Vogelesang (1989) found a similar split, although the former placed Operations Researcher with the four Managerials. □

Example 2.5: Sports: Projections
Louis Roussos asked 130 students at the University of Illinois to rank seven sports according to their preference in participating: B = Baseball; F = Football; K = Basketball; T = Tennis; C = Cycling; S = Swimming; J = Jogging. Figure 2.7 plots the first two variables. Note that these data are much more uniformly spread over the disk than in the previous example, suggesting much less agreement on overall rankings of the objects. For the horizontal variable, the value decreases with the number of people who partic-

ipate in the event: 22 play in a Football game (11 per team), 18 in Baseball, 10 in Basketball, 2 in Tennis (4 if doubles), and 1 for the other three. Variable 2 primarily separates Jogging from the rest. We see a few people who like Jogging, but most try to get as far away as possible. What support there is for Jogging comes more from the Non-team half of the data, i.e., there are more people in the sector between Jogging and Cycling/Swimming than between Jogging and Football/Baseball. □

Example 2.6. Ethnic Preferences: Projections

Beran and Walker (1990) and Walker (1993) surveyed 93 students on how close they feel to various ethnic groups: Asian, Black, Jewish, Latino/a, Native American, and White. The students ranked the groups on closeness, and categorized themselves according to the groups, resulting in sample sizes of 32, 16, 5, 12, 0, and 28, respectively. To obtain a projection, we used the combined sample. Figure 2.8 contains the result using the first two vectors. The poles are indicated by the ethnic groups' names, and the individual rankings by individual letters indicating group membership. Not surprisingly, people tend to cluster near their own pole. The horizontal variable seems to be a Black–White continuum. The Blacks are on the left side of zero, and the Whites on the right side. Asians tend to be close to the Whites and Latinos close to the Blacks, but there is some overlap. The Jewish people are closely identified with Whites. The vertical variable is interesting, linking Blacks with Whites, and Asians with Latinos. □

Remark 2.1

Multidimensional scaling includes a large variety of important and useful methods for representing all types of preference data. The field has been developed primarily by social scientists. The idea is to represent judges' and objects' relationships with each other graphically. Coombs (1964) is perhaps the landmark work in this area. The volume *Similarity and Choice* (Lantermann and Feger, 1980) contains an excellent set of papers in honor of (or by) Professor Coombs, that relates further developments. This area covers more than just ranking data, including paired comparisons and rating scales. Our book cannot do justice to the general subject without (at least) doubling in size, but we give some glimpses in this direction in Chapter 10. Note that the polytope-o-grams do

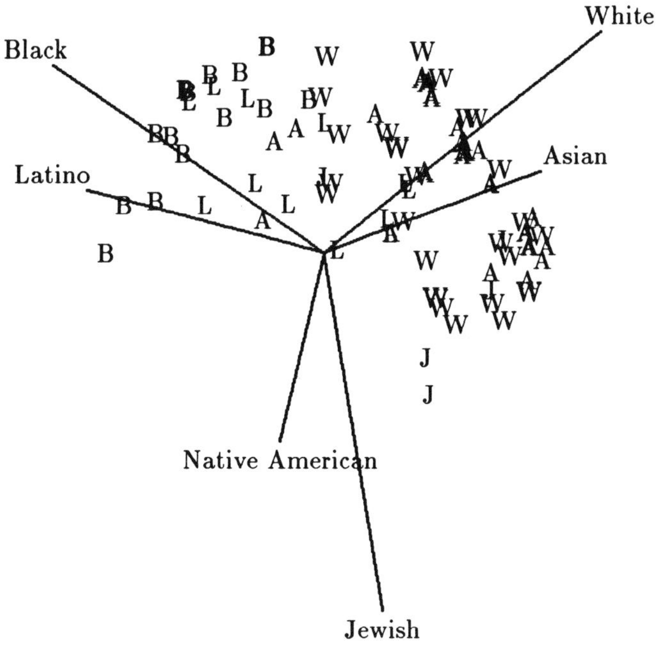

Figure 2.8 ***Ethnic Preferences: projections. The individual letters indicate the group membership of the individual rankings.***

provide a type of scaling. A judge is represented in the space according to how far the judge's ranking is from the objects' poles. □

2.3 Marginals

In the previous section, the data chose the linear combinations. Here we make systematic choices relevant for any set of data. A natural set of projections to consider are those onto the poles, extended through the center in both directions. These projections are one-dimensional, and in fact equivalent to the raw rankings for each object. Thus we can take the linear combination g for Object i to be the indicator vector $(0, \ldots, 0, 1, 0, \ldots, 0)$, where the "1" is in the ith slot. See Exercise 2.5. The projection of the polytope-o-gram yields m balls equally spaced according to the ranks given Object i, so that the frequency associated with the jth ball is the frequency that Object i has rank j. These proportions can be assembled into the *Marginals* matrix $\widehat{M}$, an $m \times m$ matrix whose

ij^{th} element is given by

$$\widehat{M}_{ij} = \frac{1}{n}\#\{y^{(k)} \mid y_i^{(k)} = j\} \equiv \sum_{k=1}^{n} I[y_i^{(k)} = j]. \tag{2.10}$$

Here, $I[A]$ for an event A is the indicator function, equaling 1 if the event occurs and 0 otherwise, and $\#B$ is the number of elements in the set B. We use the term "Marginals" since the ith row gives the observed marginal distribution of the ranks assigned to Object i, and the jth column gives the marginal distribution of objects given the rank j. Each row and column can also be represented effectively with regular histograms.

Example 2.7. Gulf War: Marginals
In January, 1991, just before the Gulf War, I asked the 21 students in my class to rank five options for the United States to pursue: I = Invade = Invade Iraq immediately with the full force of the military; K = Kuwait = Use as much force as necessary to push the Iraqis out of Kuwait, but do not make any incursions into Iraq; D = Defense = Keep the large military capability in the Gulf, but use it only if attacked by Iraqi forces; M = Multinational = Remove all but 30,000 American troops and rely on the multinational coalition to enforce the sanctions and protect Saudi Arabia for the near future; ON = Out Now = Get out now.

The options are listed from "Hawk"-est to "Dove"-est. For each option, we create the projected polytope-o-gram projected onto that option's pole. Each row in Figure 2.9 corresponds to one of these projections. It is clear that the most popular options are Kuwait and Defense, which receive the bulk of the first two rankings, followed by Multinational. The two extreme options, Invade and Out Now, are much less popular, especially the latter.

The Marginals (2.10) can be reduced further by looking at the vector of averages

$$\overline{y} = \frac{1}{n}\sum_{i=1}^{n} y^{(i)}. \tag{2.11}$$

Thus $\overline{y}_j$ is the average rank assigned to Object j. For the current data, $\overline{y}' = (3.57, 2.24, 2.33, 2.91, 3.95)$, which succinctly describes the main feature of Figure 2.9. □

Diaconis (1988, 1989) develops a hierarchy of marginal statistics, of which (2.10) is the simplest. See Section 2.6.1 for more details.

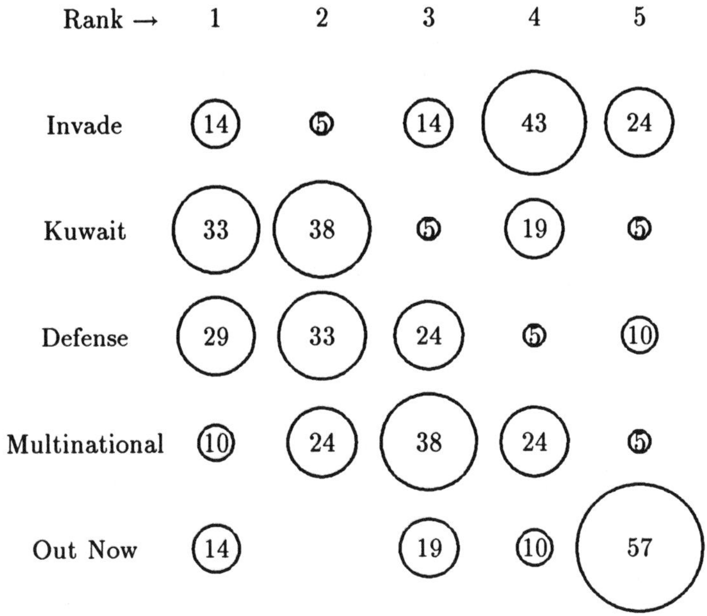

Figure 2.9 *Gulf War: projection onto the poles, i.e., the Marginals matrix. The numbers are the percentages.*

2.4 Pairs

It is natural to compare two objects based on their relative rankings. For a sample, the "Pairs" matrix, $\widehat{K}$, is an $m \times m$ matrix containing all paired comparisons, where

$$\widehat{K}_{ij} = \frac{1}{n}\#\{y^{(k)} \mid y_i^{(k)} < y_j^{(k)}\}, \tag{2.12}$$

the proportion of people who prefer Object i to Object j. See David (1988) for a comprehensive presentation of paired comparisons.

Example 2.8. Gulf War: Pairs
Table 2.2 exhibits the Pairs matrix (2.12) for the Gulf War data. The larger the entries in a row are, the more popular that row's object is. The choices have been rearranged for a purpose. Looking at the first row, Defense, we see all paired comparisons are over

Table 2.2 *Gulf War: paired comparisons. Percentage preferring row option to column option.*

	D	K	M	I	ON
Defense		52	62	71	81
Kuwait	48		76	81	71
Multinational	38	24		67	81
Invade	29	19	33		62
Out Now	19	29	19	38	

50%, i.e., Defense is preferred to each other single choice by a majority of judges. Such a selection is called the *Condorcet* choice. See Stern (1993). Note that in general there is no guarantee a Condorcet choice will exist. By contrast, the percentages in Out Now's row are all less than 50%, which would dub it the reverse Condorcet choice. In fact, we can construct a *consensus ranking*, to whit, Defense is preferred to everything, Kuwait is preferred to all but Defense, Multinational is preferred to all but the top two, and Invade is preferred only to Out Now. In other words, in the arrangement in the table everything below the diagonal is under 50%, and everything above the diagonal is over 50%. It is interesting that the consensus ranking is close to but not quite that given by the Means in Example 2.7. The difference is that the Means prefer Kuwait to Defense. □

Example 2.9. Ethnic Preferences: Condorcet Choices
Within each ethnic group in Example 2.6, the Condorcet choice is the ethnicity of the group. That is, Asian is the Condorcet choice among Asians, and Black among Blacks, etc. What is interesting are the next most popular choices. For each group there is one, i.e., a choice preferred to all but the group's ethnicity. The Asian, Jewish, and Latino groups all chose Whites; the Black group chose Latinos; and the Whites chose Native Americans. □

In addition to looking at pairs of objects, one could look at triples, quadruples, etc. McCullagh (1993a) develops a hierarchy based in *inversions*. See Section 2.6.2.

2.5 Center, spread, and distance

A common goal with rank data is to obtain a central ranking, and to assess how close the data cluster about the center. The center ranking is an element of $\mathcal{S}_m$. Although the average $\overline{y}$ is a fine candidate for describing the center, it will not be an actual rank vector unless all observed rankings are the same. It is well known (Exercise 2.6) that $\overline{y}$ is the element $z \in \Re^m$ that minimizes the sum of squared Euclidean distances of the observed vectors from z. A natural modification for obtaining a central ranking is to minimize the sum of squares over $z \in \mathcal{S}_m$. One can also use distances other than Euclidean. Many distances specialized to rank data have been proposed in the literature. Diaconis (1988) and Critchlow (1986) contain extensive discussions of a number of possibilities.

A distance is a function $d(y^*, y) \geq 0$ of y^* and $y \in \mathcal{S}_m$. We will be using distances that satisfy $d(y^*, y^*) = 0$, $d(y^*, y) = d(y, y^*) > 0$ if $y^* \neq y$. Metrics will also satisfy the triangle inequality, $d(y^*, y) \leq d(y^*, z) + d(z, y)$, but not all our distances do. For a set of data, the measure of location corresponding to d is the $\widehat{\omega}$ that minimizes the average distance

$$\overline{d}(\omega) \equiv \frac{1}{n} \sum_{i=1}^{n} d(y^{(i)}, \omega), \quad \omega \in \mathcal{S}_m. \tag{2.13}$$

Measures other than average, such as median, could be used. Note that in general the minimizer $\widehat{\omega}$ need not be unique.

The spread of the data around the center $\widehat{\omega}$ is naturally measured by $\overline{d}(\widehat{\omega})$. Normalizing the spread appropriately can yield a more interpretable coefficient. People often take $1 - 2 \cdot \overline{d}(\omega)/Max(d)$, where $Max(d) = Max\{d(y^*, y) | y^*, y \in \mathcal{S}_m\}$ is the maximum possible distance between two rank vectors. See Diaconis and Graham (1977). This coefficient is interpreted like a correlation coefficient: it ranges between -1 and 1, where "+1" means the data are as close to ω as possible, and "-1" means the data are as far from ω as possible. For some distances this is reasonable, but it is common to want "totally random" to imply that $t = 0$.

At this point it is necessary to add sampling to the pure descriptions we have presented so far. Imagine the data being generated as n independent and identically distributed replicates of $Y \sim P$, where P is a probability distribution on $\mathcal{S}_m$ represented as an $m! \times 1$

vector indexed by the rankings y:

$$P[Y = y] = P_y, \quad y \in \mathcal{S}_m. \tag{2.14}$$

For example, with $m = 3$, P is 6×1 with indices running over $\mathcal{S}_3$ in some arbitrary but fixed order, such as

$$P' = (P_{123}, P_{132}, P_{213}, P_{231}, P_{312}, P_{321}). \tag{2.15}$$

Since P is a vector of probabilities, it lies in the *simplex* $\mathcal{P}_{m!} \subset \Re^{m!}$, where

$$\mathcal{P}_K = \{x \in \Re^K \mid x_i \geq 0 \text{ for all } i, \text{ and } \sum_i x_i = 1\}. \tag{2.16}$$

We consider

$$Y^{(1)}, \ldots, Y^{(n)} \text{ independent}, \quad Y^{(i)} \sim P \text{ for each } i. \tag{2.17}$$

In most cases, one should not believe too deeply in this assumption, but it provides a useful construct for interpretation.

Now we can define "totally random" to mean each ranking is equally likely, i.e.,

$$Y \sim \mathit{Uniform}(\mathcal{S}_m) \Leftrightarrow P = u_{m!} \equiv \frac{1}{m!} 1_{m!}, \tag{2.18}$$

where for any finite set Ω,

$$W \sim \mathit{Uniform}(\Omega) \;\Rightarrow\; P[W = w] = \frac{1}{\#\Omega}, \quad w \in \Omega. \tag{2.19}$$

In a uniform permutation polytope-o-gram each ball is the same size. Uniformity, although rarely seen, is a convenient benchmark for rank data. Chapter 3 is entirely dedicated to measuring nonuniformity.

Now we are ready for our coefficient. Define

$$\delta_P(\omega) = E_P(d(Y, \omega)) \tag{2.20}$$

to be the population average distance from ω, and let $\delta_0 = \delta_P(\omega)$ for $P \sim Uniform(\mathcal{S}_m)$. Define the coefficient

$$t(\omega) = 1 - \frac{\overline{d}(\omega)}{\delta_0} \tag{2.21}$$

as in Kendall and Gibbons (1990), so that $t(\omega) = 1$ if the data are clustered at ω, and 0 if the data are uniformly distributed around ω. The minimum is $1 - Max(d)/\delta_0$, which may or may not be -1. For some distances, including Cayley, Hamming, and Ulam below, the $\overline{d}$ itself may be more intuitive than the t. (We should write δ_0 as a function depending on ω. However, in Chapter 3 and beyond, we will deal exclusively with so-called label invariant distances, for which there is no such dependence.) The population values corresponding to t and $\widehat{\omega}$ are

$$\tau_P(\omega) = 1 - \frac{\delta_P(\omega)}{\delta_0} \tag{2.22}$$

and center ω_P that minimizes $\delta_P(\omega)$ over ω.

2.5.1 Some useful distances

There are (at least) two types of distances: spatial and disorder. A spatial distance between two vectors is measured by the time it takes to travel from one vector's vertex on the permutation polytope to the other's. The distance depends on how one travels: in a straight line (Euclidean distance); along a great circle on the sphere $\mathcal{C}_m$ of (2.3); moving in unit steps parallel to the axes (Manhattan distance, or Footrule, below); stepping between connected vertices of the polytope (Kendall's distance, below). A disorder distance measures the disorder between two rankings. The canonical example is arranging books on a shelf into alphabetical order, the distance measuring the amount of work it takes to put things right. In Example 2.10 we will take this analogy literally. Some details of popular distances follow. Diaconis (1988), Chapter 6, discusses these and other distances extensively.

Spearman and Footrule. A natural set of spatial distances are the d_p distances,

$$d_p(y^*, y) = \sum_{i=1}^{m} |y_i^* - y_i|^p \tag{2.23}$$

for some $0 < p < \infty$. Take the $(1/p)^{th}$ power to turn d_p into a metric if $p > 1$. The Spearman and Footrule distances use, respectively, $p = 2$ and $p = 1$ in (2.23). Note that Spearman's distance is the square of Euclidean distance. The coefficient t_{Spear} (2.21) is famous in nonparametric statistics as *Spearman's* ρ. It can be written as the average product-moment correlation coefficient between ω and the $y^{(i)}$'s. See Exercise 2.7.

The $\overline{d}_p$ can be found directly from the Marginals matrix (2.10) as for the Hoeffding distances below. Spearman has an especially convenient expression:

$$\overline{d}_{\text{Spear}}(\omega) = \frac{m(m-1)(2m+1)}{3} - 2 \cdot \omega' \overline{y}. \tag{2.24}$$

Hoeffding. The *Hoeffding* distances have the form

$$d_{\text{Hoeff}}(y^*, y) = \sum_{i=1}^{m} a(y_i^*, y_i), \tag{2.25}$$

where a is a function on $\{1, \ldots, m\} \times \{1, \ldots, m\}$ that satisfies

$$a(i, i) = 0 \text{ and } a(i, j) = a(j, i). \tag{2.26}$$

They generalize the d_p distances. The average distance can be found by

$$\overline{d}_{\text{Hoeff}}(\omega) = \sum_{i=1}^{m} \sum_{j=1}^{m} \widehat{M}_{ij} \cdot a(j, \omega_i). \tag{2.27}$$

Kendall. Counting rather than measuring has proven very useful in nonparametric statistics and in analyzing rank data. For y^* and

y in $\mathcal{S}_m$, define a *discordant pair* to be a pair of distinct objects on which y^* and y have opposite relative preferences, that is, for $i \neq j$,

$$(y_i^* - y_j^*)(y_i - y_j) < 0. \tag{2.28}$$

The Kendall distance between two rank vectors is the number of discordant pairs,

$$d_{\text{Ken}}(y^*, y) = \sum \sum_{1 \le i < j \le m} I[(y_i^* - y_j^*)(y_i - y_j) < 0]. \tag{2.29}$$

Kendall's τ coefficient is then t_{Ken}. In addition to the spatial interpretation given above, Kendall has a disorder interpretation. Start with the books in the order indicated by y^*. Rearrange the books by sequentially interchanging well-chosen adjacent pairs. The minimum number of these interchanges needed to achieve the y order is $d_{\text{Ken}}(y^*, y)$. This interpretation was noticed as early as 1839 by O. Rodrigues.

The average Kendall distance also has a simplification. Hays (1960) shows that

$$\overline{d}_{\text{Ken}}(\omega) = \sum \sum_{\{i,j \mid \omega_i > \omega_j\}} \widehat{K}_{ij}, \tag{2.30}$$

where $\widehat{K}$ is the Pairs matrix (2.12).

Cayley. Instead of restricting to adjacent pairs, Cayley's distance, denoted d_{Cay}, counts the minimum number of arbitrary pairwise interchanges needed to bring the y^* order to the y order. Thus books far apart can be interchanged as easily as those close together.

There are two ways to calculate d_{Cay}. The direct method is to perform interchanges in y as follows. If $y_1 = y_1^*$, then count 0. If not, switch y_1 with the value y_1^* in the vector y, and count 1. Do the same with (the possibly new) y_2, then with y_3, ..., and finally with y_{m-1}. At the end, y_m will equal y_m^*. The number of interchanges is Cayley's distance. For example, suppose $y^* = (1, 4, 2, 5, 3)$ and $y = (2, 3, 4, 5, 1)$. The sequence of y's is: 1. $(1, 3, 4, 5, 2)$, 2. $(1, 4, 3, 5, 2)$, 3. $(1, 4, 2, 5, 3)$, 4. $(1, 4, 2, 5, 3)$. Interchanges occurred at steps 1, 2, and 3; hence $d_{\text{Cay}}(y^*, y) = 3$.

An interesting method, which is quicker for large m, involves counting the number of cycles in the permutation y. For given index j_1, the cycle generated by y_{j_1} consists of the set of y_{j_i}'s where $j_{i+1} = y_{j_i}$. The elements of y will be partitioned into such cycles. To find $d_{\text{Cay}}(y^*, y)$, first reorder the elements of y^* and y so that $y^* = e_m \equiv (1, \ldots, m)$. Then d_{Cay} is $m - \#$ *Cycles* in the new y. In the example above, rearranging so that $y^* = (1,2,3,4,5)$ brings y to $(2,4,1,3,5)$. The cycle starting with $y_1 = 2$ is $y_1(=2), y_2(=4), y_4(=3), y_3(=1)$. The elements y_2, y_3, and y_4 have that same cycle. $y_5(=5)$ is its own cycle. Thus there are two cycles, giving $d_{\text{Cay}} = 3$ again. See Section 3.3 for a more formal treatment.

Hamming. Letting $p \to 0$ in (2.23), we find Hamming's distance, the number of disagreements between the rank vectors:

$$d_{\text{Ham}}(y^*, y) = \#\{i | y_i^* \neq y_i\}. \tag{2.31}$$

This is more of a disorder distance. Remove the books that are in the wrong place, then replace them one by one in the correct spots. The number of replacements is (2.31). Hamming is a good distance for matching, or trying to place pegs in holes. It matters only whether the pegs are in the correct holes; there is no sense that some incorrect placings are worse than others.

Note that Hamming distance (2.31) is Hoeffding (2.25) with $a(i,j) = 0$ if $\omega_i = j$ and 1 otherwise; hence

$$\overline{d}_{\text{Ham}}(\omega) = m - \sum_{i=1}^{m} \widehat{M}_{i\omega_i}. \tag{2.32}$$

Maximum. Now let $p \to \infty$ in (2.23), or better, take $lim_{p\to\infty}$ $d_p(y^*, y)^{1/p}$. We then obtain

$$d_{\text{Max}}(y^*, y) = \max_{1 \leq i \leq m} |y_i^* - y_i|. \tag{2.33}$$

Ulam. Suppose the books can slide along the shelf. The "deletion–insertion" method of arranging chooses books one at a time, removing them and reinserting at a better place, the remaining books

Table 2.3 *Distances: arranging the books. The number of steps equals the distance between the orderings BECADF and ABCDEF.*

Steps ↓	**Kendall**	**Hamming**	**Cayley**	**Ulam**
0	BECADF	BECADF	BECADF	BECADF
1	BEACDF	⊔⊔C⊔⊔F	AECBDF	ABECDF
2	BAECDF	A⊔C⊔⊔F	ABCEDF	ABCDEF
3	ABECDF	ABC⊔⊔F	ABCDEF	
4	ABCEDF	ABCDEF		
5	ABCDEF			

possibly shifting slightly to accommodate the insertion. The minimum number of such operations is Ulam's distance. Equivalently, imagine subvectors $(y^*_{i_1}, \ldots, y^*_{i_l})$ and $(y_{i_1}, \ldots, y_{i_l})$ of y^* and y for $l \leq m$, $i_1 < \cdots < i_l$. Ulam's distance is $m - L$ for L being the largest l for which the two subvectors are in the same order. See Gordon (1979) for an example of this distance in statistics.

Example 2.10. Disorder Distances: Arranging Books
Start with six books in the order BECADF. How far is it from ABCDEF? [Here, $y^* = (4, 1, 3, 5, 2, 6)$ and $y = (1, 2, 3, 4, 5, 6)$ are the rank vectors.] The sequences of steps for some of the distances are contained in Table 2.3. The step at which the ordering is *ABCDEF* indicates $d(y^*, y)$, i.e., it is 5, 4, 3, and 2 for Kendall, Hamming, Cayley, and Ulam, respectively. □

Kruskal (1958) gives a detailed comparison of the Spearman and Kendall distances. Daniels (1950b) obtained a bound on the difference of their coefficients, $|3mt_{\text{Ken}} - 2(m+1)t_{\text{Spear}}| \leq m - 1$. Durbin and Stuart (1951) prove the inequality $d_{\text{Spear}} \geq d_{\text{Ken}}(1 + d_{\text{Ken}}/m)$. Diaconis and Graham (1977) show the following relationship among Kendall, Cayley, and Footrule: $d_{\text{Ken}} + d_{\text{Cay}} \leq d_{\text{Foot}} \leq 2d_{\text{Ken}}$. The remainder of this section presents other details of the distances, and exhibits many uses for analyzing rank data.

2.5.2 Estimating the center

In this section we address point estimation for the population center and spread, ω_P and τ_P of (2.22). The natural estimator for ω_P is the minimizer $\widehat{\omega}$ of (2.13). In principle, finding $\widehat{\omega}$ is easy: just search over all $\omega \in \mathcal{S}_m$. If the minimizer is not unique, then $\widehat{\omega}$ is the set of minimizers. The resulting estimator is at least consistent.

Theorem 2.1
If ω_P is uniquely defined, then $P[\widehat{\omega} = \omega_P] \to 1$ as $n \to \infty$. If ω_P is not unique, then $P[\widehat{\omega} \in$ (or $\subseteq$) $\omega_P] \to 1$.

Proof
For any $\omega \in \mathcal{S}_m$, the weak law of large numbers (Theorem 12.2) guarantees that

$$\overline{d}(\omega) \to \delta(\omega) \text{ in probability.} \tag{2.34}$$

Since there is only a finite number of ω's, it can be shown (Exercise 2.8) that

$$P[\max_{\omega \in \omega_P} \overline{d}(\omega) < \min_{\omega \notin \omega_P} \overline{d}(\omega) - \epsilon] \to 1, \tag{2.35}$$

where

$$2 \cdot \epsilon = \min_{\omega \notin \omega_P} \delta(\omega) - \delta(\omega_P) > 0. \tag{2.36}$$

Thus the probability that the minimizer(s) of $\overline{d}$ are in ω_P approaches 1. If ω_P is unique, then asymptotically so is $\widehat{\omega}$. □

If n and/or m is large, depending on the distance used, it may be computationally infeasible to perform a complete search for $\widehat{\omega}$, but in general there does not appear to be a one-step method for finding $\widehat{\omega}$. Instead of an exhaustive search, it is usually enough to start with a reasonable guess and then try nearby vectors obtained by interchanging two of the ranks of the original. For the Spearman distance Theorem 2.2 below implies that the center is the rank vector corresponding to $\overline{y}$, that is, $\widehat{\omega} = rank(\overline{y})$, as defined next.

Definition 2.3
For vector $z \in \Re^m$ such that $z_i \neq z_j$ if $i \neq j$, define $rank(z)$ to be the vector in $\mathcal{S}_m$ whose elements are in the same order as z: $rank(z)_i = j$ if and only if z_i is the jth smallest of $z_1, \ldots, z_m$. □

Table 2.4 *Sports: centers.*

	With Jogging		Without Jogging	
Distance ↓	**Ordering**	**100 · t**	**Ordering**	**100 · t**
Spearman	CKSBTFJ	19	CKBSTF	9
Footrule	BCSKTFJ	13	CSBKTF	7
Kendall	CSKBTFJ	14	CSKBTF	8
Hamming	BCFKTSJ	7	BSTKCF	7
Cayley	FBKTCSJ	10	SCTKBF	10
Ulam	STKBFCJ	11	CTKBFS	9
Maximum	KBFTSCJ	13	CSTKBF	13

Theorem 2.2
The maximum of $\omega'\overline{y}$ over $\omega \in \mathcal{S}_m$ is obtained by any rank vector $\widehat{\omega}$ satisfying

$$(\widehat{\omega}_i - \widehat{\omega}_j)(\overline{y}_i - \overline{y}_j) \geq 0 \tag{2.37}$$

for every i, j. If the $\overline{y}_i$'s are distinct, then the unique maximizer is $\widehat{\omega} = rank(\overline{y})$. □

Proof
Suppose $\omega \in \mathcal{S}_m$ violates (2.37), so that for some i, j, $(\omega_i - \omega_j)(\overline{y}_i - \overline{y}_j) < 0$. Let $\omega^{(0)} \in \mathcal{S}_m$ be the same as ω but with i and j interchanged: $\omega_k^{(0)} = \omega_k$ for $k \neq i, j$, $\omega_i^{(0)} = \omega_j$ and $\omega_j^{(0)} = \omega_i$. Then

$$\omega^{(0)'}\overline{y} - \omega'\overline{y} = (\omega_j - \omega_i)(\overline{y}_i - \overline{y}_j) > 0. \tag{2.38}$$

Thus any rank vector violating (2.37) can be strictly improved. Exercise 2.9 then shows that if two rankings satisfy (2.37), they yield the same $\overline{d}_{\text{Spear}}$. □

Hays (1960) has a clever procedure for searching locally for the center using Kendall's distance. In fact, if there is a consensus ranking as in Example 2.8, then that ranking is $\widehat{\omega}$ for Kendall's distance. See Exercise 2.10.

Example 2.11. Sports: Centers
Refer again to Example 2.5. What is the center of the data, i.e., what is the overall ranking of the sports? Table 2.4 exhibits modal orderings from (2.13) for seven distances, and their associated t's of (2.21). There are a number of interesting aspects to the results.

Except for Jogging, in the full data each Sport has a wide range of ranks: B: 1, 2, 4; F: 1, 3, 5, 6; K: 1, 2, 3, 4; T: 2, 4, 5; C: 1, 2, 5, 6; S: 1, 2, 3, 5, 6. Even restricting to d_p (not shown) we find orderings as diverse as SCKBTFJ, BCSKTFJ, CSKBTFJ, CKSBTFJ, and CKTBSFJ. Spearman, Footrule, and Kendall are fairly similar, the other four are quite dissimilar. Maximum and Cayley are interesting in that they rank the team sports 1, 2, 3.

When comparing the full orderings to those without Jogging, we see that Spearman, Kendall, and Footrule do not change much, but the others do. In fact, Cayley turns the order completely around, and Maximum switches the team sports from 1, 2, 3 to 4, 5, 6. The coefficients for different distances are hard to compare, but the comparisons within each distance between the cases with and without Jogging are meaningful. It is interesting that the first three distances show a drastic decrease in t when Jogging is dropped, showing much less clustering, whereas in the others, the t's stay fairly constant, suggesting that the latter are somehow more robust. □

Because of the discreteness of $\widehat{\omega}$, it is difficult to find standard errors for the $\widehat{\omega}_i$'s, or a confidence set for ω_P. Fligner and Verducci (1990) use a Bayesian approach to find ω's with high posterior probabilities based on a particular model, which we present in Section 6.4. A conservative frequentist approach is found in Remark 3.3.

2.5.3 Estimating spread, location known

If ω_P is known, it is straightforward to estimate δ or τ using $\overline{d}$ or t. These estimators are both unbiased, by definition. The variance of $d(Y, \omega_P)$ can be estimated by $\widehat{\sigma}^2$, the sample variance of the $d(y^{(i)}, \omega_P)$'s, and the standard errors of $\overline{d}$ and t are $\widehat{\sigma}/\sqrt{n}$ and $\widehat{\sigma}/(\sqrt{n} \cdot \delta_0)$, respectively.

Example 2.12. Ballots: Confidence Region for τ

Brook and Upton (1974) and Upton and Brook (1975) consider the effect on voting of the order in which candidates appear on the ballot. They collected results from a number of British elections in which the candidates are not well known; hence ballot order could reveal its effect. The data consist of ballots of three candidates, where the voters rank the candidates from 1 to 3. Only ballots for which all candidates were in the same party were used. The presumption is that the modal ranking is $\omega = (1, 2, 3)$, so that the higher a candidate is on the ballot, the higher the rank is likely to be.

Using Spearman, we calculate $\overline{d}_{\text{Spear}} = 3.521$ and $\widehat{\sigma}^2 = 8.492$. Since $n = 948$, an approximate 95% confidence interval for δ_{Spear} is (3.33, 3.71). Now $\delta_0 = 4$, so that $t_{\text{Spear}} = 0.120$, $se(t_{\text{Spear}}) =$ 0.0237, and the confidence interval is (0.073, 0.166). Thus the coefficient, Spearman's ρ, is quite small, but the interval is narrow enough to be sure the τ_{Spear} is positive.

Reworking the analysis with Hamming, we have $\overline{d}_{\text{Ham}} = 1.790$ with confidence interval (1.72, 1.86), and $t_{\text{Ham}} = 0.105$ with confidence interval (0.070, 0.140). Both distances give similar results, although the two t's are not necessarily comparable. □

2.5.4 Estimating spread, location unknown

When location is estimated, the natural impulse is to replace ω_P with $\widehat{\omega}$, and proceed by estimating the spread as above. We call this method "naive." If n is large, this approach is fine since with high probability, $\widehat{\omega} = \omega_P$. With small n, it is likely that the spread will be underestimated; hence the coefficient t would be overestimated. We use the bootstrap (Efron, 1982) to adjust the t. Define the sample vector $N = \{N_y | y \in \mathcal{S}_m\}$ of counts and the sample Probability vector $\widehat{P}$ to be the $m! \times 1$ vectors (indexed as P in 2.14) given by

$$N_y = \#\{y^{(i)} \mid y^{(i)} = y\} \text{ and } \widehat{P} = \frac{N}{n}. \tag{2.39}$$

Thus for the sample of Black Women in Example 2.1, $N' = (1, 1, 0, 0, 5, 6)$. Note that $\widehat{P}$ is a probability vector itself, $\widehat{P} \in \mathcal{P}_{m!}$ of (2.16). The idea of the bootstrap is to estimate the distribution of $\overline{d}(\widehat{\omega}) - \overline{d}(\omega_P)$ under P by the distribution of $\overline{d}(\widehat{\widehat{\omega}}) - \overline{d}(\widehat{\omega})$ under $\widehat{P}$, where the $\widehat{\widehat{\omega}}$ is the center estimated from a sample from $\widehat{P}$. The precise steps to follow are given next.

1. Find $\widehat{\omega}$ and $\overline{d}(\widehat{\omega})$ as in (2.13).
2. For $b = 1, ..., B$ (B is a large number chosen by the statistician),

   a. Randomly choose n values, $y^{(1)*}, ..., y^{(n)*}$, from the data set $\{y^{(1)}, ..., y^{(n)}\}$ *with* replacement.

b. Calculate $x_b^* = (\overline{d}_b^*(\widehat{\omega}_b^*) - \overline{d}(\widehat{\omega}))$, where $\overline{d}_b^*(\omega) = \sum d(y^{(i)*}, \omega)/\ n$ and $\widehat{\omega}_b^*$ is the minimizer of $\overline{d}_b^*(\omega)$.

3. Use $x_1^*, ..., x_B^*$ to estimate the distribution of $\overline{d}(\widehat{\omega}) - \delta(\omega_P)$.

The bootstrap estimate of $\delta(\omega_P)$ is found by solving $\overline{d}(\widehat{\omega}) - \widehat{\delta(\omega_P)} = \overline{x}^* \equiv \sum_{b=1}^{B} x_b^*/B$, which yields the estimate

$$\widehat{\delta(\omega_P)} = \overline{d}(\widehat{\omega}) - \overline{x}^* = 2 \cdot \overline{d}(\widehat{\omega}) - \overline{d}^*(\widehat{\omega}_b^*). \tag{2.40}$$

If $\overline{d}(\widehat{\omega})$ is an underestimate of $\delta(\omega_P)$, then we expect $\overline{d}^*$ to be smaller than $\overline{d}(\widehat{\omega})$. Equation (2.40) takes the estimate $\overline{d}(\widehat{\omega})$ and increases it a little. An estimate of the variance of $\overline{d}(\widehat{\omega})$ is $\sum(x_b^*)^2/B$.

To estimate a confidence interval, we need to estimate the quantities a_l and a_u that satisfy $P[a_l < \overline{d}(\widehat{\omega}) - \delta(\omega_P) < a_u] \approx 1 - \alpha$. The bootstrap estimates are $\widehat{a}_l$ and $\widehat{a}_u$ that satisfy $\#\{\widehat{a}_l < x_b^* < \widehat{a}_u\}/B \approx 1 - \alpha$. Typically one takes a_l and a_u so that there are equal areas in the tail, which suggests that a_l should be the $(\alpha/2)^{th}$ sample quantile, and a_u the $(1 - \alpha/2)^{th}$ sample quantile, of the sample of x_b^*'s. To obtain one-sided confidence intervals (or rays), set $a_l = -\infty$ and a_u to be the $(1 - \alpha)^{th}$ quantile for an upper interval, and set $a_u = \infty$ and a_l to be the αth quantile for a lower interval. The bootstrap estimate of the confidence interval is then

$$\{\delta \mid \widehat{a}_l < \overline{d}(\widehat{\omega}) - \delta < \widehat{a}_u\} = (\overline{d}(\widehat{\omega}) - \widehat{a}_u, \overline{d}(\widehat{\omega}) - \widehat{a}_l). \tag{2.41}$$

Example 2.13. Sports: Estimating τ

For the Sports data in Examples 2.5 and 2.11, we find $\widehat{\omega} = (4, 6, 2, 5, 1, 3, 7)$ using Spearman's distance, which corresponds to $\overline{d}(\widehat{\omega}) = 45.4923$. Using the direct method, as in Section 2.5.3 with $\widehat{\omega}$ in place of ω_P, we estimate the standard error of $\overline{d}$ to be 1.7765, and find the naive normal-theory 90% confidence interval for δ then to be $\overline{d} \pm 1.645 \cdot se(\overline{d})$, or (42.5700, 48.4146). For the bootstrap, we took $B = 1000$ samples from the data, and found the x_b^*'s. The average, $\overline{x}^*$, is -1.1854, suggesting a reasonably large bias in the naive method. To find $\widehat{\delta(\omega_P)}$ of (2.40), we subtract -1.1854 from 45.4923 to obtain 46.6777. For a 90% confidence interval for δ, we find the 5th and 95th percentiles of the x_b^*'s, which are $\widehat{a}_l = -3.7692$ and $\widehat{a}_u = 1.3924$. Using (2.41), we have the estimated confidence region $(45.4923 - 1.3924, 45.4923 + 3.7692) = (44.0999,$

49.2615). Thus the bootstrap shifts the confidence region higher, and also induces some asymmetry.

Now we use (2.21) and (2.22) to transform to the t's. From Table 3.3, we have $\delta_0 = m(m^2 - 1)/6 = 56$. The naive estimate of τ is $t(\widehat{\omega}) = 0.1876$, whereas the bootstrap estimate is $1 - \delta(\widehat{\omega_P})/\delta_0 = 1 - 46.6777/56 = 0.1665$. Thus the naive estimate substantially overestimates the strength of τ. The naive and bootstrapped 90% confidence intervals for τ are (0.1355, 0.2398) and (0.1203, 0.2125), again exhibiting the downward shift. □

An alternative method for estimating spread is to use *concordance* measures as in Alvo, Cabilio, and Feigin (1982). Estimation of the center is avoided by defining the spread to be $E_P(d(X, Y))$, where X and Y are independent, both distributed according to P. Then the spread is estimated by the average of the observed $d(y^{(i)}, y^{(j)})$'s. See Section 3.5 for more details.

2.5.5 *Clustering: L-centers*

As for any type of data, the center is not a very effective descriptive statistic if groups of observations are clustered about different values. Hartigan (1975) presents clustering of continuous data based on "L-means" procedure, whereby observations are grouped into L clusters by finding L means and assigning each observation to the group indexed by the closest mean. (Hartigan uses K instead of L.) The means themselves are estimated from the data to be those that minimize the sum of the distances of the observations to their group means. The analogous procedure for rank data is to find several rankings $\omega^{(1)}, ..., \omega^{(L)}$ about which the data cluster, so that we aim to minimize

$$D(\omega^{(1)}, ..., \omega^{(L)}) = \frac{1}{n} \sum_{i=1}^{n} \min_{1 \leq l \leq L} d(y^{(i)}, \omega^{(l)}). \tag{2.42}$$

For small values of n, m, and L, the minimizers in (2.42) can be found by exhaustive search. We will use Kendall's distance. For $m = 3$ or 4, exhaustive searches are feasible. For $m = 5$, we used exhaustive searches for $L = 2$ and 3. In all other situations, we randomly took a set of starting $\widehat{\omega}^{(l)}$'s and continually tried interchanging pairs of ranks until the D no longer decreased. We performed this operation several times (at least 10).

The vectors $\widehat{\omega}^{(l)}$ give an idea of where the judges cluster. The D indicates how tightly they cluster. Of course, D will decrease as the number of groups L increases. One way to adjust D to account for L is to calculate D for the uniform distribution, that is, find D_0^L defined by

$$D_0^L = \min_{\omega^{(1)},\dots,\omega^{(L)}} \frac{1}{m!} \sum_{y \in \mathcal{S}_m} \min_{1 \le l \le L} d(y, \omega^{(l)}). \tag{2.43}$$

Then a clustering analog of t is

$$t^L = 1 - D(\widehat{\omega}^{(1)}, \dots, \widehat{\omega}^{(L)})/D_0^L. \tag{2.44}$$

Again, it is not easy to find D_0^L for all but small numbers. For $L = 1$, it is δ_0. For $L = 2$ using d_{Ken}, it appears that any pair $x \in \mathcal{S}_m$ and x^c, where x^c is the reverse ranking

$$x^c = (m+1)1_m - x, \tag{2.45}$$

minimizes (2.42). In this case, letting $j = \delta_{\text{Ken},0}$, $D^2_{\text{Ken},0} = 2 \sum_{i<j} i \cdot c(i) + j \cdot c(j)$, where $c(i) = P[d(e_m, Y) = i]$ for $Y \sim Uniform(\mathcal{S}_m)$. In Section 3.4.2, an iterative procedure is given for the probabilities, so that $D^2_{\text{Ken},0}$ can be obtained exactly.

Keep in mind that the presentation here is descriptive rather than inferential. It is helpful to think of the clustering in the examples below as describing the data as it resides on a permutation polytope. Chapter 10 details a more formal, model-based approach.

Example 2.14. Living Places: Clustering

A question on the National Opinion Research Center Amalgam Survey of 1972 asked people to rank three types of areas in which to live: City over 50,000, Suburb (within 30 miles of a City), and Country (everywhere else). Table 2.5 shows the results, given in Duncan and Brody (1982), with respondents categorized by their current residence. See Thompson (1993a) for polytopes, and Diaconis (1988) for further discussion.

We are interested in clusters. Table 2.6 shows the estimated D^L_{Ken}'s of (2.42), the $D^L_{\text{Ken},0}$'s of (2.43), and t^L_{Ken} of (2.44), for the

Table 2.5 *Living Places: ranks of areas by current residence.*

Ranking (City, Suburb, Country)	Residence City	Suburb	Country	Total
(1, 2, 3)	210	22	10	242
(1, 3, 2)	23	4	1	28
(2, 1, 3)	111	45	14	170
(2, 3, 1)	8	4	0	12
(3, 1, 2)	204	299	125	628
(3, 2, 1)	81	126	152	359
Total	637	500	302	1439

Table 2.6 *Living Places: clusters.*

L	D^L_{Ken0}	City People D^L_{Ken}	City People t^L_{Ken}	Suburbanites D^L_{Ken}	Suburbanites t^L_{Ken}	Country Folk D^L_{Ken}	Country Folk t^L_{Ken}
1	3/2	1.014	0.324	0.470	0.687	0.613	0.591
2	2/3	0.363	0.456	0.202	0.697	0.119	0.822
3	1/2	0.188	0.624	0.068	0.864	0.040	0.920
4	1/3	0.049	0.853	0.016	0.952	0.003	0.991
5	1/6	0.013	0.922	0.008	0.952	0	1

three samples of people, and for $L = 1, ..., 5$. (Of course, $D^6 = 0$.) In all samples, the t's rise quite dramatically as the number of clusters increases. Except for the City People, there is little improvement going from four to five groups. What are the $\widehat{\omega}^{(l)}$'s? For all three samples, for $L = 4$, they are $(1,2,3)$, $(2,1,3)$, $(3,1,2)$, $(3,2,1)$. The objects are in the order City, Suburb, Country, so that for these vectors, Suburb is never ranked last, which conforms to the raw data. As we will see in Section 5.11, these vectors are consistent with an unfolding model: place the three objects, and all judges, on a line, and assign the ordering to each judge consistent

Table 2.7 *Living Places: percentages in L = 4 groups.*

Region	Ordering	City People	Suburbanites	Country Folk
1	(City, Sub, Ctry)	36.58%	5.20%	3.64%
2	(Sub, City, Ctry)	17.43%	9.00%	4.64%
3	(Sub, Ctry, City)	32.03%	59.80%	41.39%
4	(Ctry, Sub, City)	13.97%	26.00%	50.33%

with the order of the distances to the objects:

$$\overbrace{........City......}^{R_1}|\overbrace{........Sub}^{R_2}\overbrace{urb........}^{R_3}|\overbrace{......Country..........}^{R_4} \qquad (2.46)$$

There are four regions: judges in R_1 like the City best, then the Suburbs, and finally the Country. Those in R_2 order them Suburb, City, Country. Of course, not all vectors can be placed on this scale. Assigning each ranking to the closest region, or splitting evenly any point equidistant from two, we have the percentages assigned to each region ≡ ordering in Table 2.7. The Suburbanites and Country Folk are overwhelmingly in the right half of (2.46), i.e., they dislike the City. Suburbanites prefer the Suburbs; Country Folk the Country. City People are more City-oriented, and prefer the Suburbs to the Country, though 46% do not like the City. □

Example 2.15. Voting: Clustering
Diaconis (1988, 1989) has an extensive analysis of the voting for the 1980 election for president of the American Psychological Association (APA). Voters were asked to rank the following five candidates: A = William Bevan; B = Ira Iscoe; C = Charles Kiesler; D = Max Siegle; and E = Logan Wright. There are definite camps within the APA, and among the candidates. Candidates A and C are research psychologists, D and E are clinical psychologists, and B is a community psychologist. Diaconis' 1989 analysis reveals the split in the association, particularly by investigating the distribution of pairs of candidates receiving pairs of ranks. Most telling is the number of people who rank both A and C high, or both low, and similarly for D and E. What does clustering show?

For the $L = 1$ group, the best ordering is ACEDB, with an average distance of 4.700, and $t^1_{\text{Ken}} = 1 - 4.700/5 = 0.060$. With two groups the D^2_{Ken} is 2.837, $D^2_{\text{Ken},0} = 3.367$; hence $t^2_{\text{Ken}} = 0.157$. Although both coefficients are low, the two-group one is 2 1/2 times as large. The t^L_{Ken}'s for $L = 3$, 4, and 5 are, respectively, 0.14, 0.12, and 0.11. There does not seem much reason to go beyond two groups. The orderings, and estimated percentages, for the two groups are CABED (52.36%) and DEBAC (47.64%), exact reverse orderings. We clearly see the Researcher (AC)/Clinician (DE) split. The Researcher group appears slightly larger, which is compatible with the fact that C won the election. McCullagh (1993a) and Stern (1993) use models to reveal this same dichotomy. □

Example 2.16. Occupations: Clustering
Using the Occupation data in Example 2.4, we find the estimated t^L_{Ken}'s for $L = 1$, 2, 3, and 4 are 0.66, 0.68, 0.64, and 0.66. Thus there is a strong coefficient for one group, but more groups do not show any improvement. It is interesting that the Managerial/Technician split we saw in the polytop-o-gram in Example 2.4 is not evident here. Model based approaches in Examples 10.2 and 10.4 do reveal the split. □

Example 2.17. Sports: Clustering
For the Sports data in Example 2.5, the first four t^L_{Ken}'s are 0.14, 0.28, 0.25, and 0.25. As in the Voting data above, there is a substantial jump from one group to two, but no increase thereafter. The order for one group is (*Cycling, Swimming, Basketball, Baseball, Tennis, Football, Jogging*). For two groups, we have the orders (*Swimming, Cycling, Tennis, Basketball, Jogging, Baseball, Football*) with 53.46% and (*Baseball, Football, Basketball, Tennis, Cycling, Swimming, Jogging*) with 46.54%. There is a clear split between the people who prefer Team sports (Baseball, Football, and Basketball), and those who do not, with slightly more in the latter category. □

In Examples 2.15 and 2.17, the orders for $L = 2$ groups were close to, or exactly, reverses of each other. In these cases, the dichotomies seem real since the t^2_{Ken}'s are reasonably large, and the groups are easily interpreted. Beware, though, that if the data are uniform, then the best two-group fit has such a bipolar set of orderings. Thus it is important to refer to t before making any

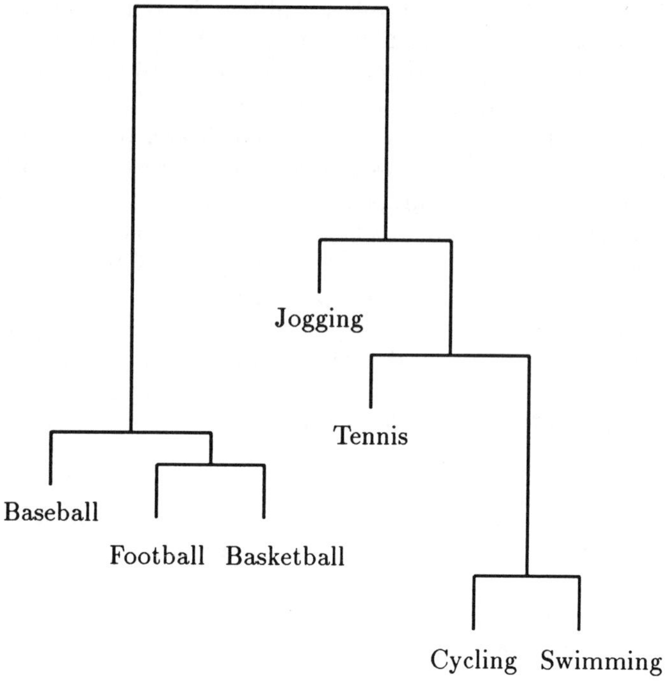

Figure 2.10 ***Sports: clustering the sports.***

strong conclusions. In Chapter 10 we look at mixture models, the model-based analog of clustering.

Example 2.18. Sports: Clustering of Objects
The projections of the polytopes reveal simultaneous clusters of judges and objects. This section considered clustering of judges. Also useful is clustering of the objects alone. We define a distance between two objects to be the squared Euclidean distance of the ranks assigned to them, i.e., the distance between Objects k and l is $(1/n)\sum_{i=1}^{n}(y_k^{(i)} - y_l^{(i)})^2$. Thus objects will be close if they are assigned similar ranks by the judges. Figure 2.10 shows a hierarchical clustering of the Sports. The idea is to start with each object in its own group, then at the first step join the two objects that are closest to each other, in this case Cycling and Swimming. Next, join the two groups that are closest, where the distance between groups of more than one object is the maximal distance between one object from each group. [This method is

called *complete linkage*. See Mardia, Kent, and Bibby (1979) for an excellent introduction.] At this step, Football and Basketball are joined. Third, Baseball joins the other two Team sports. Continue until the objects are in one large group. The vertical axis indicates the distance, groups being joined at the value of their distance. We see immediately the Team/Non-team split. Also, in the Non-team group, Cycling and Swimming are very close, then Tennis is a little bit away, and Jogging still farther. □

2.6 Linear subspaces

The summary statistics in the previous sections can be represented as sample averages of functions $z(y)$ of $y \in \mathcal{S}_m$, $\overline{z} = (1/n) \sum_{i=1}^{n} z(y^{(i)})$, where the range of z is a scalar (e.g., for distances), a vector (e.g., for Means), or a matrix (e.g., for Marginals or Pairs.) In this section we introduce an alternative representation of functions: vectors in $\Re^{m!}$. This representation is used to great effect by Alvo, Cabilio, and Feigin (1982). The Probabilities vector P in (2.14) is such a vector, where the indices run over $y \in \mathcal{S}_m$, and the value of the function at y is the value of the yth element in P, P_y. Any scalar function on $\mathcal{S}_m$ can be so represented. A function whose value is a $v \times 1$ vector is represented by an $m! \times v$ matrix X, where each column is a vector representing a scalar variable. Then the yth row contains the value of the function at y. For example, consider the function $z(y) = y$. For Object i, define the vector $x^{(i)} \in \Re^{m!}$ that picks out the rank of the object, so that $x_y^{(i)} = y_i$. Letting $X = (x^{(1)}, \ldots, x^{(m)})$, we have that the yth row is simply y'.

To connect the X notation and the $z(y)$ notation, we need the special indicator vectors ζ^y for $y \in \mathcal{S}_m$, where

$$\zeta_y^y = 1, \; \zeta_w^y = 0 \text{ for } w \neq y. \tag{2.47}$$

Thus it has a "1" in the yth slot and zeros elsewhere. Now if a matrix X and a vector $z(y)$ represent the same function, then

$$z(y) = X'\zeta^y. \tag{2.48}$$

Note that for a sample

$$N = \sum_{i=1}^{n} \zeta^{y^{(i)}} \text{ and } \widehat{P} = \overline{\zeta} \equiv \frac{1}{n} \sum_{i=1}^{n} \zeta^{y^{(i)}}; \tag{2.49}$$

hence

$$\sum_{i=1}^{n} z(y^{(i)}) = X'N \text{ and } \overline{z} = X'\widehat{P}. \tag{2.50}$$

The purpose of the vector representation is to allow us to look at the data $\widehat{P}$ as a vector in $\Re^{m!}$ and its projections on subspaces defined by the columns of X. We can then determine the effect of the particular functions on the data in much the same way that one determines effects of factors in analysis of variance, or of independent variables in linear regression. Thus we can see how much of the nonuniformity in the $\widehat{P}$ is due to the differences in the average ranks, say, as well as see where differences in $\widehat{P}$ lie after adjusting for difference in $\overline{y}$.

First, we remind the reader of some geometry and linear algebra. For a set of vectors $x^{(1)}, \ldots, x^{(v)}$ in $\Re^K$ (where for us, $K = m!$), the *Span* consists of the set of all linear combinations:

$$Span\{x^{(1)}, \ldots, x^{(v)}\} = \{\sum_{l=1}^{v} a_l x^{(l)} \mid a_1, \ldots, a_v \in \Re\}. \tag{2.51}$$

Any *subspace* is a span of a set of vectors, and vice versa. A *basis* of a subspace S is a set of linearly independent vectors that span S, where $x^{(1)}, \ldots, x^{(v)}$ are linearly independent if for no $x^{(i)}$ do there exist constants a_j, $j \neq i$, such that $x^{(i)} = \sum_{j \neq i} a_j x^{(j)}$. That is, there are no redundant vectors. There are many bases for a given subspace, but they all have the same number of elements. That number is the *dimension* of S, $dim(S)$.

Our objective is to measure the effect of the observed data due to a given subspace. The effect of any subspace is measured by the projection of the vector $\widehat{P}$ onto that subspace. For subspace $S \subset \Re^K$ and $x \in \Re^K$, the *projection of x on S* is the $\widehat{x} \in S$ that is closest to x: $||x - \widehat{x}||^2 \leq ||x - x^*||^2$, for all $x^* \in \mathcal{S}$. In fact, $\widehat{x}$ is unique. If x is already in S, then $\widehat{x} = x$. Least squares linear regression provides a way to perceive these projections. Let $X =$

Table 2.8 *Living Places: projections on Means subspace.*

			City People		Suburbanites		Country Folk	
$x^{(1)}$	$x^{(2)}$	$x^{(3)}$	$\hat{N}^{(j)}$	$\hat{R}^{(j)}$	$\hat{N}^{(j)}$	$\hat{R}^{(j)}$	$\hat{N}^{(j)}$	$\hat{R}^{(j)}$
1	2	3	136.17	73.83	6.33	15.67	−15.33	25.33
1	3	2	50.17	−27.17	−39.17	43.17	−17.00	18.00
2	1	3	192.17	−81.17	128.83	−83.83	52.00	−38.00
2	3	1	20.17	−12.17	37.83	−33.83	48.67	−48.67
3	1	2	162.17	41.83	205.83	93.17	117.67	7.33
3	2	1	76.17	4.83	160.33	−34.33	116.00	36.00

$(x^{(1)}, \ldots, x^{(v)})$, where $x^{(1)}, \ldots, x^{(v)}$ form a basis for S. Consider the linear regression equation $x = X\beta + r$, where β is $v \times 1$ and r is $v \times 1$. The least squares estimate of β is that which minimizes $||r||^2$, $\widehat{\beta} = (X'X)^{-1}X'x$, and the estimated fit of x is the projection $\widehat{x} = X\widehat{\beta}$. Thus $\mathcal{P}^S \equiv X(X'X)^{-1}X'$ is the $K \times K$ *projection matrix*:

$$\mathcal{P}^S x = \widehat{x}, \ x \in \Re^K. \tag{2.52}$$

Example 2.19. Living Places: Means Subspace
Table 2.8 shows the data vectors $N^{(j)}$ for $j \in \{City, Suburb, Country\}$, from Example 2.14 projected onto the Means subspace defined as $S_1 \equiv Span\{x^{(1)}, x^{(2)}, x^{(3)}\}$. In parentheses are the residuals $R^{(j)} = N^{(j)} - \widehat{N}^{(j)}$.

How can one understand the projections? Mallows' suggestion (Diaconis, 1989) is to find the inner product of the projections with the original vectors $x^{(i)}$ of interest, i.e., calculate the statistics using the projections of N rather than N itself. The inner products of the projections with the vectors $x^{(i)}$ are easy to interpret: they are the $n\overline{y}_i$'s. In fact, $X'N = X'\widehat{N}$. The added information in the table over simply the average rankings is contained in the residuals. These residuals are fairly large, meaning that there is quite a bit of information in the data beyond that captured by the means alone. □

By using several subspaces, one can decompose the information contained in the total sample, and assess the relative importance of the subspaces. Especially important is the difference between nested subspaces, where S_1 is nested within S_2 means $S_1 \subset S_2$. For example, the Means subspace is nested within the Pairs subspace, and within the Marginals subspace. See Exercises 2.11 and 2.12. By comparing projections, we can determine how much more the Pairs or Marginals explain than the Means, for example.

The Marginals subspace starts with m^2 vectors:

$$x_y^{(ij)} = 1 \text{ if } y_i = j, \ 0 \text{ otherwise }, \ 1 \leq i,j \leq m. \tag{2.53}$$

Thus the ij^{th} vector indicates whether Object i has rank j. Place the vectors in X with the first index running fastest, i.e.,

$$X = (x^{(11)}, x^{(21)}, \ldots, x^{(m1)}, x^{(12)}, \ldots, x^{(m2)}, \ldots, x^{(1m)}, \ldots, x^{(mm)}). \tag{2.54}$$

Then the corresponding $\overline{z}$ is the Marginals matrix $\widehat{M}$ strung out by column, i.e.,

$$X'\widehat{P} = \underline{\widehat{M}}, \tag{2.55}$$

where for any $r \times c$ matrix A,

$$\underline{A} = (a_{11}, a_{21}, \ldots, a_{r1}, \ldots, a_{1c}, \ldots, a_{rc}). \tag{2.56}$$

See (12.8). For the Pairs matrix (2.12), we define $\binom{m}{2}$ vectors

$$x_y^{(ij)} = I[y_i < y_j], 1 \leq i < j \leq m, \tag{2.57}$$

which indicates whether Object i is preferred to Object j. Let

$$X = (x^{(12)}, x^{(13)}, \ldots, x^{(1m)}, x^{(23)}, \ldots, x^{(2m)}, \ldots, x^{(m-1,m)}), \tag{2.58}$$

and for an individual ranking, set

$$\kappa(y) = X'\zeta^y, \tag{2.59}$$

the vector whose ij^{th} element is $I[y_i < y_j]$, $i < j$. Then

$$X'\widehat{P} = \overline{\kappa} = \widehat{K}^*, \tag{2.60}$$

where for any $r \times r$ matrix A,

$$A^* = (A_{12}, A_{13}, \ldots, A_{1r}, A_{23}, \ldots, A_{2r}, \ldots, A_{r-1,r}). \tag{2.61}$$

We now consider *orthogonal complements.* For any subspace S, the orthogonal complement is the set of all vectors in $\Re^K$ orthogonal to S:

$$S^\perp = \{x \mid x'w = 0 \text{ for all } w \in S\}. \tag{2.62}$$

The orthogonal complement is a subspace itself; hence x can be projected onto it. In fact, $I_K - \mathcal{P}^S$ is the projection matrix, and the residual vector $\widehat{r} = x - \widehat{x}$ is the projection. Thus any vector can be written as a sum of orthogonal vectors, one from S and one from $S^\perp$. These vectors are unique. We say $\Re^K$ is the direct sum of the subspaces:

$$\Re^K = S \oplus S^\perp. \tag{2.63}$$

Note that $dim(S) + dim(S^\perp) = dim(\Re^K) = K$, and the squared lengths of the projections sum: $||x||^2 = ||\widehat{x}||^2 + ||x - \widehat{x}||^2$. (Exercise 2.13.)

Now suppose S_0 and S_1 are two subspaces of $\Re^K$, and $S_0 \subset S_1$. Then the orthogonal complement of S_0 within S_1 is

$$S_{1\cdot 0} = S_1 \cap S_0^\perp. \tag{2.64}$$

As in (2.63), we have

$$S_1 = S_0 \oplus S_{1\cdot 0}, \; dim(S_1) = dim(S_0) + dim(S_{1\cdot 0}), \\ \text{and } ||x^{(1)}||^2 = ||x^{(0)}||^2 + ||x^{(1\cdot 0)}||^2, \tag{2.65}$$

where $x^{(a)}$ is the projection of x on S_a. The projection on S_1 of any vector can then be decomposed into the effect of S_0, and the effect of S_1 after removing the effect of S_0. Similarly, if we have a set of nested subspaces, $S_0 \subset \cdots \subset S_L \equiv \Re^K$, we can break down z into projections on the orthogonal complements of S_l in S_{l+1}:

$$\Re^K = S_0 \oplus S_{1\cdot 0} \oplus \cdots \oplus S_{L\cdot L-1}, \tag{2.66}$$

Table 2.9 *Living Places: nested projection.*

Subspace	Dim	City People SS/n	City People $100 \cdot R^2$	Suburbanites SS/n	Suburbanites $100 \cdot R^2$	Country Folk SS/n	Country Folk $100 \cdot R^2$
S_0	1	106.17		83.33		50.33	
$S_{1\cdot 0}$	2	35.89	60.87	92.02	69.55	58.60	74.28
$S_{2\cdot 1}$	2	11.86	20.12	32.78	24.78	19.77	25.00
$S_{3\cdot 2}$	1	11.21	19.01	7.50	5.67	0.57	0.72

SS = squared length of projection.

where the projections, their squared lengths, and the dimensions again sum.

Example 2.20. Living Places: Nested Projections
Continuing Example 2.19, we wish to decompose the data into interesting pieces. Start with an uninteresting subspace:

$$S_0 = Span\{1_{m!}\}, \tag{2.67}$$

which is the equiangular line. It is uninteresting since its inner product with any $\widehat{P}$ is 1, or with $\widehat{N}$ is the sample size. Note that $S_0 \subset S_1$, where S_1 is the Means subspace, since $x^{(1)} + \cdots + x^{(m)} = \binom{m+1}{2} \cdot 1_{m!}$. Then $S_{1\cdot 0}$ is the subspace of the means after taking out the constant. Exercise 2.14 shows that $S_{1\cdot 0} = Span\{x^{(i)} - c_m \mid 1 \leq i \leq m\}$.

Let S_2 be the Marginals subspace. Exercise 2.11 shows that $S_1 \subset S_2$. Finally, letting $S_3 = \Re^{m!} = \Re^6$, we have (2.66) with $L = 3$. For each data set in Table 2.5, we found the projections onto the four subspaces S_0, $S_{1\cdot 0}$, $S_{2\cdot 1}$, $S_{3\cdot 2}$. The squared lengths of the projections measure the effect of the projections; the larger the length the more information about the data in the subspace. Table 2.9 contains the squared lengths divided by the sample size. In addition, we find R^2, which is the ratio of the squared length to the sum of the squared lengths ignoring the constant subspace.

Look at the R^2's for $S_{1\cdot 0}$. They show that 61% to 74% of the differences in the observed counts is due to the differences in the average rankings of the objects. About 20–25% more is contained in the Marginals. Except for the City, there is little left unexplained by the Marginals.

The squared lengths suggest which subspaces to look at, or, as in Example 2.19, which inner products of the projections with the vectors $x^{(i)}$ to look at. In these projections the effect of the lower-ordered subspaces is conveniently removed, so that only the extra information attributed to the current subspace will be revealed. For example, consider the Country Folk. The Means subspace used the $x^{(i)}$'s in Table 2.8. Projecting $\widehat{P}$ yields $X'\widehat{P}^{(1\cdot 0)} = \overline{y} - c_m$, which in this case is $(0.88, -0.46, -0.42)'$. Thus as we saw in Example 2.19, the Country Folk dislike the city, and like the Country a bit more than the Suburbs. What about $S_{2\cdot 1}$? Now we find the $x^{(ij)'}\widehat{P}^{(2\cdot 1)}$'s from (2.53), which amounts to calculating $\widehat{M}$ of (2.10) using the projection on $S_{2\cdot 1}$. The result is

$$100 \times \widehat{M}^{(2\cdot 1)} = \begin{array}{l} \\ \text{City} \\ \text{Suburb} \\ \text{Country} \end{array} \begin{array}{c} \begin{array}{ccc} \quad 1 & \quad 2 & \quad 3 \end{array} \\ \left(\begin{array}{rrr} 14.35 & -28.70 & 14.35 \\ -10.15 & 20.31 & -10.15 \\ -4.19 & 8.39 & -4.19 \end{array} \right) \end{array}. \tag{2.68}$$

Although the City is disliked on average, there are more people who place it first or last than expected given the average rankings. Similarly, the Suburbs are more likely ranked second than expected, as is the Country but more mildly. Thus there is more to know than just the Means. □

In the next two sections we present systematic decompositions of $\Re^{m!}$ into interesting orthogonal subspaces. Section 2.6.1 discusses the spectral analysis developed by Diaconis (1988, 1989). The subspaces generalize the Marginals subspace by looking at higher-order Marginals, e.g., the number of judges giving Objects i and j ranks k and l, or in general giving sets of h objects sets of h ranks. Section 2.6.2 similarly generalizes the Pairs matrix to triples, quadruples, etc. Fundamental to this extension are the *inversions* presented by McCullagh (1993a). Other decompositions include ones based on *nested orthogonal contrasts*, which will appear in Section 5.10 and Chapter 7.

2.6.1 *Spectral decomposition*

We will give a brief introduction to Diaconis (1988, 1989), which develops a spectral theory for functions on ranked data. The theory rests on deep results from group representation theory. Here, we will try to give an idea of its use for rank statistics. Some of the discussion will be beyond the intended level of this book, but the general implications should be understandable.

In Example 2.7, the Marginals measured how many people gave each object each rank. These are first-order statistics. Second-order statistics have two types. Ordered pairs count the number of observations for which Object i has rank k and Object j has rank l. Unordered pairs count how many observations have Objects i and j ranked k and l, in either order. Third-order Marginals can be totally ordered, totally unordered, or in between. An example of an in-between ordering is Objects i and j are ranked k and l in either order, and Object a is ranked b. Fourth- and higher-order Marginals become even more complicated. There is a Marginal for each partition of m.

Definition 2.4
a) A ***partition*** of the integer λ_+ is a vector $\lambda = (\lambda_1, \ldots, \lambda_K)$ of positive integers that sum to λ_+ and are in nonincreasing order: $\lambda_1 \geq \lambda_2 \geq \cdots \geq \lambda_K$. b) An ***ordered*** partition lifts the ordering restriction. □

For any set Ω and λ a (ordered) partition with $\lambda_+ \leq \#\Omega$, let $\lambda(\Omega)$ denote the set of all ordered sets of disjoint subsets Ω with sizes given by λ. That is,

$$\begin{aligned}\lambda(\Omega) = \{(\mathcal{O}_1, \ldots, \mathcal{O}_K) \mid \mathcal{O}_k \subset \Omega,\ \mathcal{O}_k \cap \mathcal{O}_l = \emptyset \ if\ k \neq l, \\ \text{and } \#\mathcal{O}_k = \lambda_k,\ k = 1, \ldots, K\}.\end{aligned} \tag{2.69}$$

For example, if $\Omega = \{A, B, C, D, E, F\}$ and $\lambda = (2, 2, 1, 1)$, then $\lambda(\Omega)$ contains $\binom{6}{2,2,1,1} = 180$ sets of subsets, including $(\{A, B\}$, $\{C, D\}$, $\{E\}$, $\{F\})$ and $(\{C, F\}$, $\{A, E\}$, $\{B\}$, $\{D\})$.

For given λ with $\lambda_+ = m$, $C = (\mathcal{O}_1, \ldots, \mathcal{O}_K) \in \lambda(\mathcal{O})$, and $y \in \mathcal{S}_m$, define

$$C(y) = (y[\mathcal{O}_1], \ldots, y[\mathcal{O}_K]) \in \lambda(\{1, \ldots, \lambda_+\}), \tag{2.70}$$

where for any subset $\mathcal{O}^* \in \mathcal{O}$,

$$y[\mathcal{O}^*] = \{y_i \mid i \in \mathcal{O}^*\}. \tag{2.71}$$

Thus $C(y)$ groups the rankings according to the subsets in C. [In Section 5.10 we will extend the definition of $C(y)$ to cases where $\lambda_+ < m$.]

Definition 2.5
For λ a partition of m, the λ-Marginals for a set of rankings is the matrix $\widehat{M}^\lambda$ given by

$$\widehat{M}^\lambda_{C,\Gamma} = \frac{1}{n}\#\{y^{(k)} \mid C(y^{(k)}) = \Gamma\}, \tag{2.72}$$

where C ranges over $\lambda(\mathcal{O})$ and Γ ranges over $\lambda(\{1,\ldots,m\})$. □

The Marginals (2.10) use $\lambda = (m-1,1)$, i.e.,

$$\widehat{M}_{ij} = \widehat{M}^{(m-1,1)}_{C,\Gamma} \tag{2.73}$$

for $C = (\mathcal{O} - \{i\}, \{i\})$ and $\Gamma = (\{1,\ldots,m\} - \{j\}, \{j\})$, which matches the ith object with rank j. The ordered-pair Marginals have $\lambda = (m-2,1,1)$, and the unordered pairs have $\lambda = (m-2,2)$.

There is typically a great deal of redundancy in the Marginal matrices. For example, if $\lambda = (2,1,1)$ then the values for (C,Γ) $= ((\{A,B\}, \{C\}, \{D\}), (\{2,4\}, \{1\}, \{3\}))$ and $(C,\Gamma) = ((\{A,B\}, \{D\}, \{C\}), (\{2,4\}, \{3\}, \{1\}))$ are the same, being the proportion of rankings that place C first and D third. Also, there is substantial overlap in the information the matrices possess. If A and B are the most popular, then we will see the evidence in all the matrices. Thus the raw Marginals are not of direct interest. As in analysis of variance when one decomposes effects into main effects, two-way interactions, three-way interactions, etc., we would like to separate the rankings of pairs from those of individual objects.

The spectral analysis in Diaconis (1988, 1989) is exactly what we need. The observed distribution can be decomposed into effects due to the various types of Marginals, the more complicated Marginals being adjusted for the less complicated ones. Our subspaces are indexed by the partitions λ of m. For $C \in \lambda(\Omega)$ and $\Gamma \in \lambda(\{1,\ldots,m\})$, let $x^{(C,\Gamma)}$ be the indicator vector

$$x^{(C,\Gamma)}_y = I[C(y) = \Gamma]. \tag{2.74}$$

Then define the subspace

$$U^\lambda = Span\{x^{(C,\Gamma)} \mid C \in \lambda(\Omega),\ \Gamma \in \lambda(\{1,\ldots,m\})\}. \tag{2.75}$$

Many of these subspaces are contained in others, and in fact $U^{(1,1,\ldots,1)} = \Re^{m!}$, so that all are contained in that one. The goal is to find the parts of each subspace that are special to it so as to isolate the Marginals effect after taking account of the lower-order effects. First, we need to know when λ is of a higher order than μ, i.e., when is $U^\mu \subset U^\lambda$? The 0th-order subspace, $U^m = S_0$ of (2.67), is contained in all the others. The Marginals subspace spanned by the vectors in (2.53), i.e., $U^{(m-1,1)}(= S_1)$ is contained in all but U^m. The unordered pairs Marginals $U^{(m-2,2)}$ is contained in the ordered pairs $U^{(m-2,1,1)}$. A systematic way to decide is to use the *majorization* partial ordering on the partitions of m. If λ and μ are both partitions of m, then μ majorizes λ, written $\lambda \leq_M \mu$, if $\lambda_1 \leq \mu_1$, $\lambda_1+\lambda_2 \leq \mu_1+\mu_2$, $\ldots$, $\lambda_1+\cdots+\lambda_{m-1} \leq \mu_1+\cdots+\mu_{m-1}$, where components of the partitions that do not exist are assumed to be zero. Note that the sum up to the mth component is m for all partitions. See Marshall and Olkin (1979) for a complete treatment of majorization. If any of the inequalities is strict, then μ strictly majorizes λ, written $\lambda <_M \mu$. For example, $(2,2,2,1) <_M (3,2,2)$. The implication for the subspaces is given by *Young's Rule*:

$$U^\mu \subset U^\lambda \text{ if and only if } \lambda \leq_M \mu. \tag{2.76}$$

See Diaconis (1988, 1989).

The U^λ's are not completely nested unless $m \leq 4$, so that we cannot obtain the decomposition (2.66) directly. However, we can find a decomposition just as good. For each λ, we find the orthogonal complement in U^λ of all subspaces strictly contained in U^λ:

$$V^\lambda = U^\lambda \cap Span\{\cup_{\lambda <_M \mu} U^\mu\}^\perp. \tag{2.77}$$

The amazing fact is that the V^λ's provide an orthogonal decomposition of $\Re^{m!}$, i.e.,

$$\Re^{m!} = \oplus_\lambda V^\lambda, \tag{2.78}$$

where V^λ and V^μ are orthogonal if $\lambda \neq \mu$. One consequence of (2.78) is that the dimensions sum: $dim(\Re^{m!}) = m! = \sum_\lambda dim(V^\lambda)$.

Table 2.10 *Political Goals: spectral decomposition.*

	Projections on subspace λ				
	(4)	(3, 1)	(2, 2)	(2, 1, 1)	(1, 1, 1, 1)
SS/n	94.25	64.03	31.78	1.04	0.01
$100 \cdot R^2$		66.11	32.81	1.08	0.01
Dim	1	9	4	9	1
$100 \cdot R^2/Dim$		7.35	8.20	0.12	0.01

The squared lengths of the vectors similarly sum: $\|x\|^2 = \sum_\lambda \|x^\lambda\|^2$, where x^λ is the projection of x on V^λ.

The V^λ's and their dimensions can be found directly by iteratively regressing the smaller subspaces out of the larger. For small m, it is not too difficult. We illustrate with $m = 4$.

Example 2.21. Political Goals: Spectral decomposition
Recall the data in Example 2.3. The partitions of $m = 4$ are completely ordered:

$$(4) >_M (3,1) >_M (2,2) >_M (2,1,1) >_M (1,1,1,1). \tag{2.79}$$

The dimension of each U^λ can be found by starting with the $x^{(C,\Gamma)}$'s in (2.74) reduced to a basis. The dimensions are, in the order (2.79), 1, 10, 14, 23, and 24. The first and last should be obvious since $U^{(4)}$ is the constant subspace and $U^{(1,1,1,1)} = \Re^{m!}$. Linear regression can be used to find the fits, using the basis vectors as the columns in the X matrices. The projections on and dimensions of the reduced subspaces V^λ are then found by subtraction, e.g., the projection on $V^{(2,2)}$ is that on $U^{(2,2)}$ minus that on $U^{(3,1)}$. Table 2.10 contains the results.

The *Dim* row gives the dimensions of the subspaces, and the final row divides the R^2 by the dimensions, to see the information per dimension. It seems clear that the vast bulk of the information is in the $(3, 1)$ and $(2, 2)$ subspaces, that is, the first-order Marginal, and the unordered pairs Marginal. The other two subspaces do not show much at all. Table 2.11 contains the unordered pairs matrix. There are only three rows displayed since knowing the ranks assigned to two of the objects automatically implies the pair

Table 2.11 *Political Goals: unordered pairs projection.*

	Rank Pairs					
Object Pairs ↓	$\{1,2\}$	$\{1,3\}$	$\{1,4\}$	$\{2,3\}$	$\{2,4\}$	$\{3,4\}$
$\{Order, Say\}$	−175.5	73.0	102.5	102.5	73.0	−175.5
$\{Order, Prices\}$	303.5	−105.5	−198.0	−198.0	−105.5	303.5
$\{Order, Speech\}$	−128.0	32.5	95.5	95.5	32.5	−128.0

of ranks assigned to the other two. We see large positive values where the Materialist goals are 1 and 2, in some order, but also where the Post-materialist goals are 1 and 2. □

The direct approach in the example is fine for $m = 4$, but it becomes quite challenging computationally even for $m = 5$. Again, the spectral theory comes to the rescue. Diaconis (1988, 1989) gives a formula for finding the projections that do not require any regression techniques. One does need *characters* of the subspaces, a topic into which this book is not prepared to go. There is a neat formula for the dimension, based on the *Hook Length Formula.*

First, we represent each λ with its *Young Diagram*, which is a set of left-justified boxes arranged so that there are λ_1 in the first row, λ_2 in the second row, etc. The diagrams for $\lambda = (5,2,1)$ and $\lambda = (3,3,1,1)$ are shown below. Ignore for an instant the numbers in the boxes.

$\lambda = (5,2,1)$: [7][5][3][2][1]
[3][1]
[1]

$\mu = (3,3,1,1)$: [6][3][2]
[5][2][1]
[2]
[1]

Now think about the numbers. For each box in a diagram, we fill in its "hook length," which is 1 plus the number of boxes to the right in the same row plus the number of boxes below in the same column. The dimension of a subspace is a function of the product of the hook lengths:

$$dim(V^{\lambda}) = \left(\frac{m!}{\prod(Hook\ Lengths)}\right)^2. \qquad (2.80)$$

Table 2.12 *Sports: unordered pairs.* T = *Team sport,* NT = *Non-team sport.*

	Sum of Ranks					
Object Pairs ↓	**3**	**4**	**5**	**6**	**7**	**8**
$\{T,T\}$	16.2	11.0	7.8	−12.8	−13.6	−22.6
$\{T,NT\}$	−32.4	−22.0	−15.6	25.6	27.2	45.2
$\{NT,NT\}$	16.2	11.0	7.8	−12.8	−13.6	−22.6
	Sum of Ranks					
Object Pairs ↓	**9**	**10**	**11**	**12**	**13**	
$\{T,T\}$	−17.0	−0.6	11.8	7.0	12.8	
$\{T,NT\}$	34.0	1.2	−23.6	−14.0	−25.6	
$\{NT,NT\}$	−17.0	−0.6	11.8	7.0	12.8	

For λ and ν, these dimensions are $(8!/(7 \cdot 5 \cdot 3 \cdot 2 \cdot 3))^2 = (128)^2$ and $(8!/(6 \cdot 2 \cdot 3 \cdot 5 \cdot 2 \cdot 2))^2 = (56)^2$.

Example 2.22. Sports: Spectral Decomposition
The Sports data in Example 2.5 has $m = 7$, which yields 15 subspaces. The R^2/Dim values for the one-way Marginals, $\lambda = (6,1)$, and the unordered pairs Marginals, $\lambda = (5,2)$, are 4.35 and 3.30, respectively. All others are between 1.67 and 2.17, except for the last which is only 0.06. Thus the information appears in the first two subspaces. The unordered pair Marginal matrix is 21×21, so is a bit difficult to understand. However, we can see generally positive numbers when two Team or two Non-team sports are both ranked high, or both ranked low. Thus we reduce the table by adding together rows according to whether the pair of sports are both Team, both Non-team, or one of each. We also reduce the columns by summing columns if the sum of their assigned ranks are equal. The result is Table 2.12, which clearly reveals the Team/Non-team split. □

Diaconis (1988, 1989) exhibits an extensive analysis of the Voting data (Example 2.15) using the spectral decomposition. He shows that certain types of tied rankings can just as easily be handled using this approach.

2.6.2 Inversions

Just as Section 2.6.1 generalized the Marginals matrix (2.10), here we describe McCullagh's (1993a) generalization of the Pairs matrix (2.12) using *inversions*. The Pairs matrix gives pairwise information about the objects. One could also imagine triples, that is, for each triple i, j, k of objects, find the proportion of rankings that gave those three each possible relative order:

$$\widehat{K}_{(i,j,k),r} = \frac{1}{n}\#\{y^{(l)} \mid rank(y_i^{(l)}, y_j^{(l)}, y_k^{(l)}) = r\}, r \in \mathcal{S}_3. \quad (2.81)$$

See Definition 2.3. It may be that there is interaction among some sets of three objects not accounted for by the pairwise comparisons. Similarly, relative rankings of sets of four, five, ..., up to m, objects may contain information. As for the Marginals, what we really need is the effect of triples of objects after taking out the effects of the pairs, and the effect of quadruples after taking out the effects of triples, etc. To that end, we create a subspace for each set of h-tuples of objects. These subspaces are nested, so that it is easy to define the corresponding orthogonal complements. The hth subspace, $h = 1, \ldots, m$, is

$$W_h^* = Span\{x^{(s,r)} \mid s \subset \{1, \ldots, m\}, \#s = h, r \in \mathcal{S}_h\}, \quad (2.82)$$

where $x^{(s,r)}$ is the indicator vector

$$x_y^{(s,r)} = I[rank(y_{i_1}, \ldots, y_{i_h}) = r], s = \{i_1, \ldots, i_h\}, i_1 < \cdots < i_h. \quad (2.83)$$

We then define the orthogonal subspaces by $W_h = W^*_{h \cdot h-1}$, so that

$$\Re^{m!} = W_1 \oplus W_2 \oplus \cdots \oplus W_m. \quad (2.84)$$

The equality holds since $W_m^* = \Re^{m!}$. We are taking $W_1 = S_0$.

Again, we ask, "How do we find the projections?" and "What are the dimensions of the subspace?" For small m, we can do it directly by creating the vectors (2.83) and using regression as in Example 2.21. For any m, we appeal to McCullagh's (1993a) inversions. A two-object inversion for a ranking y occurs for objects i and j,

$i < j$, if $y_i > y_j$. Thus the elements below the diagonal in the Pairs matrix (2.12) contain the proportion of inversions for the pairs of objects. A three-object inversion for objects $i < j < k$ occurs if y_i, y_j, and y_k are all in the wrong spot relative to each other, that is, if $rank(y_i, y_j, y_k) = (2,3,1)$ or $(3,1,2)$. In general, an h-object inversion occurs for objects $i_1 < \cdots < i_h$ if none of the objects are in the correct relative place, i.e., if for $r = rank(y_{i_1}, \ldots, y_{i_h})$, $r \in \mathcal{I}_h$ where

$$\mathcal{I}_h = \{r \in \mathcal{S}_h \mid r_k \neq k, k = 1, \ldots, h\}. \tag{2.85}$$

Consider the subspaces of the W_h^*'s obtained by restricting to inversions,

$$T_h^* = Span\{x^{(s,r)} \in W_h^* \mid r \in \mathcal{I}_h\}, \tag{2.86}$$

and $T_1^* = W_1^*$. It can be shown (see McCullagh, 1993a, or Equation 3.53) that the number of vectors that generate T_h^* is

$$\iota_h = \binom{m}{h} \#\mathcal{I}_h = \binom{m}{h} h! \sum_{i=2}^{h} \frac{(-1)^i}{i!} \tag{2.87}$$

for $h > 1$, and $\iota_1 = 1$. McCullagh (1993a) conjectures that in fact the vectors are linearly independent, so that $dim(T_h^*) = \iota_h$. The conjecture holds for $m \leq 6$ by brute-force computing, and I strongly suspect holds in general. If it is true, then we have that

$$W_h^* = Span\{x^{(s,r)} \mid s \subset \{1, \ldots, m\}, \#s = k, r \in \mathcal{I}_k, k \leq h\}, \tag{2.88}$$

and $dim(W_h) = \iota_h$. Examples of inversions appear in Section 7.11 on models.

Both the spectral decomposition and inversion decomposition have potential to reveal the important structure in rank data. If m is too large, however, the computations become difficult, at least for the higher-order effects. To go beyond $m = 7$ for the spectral decomposition, or $m = 6$ for the inversions, appear unreasonable given my particular computing environment. As computers become more powerful and programmers more clever, these numbers will increase, but there will still be moderately small upper bounds ($m = 10$?, $m = 15$?).

2.7 Exercises

2.1. Show that for any rank vector $y \in \mathcal{S}_m$, $\sum_{i=1}^m y_i = m(m+1)/2$, $\overline{y} = (m+1)/2$, and $\sum_{i=1}^m (y_i - \overline{y})^2 = m(m^2-1)/12 = v_m$ of (2.3). Show that, as a consequence, if $U \sim Uniform\{1, \ldots, m\}$, then $E(U) = (m+1)/2$ and $Var(U) = (m^2-1)/12$.

2.2. Prove (2.5). [Find a and b so that $z_i^{(i)}(c) = a$ and $z_j^{(i)}(c) = b$ for $j \neq i$. Then $\|y - z^{(i)}(c)\|^2 = \sum_{j=1}^m (y_j - b)^2 + (y_i - a)^2 - (y_i - b)^2$. The summation depends only on c and m (not y), and expanding the final two terms yields (2.5).]

2.3. Verify (2.9). [Note that for vectors a and b, $(a'b)^2 = a'bb'a$.]

2.4. From Section 12.1.1, write $W = \sum_{i=1}^m \lambda_i \gamma^{(i)} \gamma^{(i)'}$, where $\lambda_1 \geq \lambda_2 \geq \cdots \geq \lambda_m$ are the eigenvalues of W, and the $\gamma^{(i)}$'s are the eigenvectors. Then for any vector g, $g'Wg = \sum_{i=1}^m (g'\gamma^{(i)})^2 \lambda_i$. (a) Show that since g has norm 1, and the $\gamma^{(i)}$'s are the eigenvectors, $\sum_{i=1}^m (g'\gamma^{(i)})^2 = 1$. Thus $g'Wg$ is a weighted average of the λ_i's, which is maximized when all the weight is at the maximum. Show that hence the best g is $g^{(1)} = \gamma^{(1)}$, which has $ss(g) = \lambda_1$. (b) Now restrict to g's orthogonal to $\gamma^{(1)}$. Show that the best of these is $\gamma^{(2)}$.

2.5. Show that the projection of y on the ith pole, i.e., $(y'x^{(i)} / \|x^{(i)}\|^2)\, x^{(i)}$, equals $a^{(i)} + b^{(i)} y_i$ for some vectors $a^{(i)}$ and $b^{(i)}$.

2.6. Show that $\sum_{i=1}^n \|y^{(i)} - z\|^2$ is minimized over z by $z = \overline{y}$. [Minimize each term of the form $\sum_{i=1}^n (y_j^{(i)} - z_j)^2$ separately.]

2.7. Show that $t_{\text{Spear}}(\omega)$ for Spearman's distance equals the average of the correlations between ω and the $y^{(i)}$'s, where the correlation can be written $(\omega - c_m)'(y^{(i)} - c_m)/v_m$ for c_m in (2.2) and v_m in (2.3).

2.8. Prove (2.35). [Choose any $\omega_0 \notin \omega_P$ and $\omega_1 \in \omega_P$. Then the weak law of large numbers, Theorem 12.2, shows that $\overline{d}(\omega_0) - \overline{d}(\omega_1) \to \delta(\omega_0) - \delta(\omega_P) \geq 2\epsilon > 0$. That is, letting $A(\omega_0, \omega_1)$ be the event $\{\overline{d}(\omega_0) - \overline{d}(\omega_1) > \epsilon\}$, we have that $P[A(\omega_0, \omega_1)] \to 1$. Thus $P[\cap A(\omega_0, \omega_1)] \to 1$, where the intersection is over all $\omega_0 \notin \omega_P$ and $\omega_1 \in \omega_P$, since there is only a finite number of such pairs. Verify that that limit implies (2.35).]

2.9. Show that if two rank vectors ω and $\omega^{(0)}$ both satisfy (2.37) in Theorem 2.2, then $\overline{d}_{\text{Spear}}(\omega) = \overline{d}_{\text{Spear}}(\omega^{(0)})$. [Let $a_1 < \cdots < a_K$

be the distinct values of the $\overline{y}_i$'s, and $\mathcal{O}_k = \{i \mid \overline{y}_i = k\}$. Then $\overline{y}'\omega = \sum_{k=1}^{K} a_k \sum_{j \in \mathcal{O}_k} \omega_j$. Show that (2.37) holding fixes the sets $\{i \mid \omega_i \in \mathcal{O}_k\}$ for $k = 1, \ldots, K$.]

2.10. Show that for any rank vector ω, $\overline{d}_{\text{Ken}}(\omega)$ in (2.29) is greater than or equal to $\sum\sum_{1 \leq i < j \leq m} Min\{\widehat{K}_{ij}, 1 - \widehat{K}_{ij}\}$, and that if there is a consensus ranking $\widehat{\omega}$, then $\overline{d}_{\text{Ken}}(\widehat{\omega})$ achieves that lower bound.

2.11. Show that the Means subspace, spanned by the vectors $x^{(1)}, \ldots, x^{(m)}$, where $x_y^{(i)} = y_i$, is contained in the Marginals subspace spanned by the vectors in (2.54). [Consider the vectors $\sum_{j=1}^{m} j x^{(ij)}$.]

2.12. Show that the Means subspace is contained in the Pairs subspace spanned by the vectors in (2.58). [Consider the vectors $\sum_{j \neq i} x^{(ji)}$, where $x^{(ij)} = 1_{m!} - x^{(ji)}$.]

2.13. Show that if $\mathcal{P}$ is a projection matrix, then for any x, $\|x\|^2 = \|\mathcal{P}x\|^2 + \|(I - \mathcal{P})x\|^2$. [Write $x = \mathcal{P}x + (I - \mathcal{P})x$, then expand $x'x$.]

2.14. Let S_1 be the Means subspace as in Exercise 2.11, and S_0 be as in (2.67). Show that $S_1 \perp S_0 = Span\{x^{(i)} - c_m | 1 \leq i \leq m\}$. [If $x \in S_1$, then $x = \sum_{i=1}^{m} a_i x^{(i)}$ for some a_i's. Show that if $x \perp S_0$, then it must be that $\sum a_i = 0$; hence $x = \sum a_i (x^{(i)} - c_m)$.]

CHAPTER 3

Formal Tests of Uniformity

3.1 Introduction

The uniform distribution (2.18) is central to the study of rank data, whether testing for uniformity or assessing the lack thereof. For example, in Chapter 2 the coefficients t of (2.21) and (2.44) are defined relative to the uniform: $Uniform \Rightarrow t = 0$. This chapter presents a variety of methods for testing the null hypothesis that the distribution on $\mathcal{S}_m$ is uniform:

$$H_0 : P = Uniform(\mathcal{S}_m), \tag{3.1}$$

based on the sample $Y^{(1)}, \ldots, Y^{(n)}$ as in (2.17). These are one-sample tests. Two- or many-sample tests, which determine whether several samples have the same (not necessarily uniform) distribution, appear in Chapter 4.

Nonuniformity can be measured in any number of ways. The statistics presented in Chapter 2 are all candidates. The most general statistic is the estimated Probabilities vector $\widehat{P}$ (2.39): if it is far from the constant vector $u_{m!}$ of (2.18), then we would reject uniformity. In most cases, especially if m is not small, more focused measures are indicated. Popular examples are the Means (2.11), Marginals (2.10), and Pairs (2.12). These statistics are special cases of the projections on subspaces as in Section 2.6. None (except $\widehat{P}$) is guaranteed to reveal every type of nonuniformity. For example, if the sample divides into two roughly equal groups who cluster about two diametrically opposed rankings, then the Means vector $\overline{y}$ will be close to the uniform c_m, and the Pairs' $\widehat{K}_{ij}$'s will be close to one half, neither showing the structure. If there is an *a priori* modal ranking ω of interest, then $t(\omega)$ of (2.21) is a natural statistic for testing uniformity. See Section 3.4, where we consider the distribution of the d's under uniformity.

If ω is unknown, then one can use a confidence interval technique as in Section 2.5.4 to see whether a τ of 0 is tenable. A better approach is to work directly with *diversity* or *concordance*, that is, measure how close the rankings in the sample are to each other by averaging the inter-point distances $d(y^{(i)}, y^{(j)})$. See Section 3.5.

3.2 Summary statistics

Start with v functions on $\mathcal{S}_m$ that are expected to capture interesting nonuniformities. As in Alvo, Cabilio, and Feigin (1982), and Section 2.6, these functions are represented as vectors $x^{(i)}$ in $\Re^{m!}$, and the set of these vectors is assembled in a matrix $X = (x^{(1)}, \ldots, x^{(v)})$. As discussed in Section 2.6, the constant vector is uninteresting since the elements of $\widehat{P}$ always sum to 1. Thus we will make the assumption throughout this section that the vectors are orthogonal to the constant vector, $X'1_{m!} = 0$. Equivalently,

$$S \equiv Span\{x^{(1)}, \ldots, x^{(v)}\} \perp S_0 \qquad (3.2)$$

of (2.67). Letting $\widehat{P}^S$ be the projection of $\widehat{P}$ onto S, note that the closer $\widehat{P}$ is to the uniform, the closer $\widehat{P}^S$ is to the zero vector. (Exercise 3.1.) Thus a natural statistic to consider is the squared length of the projection, $||\widehat{P}^S||^2$. Fortunately, the statistic has a convenient asymptotic distribution.

Theorem 3.1
Suppose (3.2) holds. Then as $n \to \infty$, under uniformity (3.1), $T_n^2 = nm!||\widehat{P}^S||^2 \to \chi^2_{dim(S)}$. □

The proof is given in Section 3.2.1. Thus, if n is large enough, an approximate level α test of (3.1) rejects H_0 when $T_n^2 > \chi^2_{dim(S),\alpha}$. (Here, $\chi^2_{\nu,\alpha}$ is the upper α^{th} point of a χ^2_ν, that is, $\alpha = P[\chi^2_\nu > \chi^2_{\nu,\alpha}]$.) In principle, the projection for any S can be calculated as in Section 2.6, although when m is large the vectors may become unwieldy. Fortunately, for the popular statistics there are shortcuts. In particular, we deal directly with the $\overline{z} = X'\widehat{P}$'s of (2.50). The test statistics and their asymptotic distributions under uniformity are given in Table 3.1, where

$$J_k = 1_k 1_k', \ \overline{J}_k = \frac{1}{k} J_k, \ \overline{k} = \binom{k}{2} = \frac{k(k-1)}{2}, \qquad (3.3)$$

Table 3.1 *Basic statistics for testing uniformity. Limiting distributions under uniformity are* χ^2_{df}.

Statistic	T_n^2	df
Probabilities	$nm!\|\|\widehat{P} - u_{m!}\|\|^2$	$m! - 1$
Marginals	$n(m-1)\|\|\widehat{M} - \overline{J}_m\|\|^2$	$(m-1)^2$
Means	$\frac{12n}{m(m+1)}\|\|\overline{y} - c_m\|\|^2$	$m - 1$
Pairs	$12n(\|\|\widehat{K}^* - \frac{1}{2}1_{\overline{m}}\|\|^2 - \frac{1}{m+1}\|\|\overline{y} - c_m\|\|^2)$	$\overline{m}$

$\widehat{K}^*$ is given in (2.60), and $||A||^2$ for a matrix is the sum of squares of all its elements. See also (2.2) and (2.18). The Marginals test was developed by Anderson (1959). The Means test statistic is most famous as Friedman's (1937) nonparametric statistic for testing effects in a randomized block experiment. Kendall (Kendall and Gibbons, 1990) also developed it as a measure of concordance. It will reappear in Section 3.5. The details of these test statistics fill out this section.

3.2.1 Projections

We start with a definition and lemma for distributions of quadratic forms.

Definition 3.1
For any $k \times k$ matrix Λ, let $\chi^2[\Lambda]$ denote the distribution of $\sum_{i=1}^k \lambda_i X_i^2$, where the λ_i's are the eigenvalues (Section 12.1.1) of Λ, and $X_1^2, \ldots, X_k^2$ are independent χ_1^2 random variables. □

The notation in Definition 3.1 is nonstandard, and especially should not be confused with that for noncentral chi-squares. The next result follows from Lemma 12.6 and Theorem 12.3.

Lemma 3.1
Suppose $W_n \longrightarrow N_k(0, \Sigma_0)$. *Then for symmetric* $k \times k$ *matrix* A, $W_n' A W_n \longrightarrow \chi^2[A\Sigma_0]$. □

We want to apply Lemma 3.1 to $\widehat{P}^S$. Note that (2.49) shows that $\widehat{P}$ is the sample average of independent and identically distributed random vectors $\zeta^{Y^{(i)}}$, so that we can use the central limit theorem

12.1 to find its asymptotic distribution. The next result helps us find $Cov(\zeta^Y)$.

Lemma 3.2
Suppose Z is a $k \times 1$ random vector, where $P[Z = (0, \ldots, 0, 1, 0, \ldots, 0)'] = 1/k$ no matter which of the k elements receives the "1." Then $E(Z) = u_k$ and $Cov(Z) = (1/k)H_k$, where

$$H_k = I_k - \overline{J}_k \tag{3.4}$$

for $\overline{J}_k$ in (3.3). □

Proof
Since each element of Z is distributed $Bernoulli(1/k)$, $E(Z_i) = 1/k$, giving the result for the expected value, and $Var(Z_i) = (1/k)(1 - 1/k)$. Note that by symmetry, the covariances are all equal. Since the elements of Z always sum to 1, $Var(\sum_{i=1}^{k} Z_i) = 0$, but $Var(\sum_{i=1}^{k} Z_i) = k\ Var(Z_1) + k(k-1)\ Cov(Z_1, Z_2)$; hence $Cov(Z_i, Z_j) = -1/k^2$ for $i \neq j$. The covariance result then follows. □

Remark 3.1
The matrix H_k in (3.4) is the ubiquitous *centering* matrix. For any $k \times 1$ vector w, $H_k w$ is the vector with the average of its elements subtracted from each element. In fact, H_k is the projection matrix that projects elements in $\Re^k$ to the orthogonal complement of S_0, i.e., to the set of vectors whose elements sum to 0.

Note also that if D is an $r \times c$ matrix of observations in a two-way analysis of variance, then $H_r D H_c$ is the corresponding matrix of residuals from the model with row and column effects but no interaction. □

Lemma 3.2 and the central limit theorem 12.1 imply that

$$\sqrt{n}(\widehat{P} - u_{m!}) \longrightarrow N_{m!}(0, \frac{1}{m!} H_{m!}) \tag{3.5}$$

under uniformity. From (3.5) we find the distribution of the projection $\widehat{P}^S$. Let $\mathcal{P}^S$ be the corresponding projection matrix (2.52), so that $\widehat{P}^S = \mathcal{P}^S\widehat{P}$. Multiplying both sides of (3.5) by $\mathcal{P}^S$ yields

$$\sqrt{n}\widehat{P}^S \longrightarrow N_{m!}(0, \frac{1}{m!}\mathcal{P}^S) \tag{3.6}$$

since (3.2) implies that $\mathcal{P}^S u_{m!} = 0$ and $\mathcal{P}^S H_{m!} = \mathcal{P}^S$, and $\mathcal{P}^S$ is *idempotent*, where a square matrix B is idempotent if $BB = B$. (See Section 12.1.1 and Exercises 3.1 and 3.2.)

Lemma 3.3
If $W_n \to N_v(0, B)$, *with* B *idempotent, then* $||W_n||^2 \to \chi^2_{\text{tr}(B)}$. □

Proof
By Lemma 3.1, $||W_n||^2 \to \chi^2[B]$, and by Lemma 12.2, B has eigenvalues $\lambda_1 = \cdots = \lambda_{\text{tr}(B)} = 1$ and the rest 0. Since the sum of independent χ^2's is χ^2 with degrees of freedom being the sum of those from the summands, we have tr(B) χ^2_1's yields one $\chi^2_{\text{tr}(B)}$. □

Applying Lemma 3.3 to (3.6) proves Theorem 3.1. The special cases follow.

3.2.2 Probabilities

Here we take S to be the orthogonal complement of S_0 in $\Re^{m!}$, so that $H_{m!}$ is the projection matrix. It is easy to show that $tr(H_{m!}) = m! - 1$ (Exercise 3.3), and

$$\widehat{P}^S = H_{m!}\widehat{P} = \widehat{P} - u_{m!}. \tag{3.7}$$

Thus (3.5) for Probabilities follows from Theorem 3.1.

This test can also be approached from a categorical point of view. See any basic book on categorical data, e.g., Agresti (1990) or Bishop, Fienberg, and Holland (1975). The individual y's in $\mathcal{S}_m$ are the categories, and the counts are N_y in (2.39). The null hypothesis sets the expected count in each category to $n/m!$. Then the Pearson χ^2 test statistic is

$$X^2 = \sum_{y \in \mathcal{S}_m} \frac{(N_y - n/m!)^2}{n/m!}, \tag{3.8}$$

and the likelihood ratio statistic is

$$LRS = 2 \cdot \sum_{y \in \mathcal{S}_m} N_y log(N_y/(n/m!)). \tag{3.9}$$

Both are asymptotically $\chi^2_{m!-1}$ under the null. In fact, X^2 is identical to T_n^2 for the Probabilities in Table 3.1.

3.2.3 Marginals

Let X be the matrix as in (2.54) but with columns being $x^{(ij)} - (1/m)1_{m!}$ from (2.53), the extra term needed so that $X'1_{m!} = 0$. Define $\underline{Q}^{(y)}$ via

$$X'\zeta^{(y)} = \underline{Q}^{(y)} - \frac{1}{m}1_{m!}. \tag{3.10}$$

Then $X'\widehat{P} = \overline{\underline{Q}} - (1/m)1_{m!}$, so that we need the mean and covariance of $\underline{Q}^{(Y)}$.

At this point it becomes more convenient to work with permutation matrices instead of vectors. A permutation matrix has exactly one 1 in each row and column, and zeros elsewhere. Denote the set of all $m \times m$ such matrices by $\mathcal{Q}_m$. If $Q \in \mathcal{Q}_m$, then Qw for any vector w permutes the elements of w. The sets $\mathcal{S}_m$ and $\mathcal{Q}_m$ are in one-to-one correspondence through the equation

$$Q^{(y)}e_m = y,\ Q^{(y)} \in \mathcal{Q}_m,\ y \in \mathcal{S}_m, \tag{3.11}$$

i.e., $Q_{ij} = 1$ if and only if $y_i = j$. For example, for $y' = (3, 4, 2, 1)$,

$$Q^{(y)}e_m = \begin{pmatrix} 0 & 0 & 1 & 0 \\ 0 & 0 & 0 & 1 \\ 0 & 1 & 0 & 0 \\ 1 & 0 & 0 & 0 \end{pmatrix} \begin{pmatrix} 1 \\ 2 \\ 3 \\ 4 \end{pmatrix} = \begin{pmatrix} 3 \\ 4 \\ 2 \\ 1 \end{pmatrix} = y. \tag{3.12}$$

If $Y \sim Uniform(\mathcal{S}_m)$ then $Q^{(Y)} \sim Uniform(\mathcal{Q}_m)$. Now $\underline{Q}^{(y)}$ as defined in (2.56) for $A = Q^{(y)}$ coincides with that in (3.10).

Remark 3.2
The set $\mathcal{Q}_m$ of permutation matrices is a group under regular matrix multiplication. A group $\mathcal{G}$ is a set of objects, in this case

$m \times m$ matrices, for which there is defined a multiplication, such that the set is closed under multiplication: $g, h \in \mathcal{G}$ implies $gh \in \mathcal{G}$; there exists an identity element i such that $gi = ig = g$ for all $g \in \mathcal{G}$; and for each $g \in \mathcal{G}$ there is an inverse $g^{-1} \in \mathcal{G}$ such that $gg^{-1} = g^{-1}g = i$. See, e.g., Herstein (1964). For $\mathcal{Q}_m$, the identity is I_m and $Q^{-1} = Q'$. □

It is not too hard to see that the Marginals matrix (2.10) is the average of the sample matrices:

$$\widehat{M} = \frac{1}{n}\sum_{i=1}^{n} Q^{(y^{(i)})} = \overline{Q}. \tag{3.13}$$

Thus the moments of $\widehat{M}$ under uniformity can be obtained from those of $Q \sim Uniform(\mathcal{Q}_m)$. The mean of a matrix of random variables is defined to be the matrix of the component means, as in Section 12.2. The covariance matrix is defined to be the covariance matrix for the vector obtained by stringing the columns together, i.e., $Cov(Q) \equiv Cov(\underline{Q})$.

Since each element $Q_{ij} \sim Bernoulli(1/m)$, $E(Q_{ij}) = 1/m$ and $Var(Q_{ij}) = (m-1)/m^2$. Covariances depend on cases. The distribution of a given row conforms to the Z in Lemma 3.2 with $k = m$; hence the covariance of any row is $(1/m)H_m$, as is the covariance of any given column. In Exercise 3.4 we use a trick as in the proof of Lemma 3.2 to show that $Cov(Q_{ij}, Q_{kl}) = 1/(m^2(m-1))$ if ij and kl are in different rows and columns, i.e., $i \neq k$ and $j \neq l$. Writing out the complete $m^2 \times m^2$ covariance matrix for $\underline{Q}$ will reveal certain patterns that constitute a "Kronecker product." See Section 12.2.2 for more details.

Definition 3.2
Given matrices A, $r \times s$ and B, $u \times v$, $A \otimes B$ denotes the $ru \times sv$ matrix that, when partitioned into a block matrix, has ij^{th} block $b_{ij}A$:

$$A \otimes B = \begin{pmatrix} b_{11}A & b_{12}A & \cdots & b_{1v}A \\ b_{21}A & b_{22}A & \cdots & b_{2v}A \\ \vdots & \vdots & \ddots & \vdots \\ b_{u1}A & b_{u2}A & \cdots & b_{uv}A \end{pmatrix}. \quad \square \tag{3.14}$$

The reason Kronecker products are so useful is that the standard matrix operations can be performed on the individual prodands:

$$\begin{aligned}(A \otimes B)(C \otimes D) &= AB \otimes CD;\\ (A \otimes B)' &= A' \otimes B';\\ (A \otimes B)^{-1} &= A^{-1} \otimes B^{-1},\end{aligned} \tag{3.15}$$

at least when the individual operations make sense.

The mean and covariance matrices for Q under the uniform can then be written

$$E(Q) = \overline{J}_m \text{ and } Cov(Q) = \frac{1}{m-1} H_m \otimes H_m. \tag{3.16}$$

See (3.3) and (3.4). The result for the covariance requires some manipulation. See Exercise 3.5. Collecting the results (3.13), (3.16), and the central limit theorem 12.1 gives

$$\sqrt{(m-1)n}(\widehat{M} - \overline{J}_m) \longrightarrow N_{m^2}(0, H_m \otimes H_m). \tag{3.17}$$

Now $H_m \otimes H_m$ is idempotent and $\text{tr}(H_m \otimes H_m) = (m-1)^2$. See Exercise 3.6. The Marginals result in Table 3.1 then follows from Lemma 3.3.

3.2.4 Means

The vectors of interest are $x^{(i)}$, where $x_y^{(i)} = y_i - (m+1)/2$, and the statistic $X'\widehat{P} = \overline{y} - c_m$. Equation (3.11) yields

$$E(Y) = E(Q^{(Y)} e_m) = \overline{J}_m e_m = c_m \tag{3.18}$$

since $1_m' e_m = m(m+1)/2$, and by Lemma 12.3,

$$\begin{aligned}Cov(Y) = Cov(Q^{(Y)} e_m) &= \frac{1}{m-1} H_m \otimes e_m' H_m e_m\\ &= \frac{m(m+1)}{12} H_m,\end{aligned} \tag{3.19}$$

since $e'_m H_m e_m = v_m$ of (2.3). (See Exercise 3.7.) By (3.11) and (3.13), $\overline{y} = \widehat{M} e_m$, and from (3.19),

$$\sqrt{\frac{12n}{m(m+1)}}(\overline{Y} - c_m) \longrightarrow N_m(0, H_m). \tag{3.20}$$

Hence we have the Means result in Table 3.1 by Lemma 3.3 since $tr(H_m) = m - 1$.

3.2.5 Pairs

The vectors now are $x^{(ij)} - (1/2)1_{m!}$ from (2.57), arranged as in (2.58). Thus

$$X'\widehat{P} = \widehat{K}^* - \frac{1}{2}1_{\overline{m}}. \tag{3.21}$$

See (2.60).

For $\kappa(Y)$ from (2.59), with $Y \sim Uniform(\mathcal{S}_m)$, we have that $\kappa_{ij}(Y) \sim Bernoulli(1/2)$; hence $E(\kappa(Y)) = 1/2$ and $Var(\kappa_{ij}(Y)) = 1/4$. The covariances $Cov(\kappa_{ij}(Y), \kappa_{kl}(Y))$ depend on cases. If i, j, k, l are all distinct, then they are independent. If $j \neq l$, then

$$\begin{aligned} Cov(\kappa_{ij}(Y), \kappa_{il}(Y)) &= E(I[Y_i < Y_j] \cdot I[Y_i < Y_l]) - 1/4 \\ &= P[Y_i < Y_j \text{ and } Y_i < Y_l] - 1/4 = 1/12, \end{aligned} \tag{3.22}$$

since the probability that Y_i is the smallest among Y_i, Y_j, and Y_l is 1/3. Similarly, if $i \neq k$, $Cov(\kappa_{ij}(Y), \kappa_{kj}(Y)) = 1/12$, and $Cov(\kappa_{ij}(Y), \kappa_{jk}(Y)) = -1/12$. A convenient matrix expression is

$$\Sigma_0 = Cov(\kappa(Y)) = \frac{1}{12}(I_{\overline{m}} + \Gamma_m \Gamma'_m), \tag{3.23}$$

where Γ_m is the $\overline{m} \times m$ matrix with rows indexed by ij, $1 \leq i < j \leq m$, in the same order as Z, and columns indexed by the objects 1, 2, ..., m, where the elements of Γ satisfy $g_{(ij)i} = +1$, $g_{(ij)j} = -1$, and $g_{(ij)k} = 0$ otherwise. For example, with $m = 4$,

$$\Gamma_4 = \begin{array}{c} \\ 12 \\ 13 \\ 14 \\ 23 \\ 24 \\ 34 \end{array} \begin{array}{c} \begin{array}{cccc} 1 & 2 & 3 & 4 \end{array} \\ \begin{pmatrix} 1 & -1 & 0 & 0 \\ 1 & 0 & -1 & 0 \\ 1 & 0 & 0 & -1 \\ 0 & 1 & -1 & 0 \\ 0 & 1 & 0 & -1 \\ 0 & 0 & 1 & -1 \end{pmatrix} \end{array}. \tag{3.24}$$

Thus we have that

$$W_n = \sqrt{n}(\widehat{K}^* - \frac{1}{2}1_{\overline{m}}) \longrightarrow N_{\overline{m}}(0, \Sigma_0). \tag{3.25}$$

Equation (3.23) makes it clear that Σ_0 is invertible; hence in Lemma 3.1, take $A = \Sigma_0^{-1}$. The next lemma helps to rewrite T_n^2. The proof is in Exercises 3.8 and 3.9.

Lemma 3.4
Take Γ_m as in (3.23). Then

$$\Gamma_m'(\kappa(y) - (1/2)1_{\overline{m}}) = -(y - c_m). \tag{3.26}$$

Also, for any $\alpha \neq 0$ and β,

$$(\alpha I_{\overline{m}} + \beta \Gamma_m \Gamma_m')^{-1} = a I_{\overline{m}} - b \Gamma_m \Gamma_m', \tag{3.27}$$

where $a = 1/\alpha$ and $b = \beta/(\alpha(\alpha + \beta m))$. □

The first expression shows that Γ_m takes deviations in $\widehat{K}^*$ to deviations in $\overline{y}$. Using (3.27) in (3.25) yields

$$\begin{aligned} T_n^2 = 12n(&||\widehat{K}^* - \frac{1}{2}1_{\overline{m}}||^2 \\ &- \frac{1}{m+1}(\widehat{K} - \frac{1}{2}1_{\overline{m}})'\Gamma_m\Gamma_m'(\widehat{K} - \frac{1}{2}1_{\overline{m}})), \end{aligned} \tag{3.28}$$

from which arises the result in Table 3.1 by (3.26). The statistic adjusts the deviations of the pairs with a little of those of the means. Since $A = \Sigma_0^{-1}$, the statistic (3.28) is asymptotically $\chi^2_{\overline{m}}$ under H_0. If we leave out the term with the mean, then we have Kendall's concordance statistic from Section 3.5.

Example 3.1: Tests of Uniformity
Table 3.2 collects the results of the tests applied to several data sets. The T_n^2's are the statistics in Table 3.1, "df" are the corresponding degrees of freedom, and the p is the p-value using the χ^2 approximation. The asterisks indicate there are not enough data for the χ^2 approximation to be meaningful. A "0" for the p-value indicates a value of less than 0.0001. For the Probabilities, we used the LRS from (3.9).

In the examples, most statistics are highly significant when the number of observations is large relative to $m!$, but inspecting the individual statistics can still be informative. Focus first on the Voting data. If we look at the statistics relative to their degrees of freedom, we see that the Marginals test seems to be most effective, and the Means test the least. The Marginals statistic is about 44% of the overall statistic, yet needs only 16 as opposed to 119 degrees of freedom. A similar result holds for the Occupations, although for these the Pairs test is quite competitive. The Gulf War data have only 21 observations, but we can see how much more significant the Pairs matrix is than using just the Means. For the Sports data, the Means test is not even significant. Both the Marginals and Pairs tests are quite good. We reanalyzed dropping Jogging, which many people dislike, and Jogging and Football. Significance drops sharply, suggesting that most of the nonuniformity is due to the general dislike of those sports. The Marginals test is most effective, though even it shows no significance when both sports are dropped. □

Remark 3.3. Confidence Sets for the Center
It is difficult to find directly a confidence set for ω, but it is straightforward to find a confidence region for a continuous parameter of which ω is a function, then project the region to the ω-space. We illustrate the approach for Spearman's distance, which appears to be the simplest. Let $\mu = E_P(Y)$, so that by Definition 2.3 and Theorem 2.2, $\omega = rank(\mu)$ if the elements of μ are distinct. First, find a confidence region for μ, or equivalently, $\mu_- = (\mu_1, \ldots, \mu_{m-1})'$. Let $\overline{Y}_-$ be the sample average, dropping the final term, and $\widehat{\Sigma}_-$

Table 3.2 *The tests of uniformity.*

Data Set (n, m)	Probabilities T_n^2 df p-value	Marginals T_n^2 df p-value	Means T_n^2 df p-value	Pairs T_n^2 df p-value
Voting	1717.51	748.50	30.18	189.03
(5738, 5)	119	16	4	10
	0	0	0	0
Occupations	2906.03	1874.08	84.40	1665.56
(143, 10)	3628799	81	9	45
	*	0	0	0
Gulf War	88.11	42.29	3.81	31.90
(21, 5)	119	16	4	10
	*	*	0.432	0.0004
Sports	963.12	82.98	5.28	61.02
(130, 7)	5039	36	6	21
	*	0	0.260	0
–No Jog	457.18	37.15	1.04	19.91
(130, 6)	719	25	5	15
	*	0.056	0.904	0.175
–No Jog, No Foot	181.16	22.40	0.22	6.34
(130, 5)	119	16	4	10
	*	0.131	0.994	0.786

be the corresponding sample covariance matrix. Then, assuming the true $Cov_P(Y_-)$ is nonsingular, $T^2(\mu_-) \equiv n(\overline{Y}_- - \mu_-)' \, \widehat{\Sigma}_-^{-1} (\overline{Y}_- - \mu_-)$ is asymptotically χ^2_{m-1}. Thus $B = \{\mu \mid T^2(\mu) < \chi^2_{m-1,\alpha}\}$ is an approximate $1-\alpha$ confidence region for μ; hence $C = \{rank(\mu) \mid \mu \in B\}$ is a confidence set for ω with asymptotic confidence level at least $1-\alpha$. (If μ has ties, then all ω's consistent with it are in the set C.)

We note that this procedure is available for any distance by finding a confidence region for P itself; for Hoeffding distances by finding one for $E_P(Q)$; and for Kendall by finding one for $E_P(\kappa)$. The smaller the dimension of the intermediate parameter, the less conservative the set C will be. □

Example 3.2. Voting: Confidence Region for the Center
Consider the data in Example 2.15. We invoked the above procedure to find a confidence set for the center. The vector of average ranks is $\overline{y} = (2.84, 3.16, 2.92, 3.09, 2.99)$, corresponding to a modal ranking of 15243. We use $\alpha = 0.05$, so that the cutoff point is $\chi^2_{4,.05} = 9.49$. There are only four vectors in the confidence set: 14253, 15243, 15342, 25143. The respective values of their T_n^2's are 5.73, 0, 6.14, and 8.50. With high confidence we can say that candidates A, C, and E are ranked higher than B and D, and A is higher than E. There is some overlap between A and C, C and E, and B and D. □

Example 3.3. Sports: Confidence Region for the Center
As a contrast, consider estimating the center of the Sports data in Example 2.13. Using the same $\alpha = 0.05$, we find 840 vectors in the confidence region: all 720 vectors with Jogging ranked #7, and an additional 120 vectors with Jogging ranked #6 and Football ranked #7. Hence Jogging and Football are the only ones ranked last, Jogging is never ranked higher than sixth, and otherwise all possible rankings are consistent with the data. □

The rest of this chapter presents details of the distances. The next section is useful but not crucial for understanding the subsequent material in the chapter.

3.3 Invariance and monotonicity of distances

The distances we have seen are symmetric in their arguments, $d(x, y) = d(y, x)$. Are there other invariances? The most common is label invariance, in which the distance does not change if the labels of the objects are permuted. For example, start with the objects being listed Red, Blue, Green, and two rank vectors $(1, 2, 3)$ and $(2, 3, 1)$. If one changes the listing, say to Blue, Green, Red, the vectors become $(2, 3, 1)$ and $(3, 1, 2)$. A label-invariant distance d will guarantee the distance between the vectors stays the same: $d((1, 2, 3), (2, 3, 1) = d((2, 3, 1), (3, 1, 2))$. All the distances in this book are label-invariant. Non-label-invariant distances may be reasonable if some objects are more important than others.

The formal definition is next.

Definition 3.3
A distance d is *label-invariant* if for any $Q \in \mathcal{Q}_m$ and $x, y \in \mathcal{S}_m$, $d(Qx, Qy) = d(x, y)$. □

While it is reasonable to treat all objects the same, it is not always so with the ranks. That is, a rank of "1" is inherently different than one of "5." A distance is expected to be large if the magnitude of the differences in rankings is large. Or is it? A distance is rank-invariant if it does not change upon renumbering the ranks:

Definition 3.4
A distance d is *rank-invariant* if for any Q, $Q^{(x)}$, $Q^{(y)} \in \mathcal{Q}_m$, $d(Q^{(x)}Qe_m, Q^{(y)}Qe_m) = d(Q^{(x)}e_m, Q^{(y)}e_m)$. If d is both rank- and label-invariant, it is called *bi-invariant.* □

Rank invariance captures the essential difference between a measure of spatial distance and one purely of disorder. Of the seven distances in Section 2.5.1, only Hamming and Cayley are rank invariant, and bi-invariant. Bi-invariance is intimately connected with cycles. Recall the discussion of Cayley's distance in Section 2.5.1. There is a unique partition of the objects associated with each pair of permutation vectors determined by cycles, defined next.

Definition 3.5
A cycle relative to $x, y \in \mathcal{S}_m$ is a nonempty subset $J \subseteq \mathcal{O}$ such that (a) $\{x_j | j \in J\} = \{y_j | j \in J\}$, and (b) no proper subset of J satisfies (a). □

Condition (a) says that the set of ranks given to the objects in J by x is the same set given by y; it does *not* say that x and y give the same rank to each object in J. As an illustration, let $\mathcal{O} = \{1, ..., 7\}$, and set $x = 1347625$ and $y = 2315647$. If x and y agree on an object, that object is a singleton cycle. Thus $\{2\}$ is a cycle since both vectors give Object 2 the rank 3. Similarly, $\{5\}$ is a cycle. Look at $\{4, 7\}$. Vector x gives Objects 4 and 7 the ranks 7 and 5, while y gives them 5 and 7. Though switched, the set of ranks is the same for both vectors. Thus $\{4, 7\}$ is a cycle. Finally, consider the set of Objects $\{1, 3, 6\}$. The ranks assigned by both x and y are 1, 2, and 4, though not in the same order. One can check that no proper subset of $\{1, 3, 6\}$ satisfies (a), e.g., $\{1, 3\}$ is assigned 1 and 4 by x, and 2 and 1 by y.

For any pair x and y, the cycles J_1, ..., J_G form a partition of $\mathcal{O}$, that is, $\cup J_g = \mathcal{O}$ and the J_g's are disjoint. Cycles are rank-invariant in the sense they remain the same if the ranks are changed, but they are not label-invariant. However, the number and lengths of the cycles are label-invariant. (The length of a cycle is the number of elements it has.)

Definition 3.6
The cycle distribution relative to x and y is the $m \times 1$ vector $\gamma(x, y)$, where γ_i is the number of cycles of length i. □

In the illustration above, $\gamma(1347625, 2315647) = (2, 1, 1, 0, 0, 0, 0)$. The space of possible γ's is equivalent to the partitions of the integer m, i.e., γ corresponds to the partition λ as in Definition 2.4 (a) if $\gamma_i = \{\lambda_j \mid \lambda_j = i\}$. Also, $\sum_{i=1}^{m} i\gamma_i = m$. The discussion above shows that γ is bi-invariant. The next result is even stronger.

Theorem 3.2
A function $d(x, y)$ is bi-invariant if and only if it depends on (x, y) only through $\gamma(x, y)$. □

Proof
Any function of γ is bi-invariant since γ is. Now suppose d is bi-invariant. Definitions 3.3 and 3.4 imply that for any $Q_1, Q_2 \in \mathcal{Q}_m$,

$$d(Q_1 Q^{(x)} Q_2 e_m, Q_1 Q^{(y)} Q_2 e_m) = d(Q^{(x)} e_m, Q^{(y)} e_m). \qquad (3.29)$$

First set $Q_1 = {Q^{(y)}}^{-1}$ and $Q_2 = I_m$. Then (3.29) shows

$$d(Q^{(x)} e_m, Q^{(y)} e_m) = d({Q^{(y)}}^{-1} Q^{(x)} e_m, e_m). \qquad (3.30)$$

Invoke bi-invariance again to show that

$$\begin{aligned} d({Q^{(y)}}^{-1} Q^{(x)} e_m, e_m) &= d(Q {Q^{(y)}}^{-1} Q^{(x)} Q^{-1} e_m, Q Q^{-1} e_m) \\ &= d(Q {Q^{(y)}}^{-1} Q^{(x)} Q^{-1} e_m, e_m). \end{aligned} \qquad (3.31)$$

The sequence of equalities shows that d depends only on ${Q^{(y)}}^{-1}$ $Q^{(x)} \in \mathcal{Q}_m$, and is invariant under conjugation by $\mathcal{Q}_m$. Such functions depend only on γ, as shown by Herstein (1964), pages 75 and 76. □

The two bi-invariant distributions in Section 2.5.1, Hamming and Cayley, are easily seen to be linear functions of γ:

$$d_{\mathrm{Ham}}(x,y) = m - \gamma_1(x,y) \text{ and } d_{\mathrm{Cay}}(x,y) = m - \sum_{i=1}^{m} \gamma_i(x,y). \tag{3.32}$$

It is interesting that Hamming is the only Hoeffding (2.25) distance that is bi-invariant. See Exercise 3.10. Cycles seem to capture the essence of nonspatial disorder. The longer a cycle is, the more work it is to put the elements in order.

When evaluating distances, it is natural to look for certain monotonicities, that is, one may have certain intuitive notions of when one rank vector is farther from e_m than another. Henery (1981) proposed a partial ordering on $\mathcal{S}_m$ designed to measure relative distance from e_m:

$$\begin{aligned} y <_H x \text{ if for some } i < j, y_k = x_k \text{ for } k \neq i, j; \\ y_i < y_j; \text{ and } x_i > x_j. \end{aligned} \tag{3.33}$$

See Block, Chhetry, Fang, and Sampson (1990, 1991, 1992) for this and other partial orderings. Thus x and y are equal except for a pair of indices for which y is in the right order, and x is not. Transitivity is used to extend the ordering, so that $y <_H x$ if there is a sequence of vectors such that $y <_H z^{(1)} <_H z^{(2)} <_H \cdots <_H z^{(k)} <_H x$. Which distances are (strictly) monotone in $<_H$, i.e., satisfy $d(e_m, y) \leq (<) d(e_m, x)$ whenever $y <_H x$? Take $m \geq 3$. It is easy to see that Hamming and Cayley are not monotone, which is a consequence of their rank invariance. Ulam for $m = 3$ and Maximum are monotone, but not strictly so. When $m \geq 4$, Ulam is not even monotone. For example, when $m = 4$, $(2,1,4,3) <_H (2,3,4,1)$, but $d_{\mathrm{Ulam}}(e_m, (2,1,4,3)) = 2 > d_{\mathrm{Ulam}}(e_m, (2,3,4,1)) = 1$. For the d_p distances, (strict) monotonicity holds if and only if for all $i < j$ and $k < l$,

$$|k-i|^p + |l-j|^p \leq (<) |l-i|^p + |k-j|^p. \tag{3.34}$$

When $p < 1$, (3.34) fails for $(i,j,k,l) = (1,2,2,3)$. If $p \geq 1$, (3.34) with $\leq$ is always satisfied since $|a-b|^p$ is convex in (a,b); strictly so if $p > 1$, showing strict monotonicity in that case. When $p = 1$, Footrule, (3.34) is not always strict: again take $(i,j,k,l) = (1,2,2,3)$. It can be shown that Kendall is strictly monotone. In fact, Kendall, or any strictly monotone function of Kendall, is the coarsest distance that is strictly monotone. (Coarsest meaning taking on the fewest distinct values.) Thus of the standard distances, only Kendall and Spearman are strictly monotone.

Henery's monotonicity is not relevant to bi-invariant distances. For them, long cycles are associated with large distances. It is easier to rearrange two 2-cycles than one 4-cycle, which suggests the following monotonicity. Suppose (y, e_m) has cycles $J_1, \ldots, J_k$, and (x, e_m) has cycles $J_1, \ldots, J_{k-2}$ and $J_{k-1} \cup J_k$. Then we say that $y <_C x$. (The C stands for Cayley.) Thus y is closer to e_m than x is if y and x have the same cycles, except that one of the cycles in x is split in y. The ordering is extended by transitivity, again. Hamming is monotone $[<_C]$, but not strictly, since $m - d_H$ is the number of cycles of length 1. Cayley's is strictly monotone since $m - d_{\text{Cay}}$ is the number of cycles. Cayley's is the coarsest strictly monotone distance. □

Remark 3.4

Most of the statistics of interest outside of those based on distances are not label- or rank-invariant. However, it may be that the corresponding subspaces, hence test statistics, are. Given a function vector $x \in \Re^{m!}$ as in Section 2.6, so that its elements are x_y for $y \in \mathcal{S}_m$, the permutation matrix $Q \in \mathcal{Q}_m$ permutes the labels of x by permuting the labels of its indices. That is, the permuted vector $u = L(Q, x) \in \Re^{m!}$ is given by $u_w = x_y$, where $w = Qy$. A subspace S is label-invariant if $S = \{L(Q,x) \mid x \in S\}$ for each $Q \in \mathcal{Q}_m$. Similarly, Q permutes the ranks of x via $u = R(Q,x)$ if $u_w = x_y$ where $w = Q^{(y)}Qe_m$, and S is rank-invariant if $S = \{R(Q,x) \mid x \in S\}$ for each $Q \in \mathcal{Q}_m$. The Marginal subspaces U^λ (2.75) and V^λ (2.77) are bi-invariant, i.e., both label- and rank-invariant. Diaconis (1988, 1989) shows that the decomposition (2.78) is close to a decomposition of $\Re^{m!}$ into bi-invariant irreducible subspaces. A bi-invariant subspace is irreducible if there is no other subspace strictly contained in it that is also bi-invariant. Each of the $V^{(\lambda)}$'s can be split into h_λ irreducible subspaces, each with dimension h_λ, where h_λ is the square root of the dimension in (2.80). This decomposition is the key to the spectral analysis mentioned in Section 2.6.1. □

3.4 Distance from a fixed vector

Recall Example 2.12. There, it is hypothesized that voters are more likely to vote for people near the top of the ballot. The ranking $(1, 2, 3)$ is a predefined target, as opposed to many other examples where there is no such target. Another, classic, example is the Draft Lottery data.

Example 3.4. Draft Lottery
In 1969, the U.S. Government instituted a lottery system for inducting young men into the military. A unique number from 1 to 366 was randomly assigned to each day of the year, and each young man born in 1950 or earlier received the number corresponding to his birthday. People were then to be drafted in order of the numbers. The randomization was a national television event, watched with special interest by the people in the pool. The days of the year were placed in capsules in a large bowl, and drawn one by one. September 14 was first, bestowing everyone born on that day with #1. Draws continued until all days had been chosen, the last being June 8. (Historical note: President Clinton's birthday, August 19, was drawn 311th.)

The purpose of the public randomization was to convince everyone of its fairness. Much of the impetus behind the lottery was the perceived lack of fairness in the previous system. After the drawing, clever statisticians went to work. See Fienberg (1971) and Diaconis (1988), pages 93–4 and 102–3. A plot of the days of the year versus lottery numbers does not reveal anything; however, plotting the average number per month versus month shows a distinct downward trend. See Figure 3.1, where the months' names are plotted at their average lottery numbers. Numerous statistical tests proved that indeed the lottery was not as "random" as hoped. Why not? After all, the capsules were mixed up in the bowl and drawn randomly. A closer look at the process will explain it. Before the drawing, each capsule was placed in one of twelve boxes according to its month. The capsules were sequentially placed in a larger box, first the January capsules, then February, etc. Each time, the capsules were swept towards one end of the box and mixed. The box was carried upstairs to await the drawing, then at the appointed hour carried back down. The capsules were poured

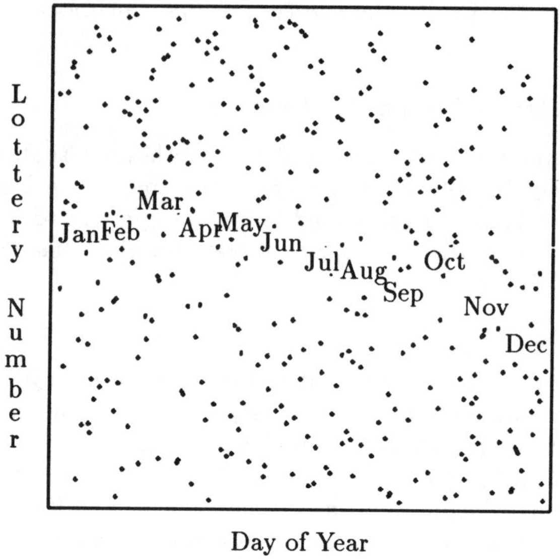

Figure 3.1 *The 1969 Draft Lottery.*

from one end of the box (which one remains a mystery) into the ultimate bowl, where they were stirred some more. During the drawing, capsules were picked at random, but more likely than not from near the top. If each mixing had been extremely thorough, the process would have approached total randomness. But there is enough danger in there being some trend within the bowl coupled with the non-uniform choices from the bowl that a bias would result. It did.

Mathematically, the list of lottery numbers is $Y \in \mathcal{S}_{366}$, where Y_i is the number for the ith day of the year. The problem is to test whether Y is uniform versus close to a trend, i.e., close to e_m, or e_m^c. [Recall (2.45), so that $e_m^c = (m, \ldots, 1)$.] We take a two-sided hypothesis since we do not know *a priori* out of which end of the box the capsules were poured. Any of the distances can now be used as a test statistic. This example will be continued in Example 3.5. □

If the number of vectors n is large, then $\overline{d}(\omega)$ (2.13) and $t(\omega)$ (2.21) are approximately normal under uniformity. Thus we have $\sqrt{n}(\overline{d}(\omega) - \delta_0)/\sigma_0 \rightarrow N(0,1)$, where δ_0 and $\sigma_0^2 \equiv Var_0(d)$ are found in Table 3.3 for five of the distances. See also (3.40), (3.42), and (3.43) below. When there is only $n = 1$ rank vector, as in Example 3.4, we have to look more carefully at the distribution of $d(\omega, Y)$ under uniformity, or develop another kind of asymptotic result, where the number m of objects is large. Our next task is to find the distributions and moments, exact and asymptotic, for the distances. For small m, the exact distribution of any distance is easily enumerated. The task quickly becomes daunting as m increases. In Section 3.4.3 we look at asymptotic distributions.

3.4.1 Means and variances

Start with Hoeffding distances. The linearity in (2.25) can be written

$$d_{\text{Hoeff}}(x, y) = \text{tr}(Q^{(x)} \Delta Q^{(y)\prime}), \tag{3.35}$$

where Δ is the $m \times m$ matrix with elements $\Delta_{ij} = a(i, j)$. See (3.11). Useful expressions are available for the Δ's for Hamming, Footrule, and Spearman:

$$\begin{aligned} \Delta_{\text{Ham}} = J_m - I_m; \quad \Delta_{\text{Foot}} &= e_m 1_m' + 1_m e_m' - 2 \cdot F_m F_m; \\ \Delta_{\text{Spear}} &= f_m 1_m' + 1_m f_m' - 2 \cdot e_m e_m', \end{aligned} \tag{3.36}$$

where $f_m' = (1, 2^2, 3^2, \ldots, m^2)$ and F_m is the $m \times m$ matrix with $F_{m,ij} = 1$ if $i + j > m$ and $F_{m,ij} = 0$ otherwise. See Exercise 3.11.

By (3.35), if $Q \sim \mathit{Uniform}(\mathcal{Q}_m)$,

$$\begin{aligned} E(d_{\text{Hoeff}}(Qe_m, e_m)) &= E(\text{tr}(Q\Delta)) = \text{tr}(E(Q)\Delta) \\ &= \text{tr}(\overline{J}_m \Delta) = -\text{tr}(H_m \Delta) = m \cdot a_{..}, \end{aligned} \tag{3.37}$$

where $a_{..}$ is the average of the $a(i, j)$'s. The fourth equality follows since the diagonal elements of Δ are 0. Since $Cov(Q) = (m - 1)^{-1} H_m \otimes H_m$, Lemma 12.5 shows that

$$\begin{aligned} Cov(d(Qe_m, e_m)) &= \frac{1}{m-1} \text{tr}(\Delta H_m \Delta H_m) \\ &= \frac{1}{m-1} \text{tr}((H_m \Delta H_m)^2), \end{aligned} \tag{3.38}$$

the second equation following from $H_m H_m = H_m$. As in Remark 3.1, the final trace is the sum of squares of residuals of the $a(i, j)$'s.

Thus the covariance in (3.38) can be calculated from the $a_{i\cdot} = a_{\cdot i}$'s where $a_{i\cdot} = (1/m)\sum_{j=1}^{m} a(i,j)$ and $a_{\cdot\cdot}$, i.e.,

$$Cov(d(e_m, Qe_m)) = \frac{||\Delta||^2 - 2 \cdot m \cdot \sum_{i=1}^{m} a_{i\cdot}^2 + m^2 \cdot a_{\cdot\cdot}^2}{m-1}. \tag{3.39}$$

Consider Spearman distance. From (3.36), since $H_m 1_m = 0$, $-\text{tr}(H_m \Delta_{\text{Spear}}) = 2 \cdot e_m' H_m e_m = 2v_m$ of (2.3); hence (3.37) yields the mean in Table 3.3. For the variance, $\text{tr}((H_m \Delta H_m)^2) = 4$ $\text{tr}((H_m e_m e_m' H_m)^2)$, which rearranges to $4(e_m' H_m e_m)^2$. Dividing by $m-1$ gives the result. The Footrule can be handled similarly, although it is a bit more complicated. See Diaconis and Graham (1977) and Exercise 3.12. Hamming is quite easy. Since $\Delta_{\text{Ham}} = J_m - I_m$ and $H_m J_m = 0$, the mean is $-\text{tr}(H_m(J_m - I_m)) = \text{tr}(H_m) = m-1$, and the variance is $\text{tr}(H_m(J_m - I_m)H_m(J_m - I_m))/(m-1) = \text{tr}(H_m)/(m-1) = 1$. The d_p distances for other p do not all have tractable answers, but it can be shown that as $m \to \infty$,

$$\begin{aligned} \frac{1}{m^{p+1}} E(d_p) &\to \frac{2}{(p+1)(p+2)} \quad \text{and} \\ \frac{1}{m^{2p+1}} Var(d_p) &\to \frac{1}{(2p+1)(p+1)} + \frac{4}{(p+1)^2(p+2)^2} \\ &\quad - \frac{4}{(p+1)^2}\left(\frac{1}{2p+3} + Beta(p+2, p+2)\right) \equiv v(p), \end{aligned} \tag{3.40}$$

where $Beta(a,b)$ is the beta function $\int_0^1 x^{a-1}(1-x)^{b-1}\,dx$.

For Kendall (2.29), note that by (2.59),

$$d_{\text{Ken}}(e_m, y) = \overline{m} - \sum\sum_{i<j} \kappa_{ij}(y) = \overline{m} - 1_{\overline{m}}' \kappa(y). \tag{3.41}$$

Thus by (3.23) and thereabove, $E(d_{\text{Ken}}) = \overline{m}/2$ and $Var(d_{\text{Ken}}) = 1'\Sigma_0 1 = (1/12)(\overline{m} + 1_{\overline{m}}' \Gamma_m \Gamma_m' 1_{\overline{m}})$. Now $1_{\overline{m}}' \Gamma_m$ is the vector of sums of the columns of Γ_m. The jth column has $(j-1)$ -1's and $(m-j)$ +1's; hence its sum is $(m-j)-(j-1) = -2(j-(m+1)/2)$. Thus $1_{\overline{m}}' \Gamma_m \Gamma_m' 1_{\overline{m}}$ is the sum of squares of those, which by (2.3) is $4v_m$. The result in Table 3.3 then follows.

We will defer Cayley's to the next section, after developing a convenient decomposition. Ulam's distance is difficult to handle

Table 3.3 *Means and variances of five distances.*

	Mean	**Variance**
Spearman	$m(m^2-1)/6$	$m^2(m-1)(m+1)^2/36$
Footrule	$(m^2-1)/3$	$(m+1)(2m^2+7)/45$
Kendall	$m(m-1)/4$	$m(m-1)(2m+5)/72$
Hamming	$m-1$	1
Cayley	$m-\sum_{i=1}^{m} 1/i$	$\sum_{i=1}^{m}(1/i)(1-1/i)$

analytically. Kerov and Vershick (1985) show that

$$\frac{E(d_U)-m}{\sqrt{m}} \to -2. \tag{3.42}$$

We do not know of convenient formulas for the Maximum, either, but from (3.67) we have the asymptotics

$$\frac{E(d_{\text{Max}})-m}{\sqrt{m}} \to -\frac{\sqrt{\pi}}{2} \text{ and } \frac{Var(d_{\text{Max}})}{m} \to 1-\frac{\pi}{4}. \tag{3.43}$$

3.4.2 Exact distributions

For Hamming, Kendall, and Cayley, the exact distribution can be described using simple recursions. The same is true for Maximum, but only for values exceeding $m/2$. The first three derivations are found in Feller (1968). The other distances do not appear to have comparable simplifications. Ury and Kleinecke (1979) contains the null distribution of the Footrule, (2.23) with $p = 1$, for $m \leq 15$, and Baer and Brock (1968) have those for Ulam for $m \leq 36$.

Both Kendall and Cayley can be written as sums of independent variables. Diaconis (1988), pages 122–123, indicates deep reasons for such decomposition for these distances and other distances on groups. We start with Kendall. Write d_{Ken} in (2.29) as

$$d_{\text{Ken}}(e_m, y) = \sum_{i=1}^{m-1} U_i, \quad U_i = \sum_{j=i+1}^{m} I[y_i > y_j]. \tag{3.44}$$

The variable U_i is the number of objects listed below the ith component in the vector that have ranks smaller than y_i. In Section 5.10 we will see that these U_i's represent a set of orthogonal contrasts. An alternative explanation in terms of multistage ranking is given in Fligner and Verducci (1988) and Section 5.6.2. The key property of these variables is their independence, which will be seen in Lemma 5.1.

Lemma 3.5
If $Y \sim Uniform(\mathcal{S}_m)$, then the $U_1, \ldots, U_{m-1}$ are independent, and $U_i \sim Uniform(\{0, \ldots, m-i\})$. □

The mean and variance of d_{Ken} can be found as the sums of the means and variances of the U_i's. See Exercise 3.13.

To motivate the formula for Kendall, let $Kendall_k$ denote the distribution of $d_{\text{Ken}}(e_m, Y)$ for $Y \sim Uniform(\mathcal{S}_k)$. Then by Lemma 3.5,

$$\begin{aligned} V \sim \ Kendall_{m-1} \text{ and } U \sim \ Uniform(\{0, \ldots, m-1\}), \\ \text{independent}, \Rightarrow W = V + U \sim \ Kendall_m. \end{aligned} \tag{3.45}$$

We have the following convolution result. See Exercise 3.14.

Lemma 3.6
Suppose V and U are independent with spaces $\{0, ..., a\}$ and $\{0, ..., b\}$ respectively, and $W = V + U$. Then for $0 \leq w \leq a + b$,

$$P[W = w] = \sum_{u=\max(0,w-a)}^{\min(w,b)} P[U = u] P[V = w - u]. \quad \Box \tag{3.46}$$

Now the distribution for $Kendall_m$ can be easily found from that of $Kendall_{m-1}$ by using Lemmas 3.5 and 3.6. Here, $a = \overline{m-1}$, $b = m - 1$, and $P[U = j] = 1/m$ for $0 \leq j \leq m - 1$; hence from (3.45),

$$P[W = w] = \frac{1}{m} \sum_{u=\max(0,w-\overline{m-1})}^{\min(w,m-1)} P[V = w - u], \quad w = 0, \ldots, \overline{m}. \tag{3.47}$$

To illustrate, let $\eta^{(k)}$ be the vector of probabilities for $Kendall_k$, so that $\eta_i^{(k)} = P[Kendall_k = i]$ for $0 \le i \le \overline{m}$. Since $Kendall_2$ takes on the two values 0 and 1 with equal probability, $\eta^{(2)} = (1,1)/2$. For $Kendall_3$, (3.46) is applied with $z = 0, ..., 3$:

$$\eta^{(3)} = \frac{1}{3}(\eta_0^{(2)}, \eta_0^{(2)} + \eta_1^{(2)}, \eta_0^{(2)} + \eta_1^{(2)}, \eta_1^{(2)}) = \frac{1}{6}(1,2,2,1). \quad (3.48)$$

Convolving $\eta^{(3)}$ with $Uniform(\{0,1,2,3\})$ yields $\eta^{(4)} = (1, 1+2, 1+2+2, 1+2+2+1, 2+2+1, 2+1, 1)/24 = (1,3,5,6,5,3,1)/24$. Thus at the kth step, the factor outside is $1/k!$, and the ith element of the vector is obtained from the previous vector by summing the k elements starting from i and working backwards.

Cayley has a similar recursion. Given y, imagine calculating $d_{\mathsf{Cay}}(e_m, y)$ by first making the interchange (if needed) to set $y_m = m$, then $y_{m-1} = m-1$, etc. Recall that Cayley's distance is the number of such interchanges necessary. We define the vector Y^{-m} from Y as the first $m-1$ elements of Y *after* possible interchanging to place m in the last slot: Let k be the object ranked m, $Y_k = m$. If $k = m$, then Y^{-m} is the vector of the first $m-1$ elements of Y. If $k \ne m$, then $Y_i^{-m} = Y_i$ for $i \ne k, m$, and $Y_k^{-m} = Y_m$. Note that Y^{-m} is in $\mathcal{S}_{m-1}$. The key fact is the following, shown in Exercise 3.15.

Lemma 3.7
If $Y \sim Uniform(\mathcal{S}_m)$ and Y^{-m} is defined as above, then Y_m and Y^{-m} are independent, $Y^{-m} \sim Uniform(\mathcal{S}_{m-1})$, and $Y_m \sim Uniform$ $(\{1,\ldots,m\})$. □

Define $Cayley_k$ as above for $Kendall_k$, except using d_{Cay}. Write

$$d_{\mathsf{Cay}}(Y, e_m) \equiv W = V + U \equiv d_{\mathsf{Cay}}(Y^{-m}, e_{m-1}) + I(Y_m \ne m). \quad (3.49)$$

Lemma 3.7 shows that V and U are independent. Now invoke Lemma 3.6, where $a = m-1$, $U \sim Bernoulli((m-1)/m)$, and $b = 1$, which gives

$$P[W = k] = \frac{1}{m}P[V = k] + \frac{m-1}{m}P[V = k-1], \; k = 0, \ldots, m, \quad (3.50)$$

and $P[V = -1] = P[V = m] = 0$.

Lemma 3.7 itself can be iterated, so that letting $U_{m-1} = I(Y_{m-1}^{-m} \neq m-1)$, $U_{m-2} = I((Y_{m-1}^{-m})_{m-2}^{-(m-1)} \neq m-2)$, ..., we have that

$$Cayley_m = U_2 + \cdots + U_m, \tag{3.51}$$

where the U_k's are independent, $U_k \sim Bernoulli((k-1)/k)$. (Note that $U_1 \equiv 0$.) It is now easy to find the moments of d_{Cay} for Table 3.3. See Exercise 3.16. We note that $\#\{y|d_{\text{Cay}}(e_m, y) = k\}$ is the absolute value of a Stirling number of the first kind. See, e.g., Abramowitz and Stegun (1972), Section 24.1.3.

Next is Hamming. We start by finding the probability that there are no matches between e_m and y, i.e., that $d_{\text{Ham}}(e_m, y) = m$. (Thus y is an m-object inversion as in 2.85.) We will use the inclusion–exclusion argument as in Feller (1968), pages 100–106. Let $A_i = \{y_i = i\}$, and B_i be the complement $\{y_i \neq i\}$. We wish to find $P[\cap B_i] \equiv 1 - P[\cup A_i]$. Now we expand the last probability:

$$\begin{aligned} P[\cup A_i] = \sum_i P[A_i] - \sum\sum\nolimits_{i<j} P[A_i \cap A_j] \\ + \sum\sum\sum\nolimits_{i<j<k} P[A_i \cap A_j \cap A_k] \\ \pm \cdots (-1)^{m+1} P[A_1 \cap \cdots \cap A_m]. \end{aligned} \tag{3.52}$$

Each individual probability depends only on the number of sets being intersected. With r sets, the probability is $(m(m-1)\cdots(m-r+1))^{-1}$. In the rth summation, there are $\binom{m}{r}$ terms; hence the sum is $1/r!$. Thus taking the complement in (3.52), we have

$$P[0 \text{ matches}] = 1 - P[\cup A_i] = \frac{1}{2!} - \frac{1}{3!} \pm \cdots \frac{(-1)^m}{m!} \equiv E_m, \tag{3.53}$$

and we set $E_0 = 1$ and $E_1 = 0$. Now consider the probability there are exactly k matches. There are $\binom{m}{k}$ ways to choose which of the A_i's hold, and each set has the same probability. That is,

$$\begin{aligned} P[d_{\text{Ham}} = m-k] &= P[\text{Exactly } k\ A_i\text{'s occur}] \\ &= \binom{m}{k} P[A_1 \cap \cdots \cap A_k \cap B_{k+1} \cap \cdots \cap B_m]. \end{aligned} \tag{3.54}$$

By successively conditioning on the A_i's, we have that

$$P[d_{\text{Ham}} = m - k] = \binom{m}{k} \frac{1}{m} \frac{1}{m-1} \cdots \frac{1}{m-k+1} \times P[B_{k+1} \cap \cdots \cap B_m | A_1 \cap \cdots \cap A_k]. \tag{3.55}$$

Given that the first k elements of y match, the other elements are equally likely to be in any order; hence the conditional probability none of them match is E_{m-k} in (3.53). Putting (3.53), (3.54), and (3.55) together yields

$$P[d_{\text{Ham}} = m - k] = \frac{1}{k!} E_{m-k}. \tag{3.56}$$

Finally, turn to the Maximum. We will find the distribution exactly for values exceeding $m/2$. For moderate to large m, the vast majority of the mass is above $m/2$. Consider $P[d_{\text{Max}} \geq m-k]$. Note that in order for $\max\{|y_i - i|\} \geq m - k$, we must have either $y_1 \geq m - k + 1$, $y_2 \geq m - k + 2$, ..., or $y_k \geq m$, or $y_m - k + 1 \leq 1$, ..., or $y_m \leq k$. That is, letting $C_i = A_i = \{y_{k-i+1} \geq m - i + 1\}$ and $C_{k+i} = B_i = \{y_{m-k+1} \leq i\}$, we have

$$P[d_{\text{Max}} \geq m - k] = P[\cup_{i=1}^{2k} C_i]. \tag{3.57}$$

Now expand the last probability using (3.52). There are $2k$ terms, the rth one being $(-1)^{r+1} v_r^k$, where

$$v_r^k = \sum \cdots \sum_{1 \leq i_1 < i_2 < \cdots < i_r \leq m} P[\cap_{j=1}^r C_{i_j}]. \tag{3.58}$$

For integers $a + b = r$, $0 \leq a, b \leq k$, the summation will contain $\binom{a+b}{a}$ probabilities of intersections of a of the A_i's and b of the B_i's. It can be shown that if $k \leq m/2$, then for $1 \leq j_1 < \cdots < j_a \leq k$

and $1 \leq l_1 < \cdots < l_b \leq k$,

$$P[A_{j_1} \cap \cdots \cap A_{j_a} \cap B_{l_1} \cap \cdots \cap B_{l_b}] = \frac{i_1}{m}\frac{i_2-1}{m-1} \cdots \frac{i_k-(k-1)}{m-(k-1)} \times \frac{j_1}{m-k} \cdots \frac{j_l-(l-1)}{m-k-l+1}. \tag{3.59}$$

The sum of all such terms (3.59) with given a and b is $u_a^k u_b^k/(m)_r$, where

$$u_a^k = \sum \cdots \sum\nolimits_{1 \leq i_1 < \cdots < i_a \leq k} i_1(i_2-1)(i_3-2)\cdots(i_a-a+1). \tag{3.60}$$

Thus from (3.58),

$$v_r^k = \sum \sum\nolimits_{a+b=k, 0 \leq a,b \leq k} u_a^k u_b^k/(m)_k, \tag{3.61}$$

and

$$P[d_{\text{Max}} \geq m-k] = \sum_{r=1}^{2k} (-1)^{r+1} v_r^k. \tag{3.62}$$

3.4.3 Asymptotics as $m \to \infty$

The asymptotics in previous sections were relatively easy since we had the number of independent replicates approaching infinity. Here we will deal with the case that there is just one ranking, but the number of objects is increasing. It is more difficult since the components of Y are not independent.

For the Hoeffding distances, we have the famous theorem by Hoeffding (1951). We can get away with a weaker version of his Theorem 4, the condition given by his Equation (12).

Theorem 3.3: Hoeffding
If for Hoeffding distance (2.25), with $\Delta_{ij} = a(i,j)$,

$$\frac{\max\{(H_m \Delta H_m)^2\}}{Var(d)} \to 0 \tag{3.63}$$

as $m \to \infty$, then $(d-\delta_0)/\sqrt{Var(d)} \to N(0,1)$. □

The quantities d, δ_0, H_m, and Δ depend on m. The "max$\{A\}$" in (3.63) means the maximum over the elements of the matrix A.

Consider the d_p distances (2.23) for $0 < p < \infty$. Equation (3.40) gives the asymptotic variance. Exercise 3.17 shows that

$$\max\{(H_m \Delta_p H_m)^2\} \leq (m-1)^p \text{ and } v(p) > 0; \tag{3.64}$$

hence (3.63) follows from (3.64). Thus Theorem 3.3 applies to the d_p distances for $p > 0$. Hamming ($p = 0$) does not work here. See Exercise 3.18.

Kendall and Cayley both make use of their decompositions into independent components. We will use the following corollary of the Lindeberg–Feller theorem. See, e.g., Serfling (1980), page 32.

Theorem 3.4
Suppose $X_i, i = 1, 2, \ldots$ are independent, and let $W_m = \sum_{i=1}^m X_i$. If for some $\nu > 2$,

$$\frac{\sum_{i=1}^m E|X_i - E(X_i)|^\nu}{Var(W_m)^{\nu/2}} \to 0 \text{ as } m \to \infty, \tag{3.65}$$

then $(W_m - E(W_m))/\sqrt{Var(W_m)} \to N(0,1)$. □

Kendall's can be represented as in Theorem 3.4 by taking $X_i = U_{m-i} \sim Uniform(\{0, 1, ..., i-1\})$ from Lemma 3.5. Thus we have $|X_i - E(X_i)| < i - 1$. The numerator in (3.65) is of order $m^{\nu+1}$, while the denominator is of order $m^{3\nu/2}$ by Table 3.3; hence for $\nu > 2$, (3.65) holds. The same can be done for Cayley. See Exercise 3.19.

As mentioned above, Hoeffding's Theorem 3.3 does not apply to Hamming, but Hamming is easy to work with directly. Consider (3.56). The term E_{m-k}, by (3.53), consists of the first $m - k + 1$ terms in the Taylor expansion of $\exp(-1)$; hence fixing k yields

$$P[d_{\text{Ham}} = m - k] \to \frac{\exp(-1)}{k!}, \quad k = 0, 1, 2, ..., \tag{3.66}$$

which is the *Poisson(1)* distribution. (The *Poisson* (η) distribution has probabilities $P[X = k] = \exp(-\eta)\ \eta^k/k!$ for $k = 0, 1, 2, \ldots$.) See Feller (1968), page 108.

For Maximum we quote the following result from Marden (1994):

$$\frac{m - d_{\text{Max}}}{\sqrt{m}} \to W, \ W^2 \sim \mathit{Exponential}(1). \tag{3.67}$$

Example 3.5. Draft Lottery: Distance Tests
We now have enough to test whether the Draft Lottery in Example 3.4 was fair. We calculated the Z score, $Z = (d(e_{366}, y) - \delta_0)/\sqrt{Var(d(e_{366}, Y))}$, for the various distances under the assumption that the lottery is fair, i.e., $Y \sim \mathit{Uniform}(\mathcal{S}_m)$. See Table 3.3. Table 3.4 contains the Z's for the seven main distances. Only Spearman and Kendall are symmetric in the sense that $d(e_m, y) - \delta_0 = -(d(e_m^c, y) - \delta_0)$; hence it may be that the others can capture the trend better using e_m^c than e_m. This is true for Maximum and Ulam in 1969, but Footrule actually is worse.

Footrule, Spearman, Kendall, and Cayley are approximately normal for large m under uniformity, so that the p-value can be found from Z using the normal distribution. For Hamming, using the Poisson approximation, the one-sided p-values for the Hamming values in Table 3.4 are 0.2642, 0.2642, 0.2642, and 0.0190; for Maximum, using (3.57) (3.67 with continuity correction), they are 0.1435 (0.1425), 0.1098 (0.1093), 0.0684 (0.0787), and 0.3084 (0.3039); and for Ulam, using 10000 simulations, they are 0.2411, 0.0594, 0.0263, 0.3916.

We calculated the Z for the d_p distances, also. They are most significant (with Z hovering around 4.3) near $p = 1$ and 2, i.e., Footrule and Spearman, the Z's slowly decreasing as p increases, but becoming starkly insignificant as $p \to 0$. At $p = 0$ we have Hamming, which shows no significant trend. We see that Kendall is quite close to Footrule and Spearman. Maximum and Ulam show a barely significant trend. Cayley is like Hamming in not detecting a trend. Thus Kendall, Spearman, Footrule, and the d_p distances show clearly that the lottery was not fair. There is a significant trend downward, so that dates earlier in the year tended to have higher numbers. This fact suggests that the capsules were poured into the bowl with January first.

For the 1970 lottery, randomization was approached much more carefully. See Fienberg (1971). Before any public mixing was performed, two computer-generated random permutations were used to reorder both the days of the year and the lottery numbers. Then for the televised randomization, there were two bowls, one with the

Table 3.4 *Draft Lottery: Z statistics.*

	1969			
Distance ↓	$d(e_m, y)$	Z	$d(e_m^c, y)$	Z
Footrule	51292	4.49	39276	−3.64
Spearman	10015392	4.31	6327116	−4.31
Kendall	38369	4.25	28426	−4.25
Hamming	364	−1.00	364	−1.00
Cayley	359	−0.24	360	0.22
Ulam	335	1.40	328	−1.82
Maximum	362	1.48	335	−1.68
	1970			
Distance ↓	$d(e_m, y)$	Z	$d(e_m^c, y)$	Z
Footrule	44218	−0.13	44826	0.28
Spearman	7988976	−0.27	8219944	0.27
Kendall	32883	−0.29	33547	0.29
Hamming	363	−1.00	361	−3.00
Cayley	359	0.22	357	−0.69
Ulam	328	−1.38	331	0.01
Maximum	356	0.89	354	0.66

date capsules, the other with the lottery number capsules. Capsules were drawn in pairs, one from each bowl, and the lottery number was assigned to the date. Since the capsules were already thoroughly randomized, it did not matter to the actual fairness of the lottery how well the capsules were physically mixed, or how uniformly the capsules were drawn from the bowls. Of course, the public mixing was important to convince the country that the lottery was fair. Statistically, the 1970 lottery shows nothing significant, as expected, no matter which distance is used, except slightly for Hamming with e_m^c. □

Example 3.6. Ballots: Distance Tests
Recall Example 2.12, where $\omega = (1, 2, 3)$. The Z's for d_p peak at -6.53 around $p = 1/4$, and slowly tail off as p increases to -3.71 at $p = 10$. The Z's for the other distances, in the order *H, F, S, K, M, C, U* are −6.46, −6.19, −5.21, −5.39, −6.19, −5.91, and −5.12. (When $m = 3$, Footrule and Maximum are the same distance.)

Notice that in Example 3.5, where m is large, there is a great difference in the ability of the distances to show significance. By contrast, the present example has a very small m, and the Z's are all between -5 and -6.5, except for d_p when p is large. Also, while in the previous example Footrule, Kendall, and Spearman were much better than the others, here Hamming is the best, followed by Footrule $\equiv$ Maximum and Cayley. □

3.5 One-sample diversity and concordance

When there is *a priori* knowledge of the center ω, then $\overline{d}(\omega)$ in (2.13) is a reasonable measure of spread. Alvo, Cabilio, and Feigin (1982) show how spread can be measured directly, avoiding the search for $\widehat{\omega}$, by averaging the distances between all pairs of observations:

$$\widehat{\alpha} = \frac{1}{n^2}\sum_{i=1}^{n}\sum_{j=1}^{n} d(y^{(i)}, y^{(j)}). \tag{3.68}$$

We are still assuming that d is symmetric in its arguments, and $d(y,y) = 0$, so that the sum need be only over $i < j$. [Alvo, Cabilio, and Feigin actually define the diversity as $(n/(n-1))\widehat{\alpha}$.] For small n or m, the computations are not difficult. For larger samples, they can become tedious. Fortunately, there are simpler formulas for many of the common distances. If d is a Hoeffding distance, then using the matrix notation (3.11) for the rank vectors, we have from (3.13), (3.35), and (3.68) that

$$\widehat{\alpha}_{\text{Hoeff}} = \frac{1}{n^2}\sum_{i=1}^{n}\sum_{j=1}^{n} tr(Q_i \Delta Q'_j) = tr(\widehat{M}\Delta\widehat{M}'). \tag{3.69}$$

For Spearman, use (3.36). Since each row and column of $\widehat{M}$ adds to 1, $\widehat{M}1_m = \widehat{M}'1_m = 1_m$, so that

$$\begin{aligned}\widehat{\alpha}_{\text{Spear}} &= tr(\widehat{M}\Delta_{\text{Spear}}\widehat{M}') = 2\cdot 1'_m f_m - 2(\widehat{M}e_m)'(\widehat{M}e_m) \\ &= \frac{m(m+1)(2m+1)}{3} - 2\|\overline{y}\|^2\end{aligned} \tag{3.70}$$

since $\widehat{M}e_m = \overline{y}$. For Footrule, use $\Delta_{\text{Foot}} = 1_m e'_m + e_m 1'_m - 2F_m F_m$ from (3.36) to derive

$$\widehat{\alpha}_{\text{Foot}} = tr(\widehat{M}\Delta_{\text{Foot}}\widehat{M}') = m(m+1) - 2\sum_{i=1}^{m}\sum_{j=1}^{m}(\sum_{k=m-j+1}^{m} \widehat{M}_{ij})^2. \tag{3.71}$$

See Exercise 3.21. Hamming has $\Delta_{\text{Ham}} = J_m - I_m$; hence

$$\widehat{\alpha}_{\text{Ham}} = tr(\widehat{M}\Delta_{\text{Ham}}\widehat{M}) = tr(\widehat{M}J_m\widehat{M}') - tr(\widehat{M}\widehat{M}') = m - ||\widehat{M}||^2. \tag{3.72}$$

For Kendall (2.29), Exercise 3.21 shows that

$$d_{\text{Ken}}(x,y) = \frac{\overline{m}}{2} - 2\cdot(\kappa(x) - \frac{1}{2}1_{\overline{m}})'(\kappa(y) - \frac{1}{2}1_{\overline{m}}), \tag{3.73}$$

for κ in (2.59). Thus

$$\begin{aligned} \widehat{\alpha}_{\text{Ken}} &= \frac{\overline{m}}{2} - \frac{2}{n^2}\sum_{i=1}^{n}\sum_{j=1}^{n}(\kappa(x^{(i)}) - \frac{1}{2}1_{\overline{m}})'(\kappa(x^{(j)}) - \frac{1}{2}1_{\overline{m}}) \\ &= \frac{\overline{m}}{2} - 2||\widehat{K}^* - \frac{1}{2}1_{\overline{m}}||^2 \end{aligned} \tag{3.74}$$

since $\widehat{K}^* = \overline{\kappa}$.

We have seen some of these statistics already. The $\widehat{\alpha}_{\text{Spear}}$ for Spearman is equivalent to the statistic for the Means test in Table 3.1. It was introduced by Kendall. Hamming's $\widehat{\alpha}_{\text{Ham}}$ is equivalent to the Marginals statistic. The $\widehat{\alpha}_{\text{Ken}}$ for Kendall was proposed by Ehrenberg (1952). Note that this statistic is close to that for the Pairs statistic of Table 3.1. The latter includes an extra term with the average ranks.

The other distances, Cayley, Ulam, and Maximum, do not appear to have a similar simplification. For any distance, we have

$$\widehat{\alpha} = \widehat{P}'D\widehat{P}, \tag{3.75}$$

where D is the $m! \times m!$ matrix with $D_{xy} = d(x,y)$, the indices running over $\mathcal{S}_m$ in the same order as for $\widehat{P}$. See Alvo, Cabilio, and Feigin (1982).

The first two moments of $\widehat{\alpha}$ under uniformity are obtained from the corresponding moments of $d(e_m, Y)$. Since d is label-invariant, for fixed $x \in \mathcal{S}_m$, $d(x, Y) = d(Q^{(x)}e_m, Q^{(Y)}e_m) = d(e_m, Q^{(x)\prime} Q^{(Y)}e_m)$. Since $Q^{(Y)} \sim Uniform(\mathcal{S}_m)$ implies that $QQ^{(Y)} \sim Uniform\ (\mathcal{S}_m)$ for any fixed $Q \in \mathcal{Q}_m$, $E(d(x,Y)) = E(d(e_m, Y)) = \delta_0$ for any $x \in \mathcal{S}_m$. Thus $E(d(X,Y)) = E[E(d(X,Y) \mid X)] = \delta_0$, from which we immediately have

$$E(\widehat{\alpha}) = \frac{n-1}{n} E(d(X,Y)) = \frac{n-1}{n}\delta_0 \qquad (3.76)$$

since $d(y^{(i)}, y^{(i)}) = 0$. For the variance of $\widehat{\alpha}$, we show that the covariances between distinct terms in the summation are 0. If i, j, k, l are all distinct, then $d(Y^{(i)}, Y^{(j)})$ and $d(Y^{(k)}, Y^{(l)})$ are clearly independent. If $j = l$, but i, j, k are distinct, then

$$\begin{aligned} &Cov(d(Y^{(i)}, Y^{(j)}), d(Y^{(k)}, Y^{(j)})) \\ &\quad = E[Cov(d(Y^{(i)}, Y^{(j)}), d(Y^{(k)}, Y^{(j)})|Y^{(j)})] \\ &\qquad + Cov[E(d(Y^{(i)}, Y^{(j)})|Y^{(j)}), (E(d(Y^{(k)}, Y^{(j)})|Y^{(j)})] \end{aligned} \qquad (3.77)$$

The covariance within the expectation is 0 since, given $Y^{(j)}$, the two distances are independent. The two expectations within the covariance are both δ_0; hence the covariance is 0. Only the variance terms in $\widehat{\alpha}$ remain, and these are $Var(d(e_m, Y))$ using the same reasoning as for the means. Thus,

$$Var(\widehat{\alpha}) = \frac{2(n-1)}{n^3} Var(d(e_m, Y)). \qquad (3.78)$$

As in Section 2.5, it is usually easier to interpret $\widehat{\alpha}$ by normalizing to mimic a correlation coefficient. Since there is no particular sense of direction for these concordances, the normalization more closely resembles the R^2 in a regression: $R^2 = 1$ indicates maximal concordance, i.e., everybody has the same rank ($\widehat{\alpha} = 0$), and R^2 of 0 will indicate total non-concordance, which we take to mean uniformity for $\widehat{P}$ in (2.39), $\widehat{P} = u_{m!}$, so that $\widehat{\alpha} = u'_{m!} D u_{m!} = \delta_0$. See Exercise 3.22. Thus

$$R^2 = 1 - \frac{\widehat{\alpha}}{\delta_0} \qquad (3.79)$$

is the measure of concordance.

It is important to point out that the coefficient is often normalized by the maximum,

$$Max\{\alpha\} \equiv \sup_{P \in \mathcal{P}_{m!}} P'DP, \tag{3.80}$$

where $\mathcal{P}_{m!}$ is the simplex (2.16), so that $0 \leq R^2 \leq 1$. In many cases, $\delta_0 = Max\{\alpha\}$ as we will see in Remark 3.5. It is generally not true, however, for Cayley or Ulam. See Exercise 3.23. Thus it is possible for R^2 in (3.79) to be negative. The main reason we are using δ_0 is (as in Section 2.5) to have "0" denote uniformity. Another practical reason is that it is typically easier to figure out δ_0 than $Max\{\alpha\}$.

To use $\widehat{\alpha}$ or R^2 as a test statistic, we need to find the asymptotic distribution under uniformity. As noted above, we have done this for Spearman and Hamming. The next theorem, from Alvo, Cabilio, and Feigin (1982), gives the result for the general case (3.75), and the subsequent one specializes to Hoeffding distances.

Theorem 3.5
Under uniformity, for $\widehat{\alpha}$ in (3.75), $m!n(\widehat{\alpha} - \delta_0) \rightarrow \chi^2[H_{m!}D]$. □

Proof
From (3.5) and Lemma 3.1, we have that $m!n(\widehat{P} - u_{m!})'\ D\ (\widehat{P} - u_{m!}) \rightarrow \chi^2[H_{m!}D]$. Expanding yields $(\widehat{P} - u_{m!})'D(\widehat{P} - u_{m!}) = \widehat{\alpha} - 2u'_{m!}D\widehat{P} + u'_{m!}Du_{m!}$. Each row and column of D contains the same values, but in a different order, since d is label-invariant. Thus $u'_{m!}D = \delta_0 1'_{m!}$, $u'_{m!}D\widehat{P} = \delta_0 1'_{m!}\widehat{P} = \delta_0$, and $u'_{m!}Du_{m!} = \delta_0$. Thus the quadratic form equals $\widehat{\alpha} - \delta_0$, proving the result. □

Theorem 3.6
For Hoeffding distance (2.25) and (3.69), $(m-1)n(\widehat{\alpha}_{\text{Hoeff}} - \delta_0) \rightarrow \sum_{i=1}^{m-1} C_i^2$, where C_i^2's are independent $\chi^2[H_m\Delta]$'s. □

See Exercise 3.24. We note that, as will be seen in Remark 3.5, the matrices $H_{m!}D$ and $H_m\Delta$ are often non-negative definite (Section 12.1), which means their eigenvalues are less than or equal to 0. Hence the statistics in the above two theorems will be negative. The results for Spearman and Hamming appear in Sections 3.2.4 and 3.2.3, respectively. Footrule can be found using Theorem 3.5, but there does not appear to be a simple description of the eigenvalues of $H\Delta_{\text{Foot}}$. They are the same as the eigenvalues of $H_m \Delta_{\text{Ham}} H_m = H_m F_m F_m H_m$. Kendall can be handled directly. See Alvo, Cabilio, and Feigin (1982), Theorem 4.3. By Lemma 3.1,

$$n\|\widehat{K}^* - \frac{1}{2}1_{\overline{m}}\|^2 \longrightarrow \chi^2[\frac{1}{12}(I_{\overline{m}} + \Gamma_m \Gamma_m')]. \tag{3.81}$$

Thus by (3.74),

$$n(3\overline{m} - 6\widehat{\alpha}_{\text{Ken}}) \longrightarrow \chi^2[I_{\overline{m}} + \Gamma_m \Gamma_m'] = (m+1)\chi^2_{m-1} + \chi^2_{\overline{m-1}}, \tag{3.82}$$

where the two χ^2's are independent. The final equation follows from the next lemma.

Lemma 3.8
The eigenvalues of $\alpha I_{\overline{m}} + \beta\Gamma_m\Gamma_m'$ consist of $(m-1)$ values of $\alpha + m\beta$ and $\overline{m-1}$ values of α. □

Proof
The eigenvalues of $\Gamma_m\Gamma_m'$ are the same as those of $\Gamma_m'\Gamma_m$, which equals mH_m. We know that H_m has $(m-1)$ 1's and one 0 for its eigenvalues; hence $\beta\Gamma_m\Gamma_m'$ has $(m-1)$ values of $m\beta$ and the other $\overline{m-1}$ values are 0. To each of these eigenvalues we can add α since $\alpha I_{\overline{m}}$ has $\overline{m}$ values of α, proving the result. □

Example 3.7. Concordances
Table 3.5 calculates the $100 \times R^2$ and its standard error $(100 \times \sqrt{Var(R^2)})$ for a few familiar data sets. The expected value of R^2 under uniformity is $1/n$. From previous examples, we know that the Voting data have small but very significant effects and the Occupations data have quite large effects. The concordances show the same. For the Voting data, The R^2's are generally around 1%, but relative to their standard errors that is very high. On the other hand, the Occupations data have R^2 of almost 66% for Spearman, and over 52% for Kendall. Looking at the Sports data, we know that most of the nonuniformity comes from Jogging, and not as much from Football. The set without Jogging and Football has very low R^2's, most not even significant judging from the standard errors. Adding Football boosts the R^2's somewhat, and including all sports pumps some of the R^2's to around 4%, small but signif-

Table 3.5 *Concordances.*

Data Set → **Distance ↓**	**Voting** **100 · R^2**	 **se**	**Occupations** **100 · R^2**	 **se**	**Sports** **100 · R^2**	 **se**
Spearman	0.658	0.012	65.582	0.328	4.740	0.442
Footrule	0.894	0.008	47.589	0.212	3.485	0.293
Kendall	0.548	0.010	52.349	0.245	3.905	0.344
Hamming	0.815	0.006	16.180	0.110	1.773	0.181
Cayley	1.263	0.008	22.977	0.164	2.692	0.256
Ulam	0.571	0.008	28.124	0.156	2.787	0.244
Maximum	1.029	0.008	38.032	0.174	4.426	0.248

Data Set → **Distance ↓**	**Sports: No Jog** **100 · R^2**	 **se**	**Sports: No Jog, Foot** **100 · R^2**	 **se**
Spearman	0.957	0.485	0.214	0.542
Footrule	1.144	0.326	0.651	0.373
Kendall	0.978	0.384	0.305	0.442
Hamming	1.143	0.217	1.077	0.271
Cayley	2.002	0.299	1.868	0.361
Ulam	1.194	0.286	0.635	0.346
Maximum	3.393	0.281	2.271	0.331

icant. Basically, there are no new surprises, but the concordances are highly capable of revealing nonuniformity.

An interesting aspect is to see which concordances seem to do best, in the sense of having a high R^2, within data sets. It is not especially meaningful to compare the R^2's directly since they are not on the same scale: a unit $(i-j)^2$ for Spearman and one $I[i \neq j]$ for Hamming are noncomparable. One rough way to assess the relative effectiveness of a distance is to use $R^2/se(R^2)$. For the Voting data, all the distances have large ratios, but the best are Hamming, Maximum, and Cayley. Spearman and Kendall are worst. In the Occupations data, Spearman, Footrule, Kendall, and Maximum are fairly similar, with ratios around 200; Ulam is almost as good; and Hamming and Cayley are worst. Turning to the Sports data, with the complete data, all distances hover around 10 except for Maximum which has a ratio of about 18. When Jogging is dropped, Maximum is best with 12, but Hamming and Cayley (5.3, 6.7) are next best, followed by Ulam (4), Footrule

(3.5), Kendall (2.5), and Spearman (2). Dropping Football too does not change the relative order of effectiveness, although now Ulam, Spearman, Kendall, and Footrule are even less effective.

Tentative conclusions (these are only three data sets, after all) are that when there is strong agreement around a single order, as in Occupations, the Spearman/Footrule/Kendall/Maximum group does best. If there is distinct lack of uniformity, but not in the direction of clustering about a single ranking, then Hamming and Cayley are quite effective, but interestingly so is Maximum. Recall Section 3.3. The distances that satisfy the Henery monotonicity are good in the former case, and the bi-invariant distances are good in the latter, as might be expected, at least in retrospect. □

Remark 3.5. The Maximum of α

If $\widehat{\alpha}$ in (3.75) is maximized by the uniform $\widehat{P} = u_{m!}$, then we know $Max\{\alpha\} = \delta_0$. Feigin and Alvo (1986) give conditions under which the uniform does maximize $\widehat{\alpha}$, which we now describe. For a general distance, set $G = \delta_0 J_{m!} - D$. By construction, $u'_{m!} G u_{m!} = 0$. If G is positive semi-definite (Section 12.1), then for any other probability vector $\pi \in \mathcal{P}_{m!}$ of (2.16), $\pi' G \pi \geq u'_{m!} G u_{m!}$. But $\pi' J_{m!} \pi = 1$ for any probability vector; hence $u'_{m!} D u_{m!} \geq \pi' D \pi$, so that (3.75) is maximized by $u_{m!}$. Since $1'_{m!} G = 0$, $H_{m!} G H_{m!} = G$; hence G is positive semi-definite if and only if $H_{m!} D H_{m!}$ is negative semi-definite. It is not always easy to determine whether $H_{m!} D H_{m!}$ is negative semi-definite, but as usual there is a simpler formula for Hoeffding distances. The argument is basically the same, but we let $G = \delta_0 J_m - \Delta$, so it is now $m \times m$, and look at $\mathrm{tr}(MGM')$ where $M = E_P(Q)$. Then $\overline{J}_m G \overline{J}_m = 0$ and $M J_m M' = J_m$; hence if G is positive semi-definite, $\mathrm{tr}(M \Delta M') \leq \mathrm{tr}(\overline{J}_m \Delta \overline{J}_m)$ for all M. Equivalently, it is enough if $H_m \Delta H_m$ is negative semi-definite. It is easy to show that negative semi-definiteness holds for Spearman, Footrule, and Hamming: the $H_m \Delta H_m$'s are, respectively, $-2 H_m e_m e'_m H_m$, $-2 H_m F_m F_m H_m$, and $-H_m$. The result for Kendall is also immediate since clearly $\widehat{\alpha}_{\mathrm{Ken}}$ (3.74) is maximized when all $\widehat{K}_{ij} = 1/2$.

3.6 Summary

This chapter covered several tests of uniformity, but which one(s) should be used? It depends mainly on what non-uniformity one is interested in or suspects, but also has to be guided by sample size, number of objects, and ease of implementation. All the statistics are easy to calculate, although some of the distance and concordance measures, especially Cayley, Ulam, and Maximum, may be tedious when the numbers of objects and observations are large. The asymptotic distributions of the Probabilities, Means, Marginals, and Pairs tests are simple χ^2's. The distances all have convenient asymptotics, too. Some trouble occurs with some of the concordance tests. Spearman and Hamming coincide with Means and Marginals, so they are fine. Kendall is asymptotically a linear combination of two χ^2's which is not too hard to handle, and pointwise close to the Pairs test. The other concordances require more extensive linear combinations based on finding eigenvalues of a particular matrix.

If one has a fixed target rank vector, and is interested to see the extent to which the data cluster near it, the distance tests are indicated. I would otherwise tend to concentrate on the four general test statistics, using whichever the data can bear: If $n \geq 5m!$, the Probabilities test; if $n \geq 5m^2$ the Marginals test; if $n \geq 3m^2$ the Pairs test, and if $n \geq 3m$, the Means test. These recommendations are very conservative, based on having about five observations per degree of freedom. The approximations should work for lower factors, and in any case it is straightforward to simulate the distributions. (See Exercise 3.25.) The distance tests work even for $n = 1$, and for Kendall, Hamming, Cayley, and Maximum, exact formulas for the densities are available.

3.7 Exercises

3.1. Show that if the subspace S is orthogonal to $S_0 = Span\{1_{m!}\}$, then $\mathcal{P}^S u_{m!} = 0$, where $\mathcal{P}^S$ is the projection matrix for S.

3.2. Using the setup in the previous problem, show that $\mathcal{P}^S H_{m!} = \mathcal{P}^S$.

3.3. Show that $\text{tr}(H_K) = K - 1$.

3.4. Let $Q \sim Uniform(\mathcal{S}_m)$ for $m > 3$. Write

$$0 = Var(\sum_{i=1}^{m}\sum_{j=1}^{m} Q_{ij}) = a_m Var(Q_{11}) + b_m Cov(Q_{12}, Q_{13}) \\ + c_m Cov(Q_{13}, Q_{23}) + d_m Cov(Q_{12}, Q_{34}).$$

Show that $a_m = m^2$, $b_m = c_m = m^2(m-1)$ and $d_m = m^2(m-1)^2$; hence $Cov(Q_{12}, Q_{34}) = 1/(m^2(m-1))$.

3.5. Verify the equation for $Cov(Q)$ in (3.16).

3.6. Show that if A and B are idempotent, then so is $A \otimes B$. For any square matrices, show that $\text{tr}(A \otimes B) = \text{tr}(A) \times \text{tr}(B)$.

3.7. Show that $e_m' H_m e_m = v_m$ of (2.3). [Recall Exercise 2.1.]

3.8. Verify (3.26).

3.9. Verify (3.27). [Show that $\Gamma_m' \Gamma_m = mH_m$ and $\Gamma_m H_m = \Gamma_m$.]

3.10. Hoeffding distance d_{Hoeff} is bi-invariant if $d_{\text{Hoeff}}\,(Q^{(y)}e_m, e_m) = d_{\text{Hoeff}}\,(Q'Q^{(y)}Qe_m, e_m)$ for all $Q \in \mathcal{S}_m$. Thus

$$\text{tr}(Q^{(y)}\Delta) = \text{tr}(Q^{(y)}Q\Delta Q') = E(\text{tr}(Q^{(y)}Q\Delta Q')) = \text{tr}(Q^{(y)}B),$$

where $B = E(Q\Delta Q')$, and the expected value is over $Q \sim$ *Uniform* $(\mathcal{Q}_m)$. Show that B_{kk} is the average of the diagonal terms of Δ, i.e., 0, and B_{kl} for $k \neq l$ is the average of the off-diagonal terms, i.e., $b \equiv \sum_{1 \le i < j \le m} a_{ij} / \binom{m}{2}$. Thus $B = b(J_m - I_m)$; hence $d_{\text{Hoeff}} = b \cdot d_{\text{Ham}}$.

3.11. Verify (3.36). Show that for non-negative a and b, $|a - b| = a + b - 2\min\{a, b\}$, and $(F_m F_m)_{ij} = \min\{i, j\}$.

3.12. Verify the variance formula in Table 3.3 for the Footrule:

(a) Recall that $\sum_{i=1}^m i^k$ equals $m(m+1)/2$, $m(m+1)(2m+1)/6$, $m^2(m+1)^2/4$ and $m(m+1)(2m+1)(3m^2+3m-1)/30$ for $k = 1, 2, 3, 4$, respectively.

(b) Use (3.39), where $a_{ij} = |i-j|$. Show that $a_{\cdot\cdot} = 2\sum_{k=1}^{m-1} k(m-k)/m^2 = (m-1)(m+1)/3m$ and $||\Delta||^2 = 2\sum_{k=1}^{m-1} k^2\,(m-k) = m^2(m^2-1)/6$.

(c) Show that $a_{i\cdot} = \sum_{j=1}^m |i-j|/m = (m+1)/2 + i^2/m - (m+1)i/m$. [Note that $\sum_{j=1}^m |i-j| = \sum_{j=1}^{i-1} j + \sum_{j=1}^{m-i} j$.] Use Part (a) to find $\sum_{i=1}^m a_{i\cdot}^2 = (m+1)(m-1)(7m^2-8)/60m$.

(d) Place the results in Parts (b) and (c) into (3.39) to obtain the result.

3.13. By Lemma 3.5 and (3.44), under uniformity we have that $E(d_{\text{Ken}}\,(Y, e_m)) = \sum_{i=1}^{m-1} E(U_i)$ and $Var(d_{\text{Ken}}(Y, e_m)) = \sum_{i=1}^{m-1} Var(U_i)$. Use the results from Exercise 2.1 to verify the results for Kendall's distance in Table 3.3.

3.14. Verify (3.46).

3.15. Prove Lemma 3.7. [Show that the space of (Y^{-m}, Y_m) is $\mathcal{S}_{m-1} \times \{1, \ldots, m\}$, which has $m!$ elements. Thus $\mathcal{S}_m$ and $\mathcal{S}_{m-1} \times \{1, \ldots, m\}$ are in one-to-one correspondence, showing that uniformity on one space implies uniformity on the other. Since the latter space in a product space, Y^{-m} and Y_m are independent.]

3.16. Use (3.51) to verify the results in Table 3.3 for Cayley's distance.

3.17. Verify (3.64). [The first inequality follows since each element in Δ_p is at most $(p-1)^2$, and by the second paragraph in Remark 3.1, the values of the squared residuals can be no larger. For the second inequality, write

$$Beta(p+2, p+2) = \frac{p(p+1)}{(p+2)(p+3)} Beta(p, p+4),$$

and $Beta(p, p+4) < Beta(p, 4) = 6/(p(p+1)(p+2)(p+3))$. Thus $v(p)$ is bounded below by the expression in (3.40) with $Beta(p+2, p+2)$ replaced by the bound just obtained. After much algebra, the resulting lower bound for $v(p)$ is

$$\frac{p^2(2p^4 + 17p^3 + 59p^2 + 127p + 107)}{(2p+1)(p+1)^2(p+2)^2(2p+3)(p+3)^2}.]$$

3.18. Show that the limit in (3.63) for Hamming's distance is 1.

3.19. Prove that (3.65) holds for Cayley's distance for $\nu \geq 3$. [Use $X_i = U_i$ from (3.51) to show that $E|X_i - E(X_i)| \leq 1/i^\nu + 1/i$, and use the variance formula from Table 3.3.]

3.20. Let $\mathcal{O}_1$ be the cycle that contains Y_1, so that for some set of distinct $i_1, \ldots, i_k$ in $\{2, \ldots, m\}$, $Y_1 = i_1$, $Y_{i_1} = i_2$, ..., $Y_{i_k} = 1$, where $\#\mathcal{O}_1 = k+1$. For any fixed set of i_j's, $P[Y_1 = i_1, \ldots, Y_{i_k} = 1] = 1/(m(m-1)\cdots(m-k)$. There are $(m-1)(m-2)\cdots(m-k)$ such sets with k elements; hence $P[\#\mathcal{O}_1 = k+1] = 1/m$.

3.21. Verify (3.71), (3.72), and (3.73). [For the last one, show that

$$(I[y_i < y_j] - \frac{1}{2})(I[x_i < x_j] - \frac{1}{2}) = -\frac{1}{2}(I[(y_i - y_j)(x_i - x_j) < 0] - \frac{1}{2}).]$$

3.22. Show that for any distance matrix, $u'_{m!}Du_{m!} = \delta_0$.

3.23. Show that when $m = 3$, for Cayley and Ulam distances, the $Max\{\alpha\}$ in (3.80) is not attained at $P = u_{m!}$. [Choose P to have all its mass uniformly distributed on the points 123, 312 and 231.]

3.24. Prove Theorem 3.6. [First show that $\widehat{\alpha} - \delta_0 = \text{tr}(\widehat{M} - \overline{J})\Delta(\widehat{M} - \overline{J})' = \underline{(\widehat{M} - \overline{J})'}(I_m \otimes \Delta)\underline{(\widehat{M} - \overline{J})}$. Then use (3.17), Lemma 3.1 and Lemma 12.4.]

3.25. Explain how to simulate from the $Uniform(\mathcal{S}_m)$ distribution by first simulating Cayley's $m - 1$ Bernoulli's in (3.51), and using them to construct the vector $y \in \mathcal{S}_m$ by reversing the steps above (3.51).

CHAPTER 4

Comparing Populations of Judges

4.1 Introduction

The most interesting feature of the data in Example 2.1 is the difference in the rankings given by the two groups of women. This chapter deals with formal tests of such differences. The tools used here are analogs of those in Chapter 3 for testing uniformity. In later chapters we explore modeling the differences.

We will consider two general approaches for testing the equality of populations of judges, both neatly presented in Feigin and Alvo (1986). The multivariate analysis of variance approach chooses a statistic z that captures an interesting aspect of the populations, then compares the average z's for the populations. The other approach assesses agreement or diversity among judges using a distance on rankings, measuring the relative agreement within samples to that between samples.

Throughout this chapter, we assume L populations of judges, and from population l we have a sample of $n^{(l)}$ observations,

$$Y^{(l1)}, \ldots, Y^{(ln^{(l)})}. \tag{4.1}$$

Let $n = n^{(1)} + \cdots + n^{(L)}$. The formal null hypothesis that the populations are equal is expressed

$$H_0^D : P^{(1)} = \cdots = P^{(L)} \text{ versus } H_A^D : P^{(k)} \neq P^{(l)} \text{ for some } k \neq l, \tag{4.2}$$

where $P^{(l)}$ is the Probability vector for the lth population. The "D" in the superscript of the hypotheses denotes "Difference."

The immediately obvious test, at least when n is large relative to $m!$, is the χ^2 test of homogeneity. Thus an $L \times m!$ contingency table is constructed, where the rows indicate the populations and the columns the possible values of $y \in \mathcal{S}_m$. The lk^{th} cell contains the number of observations from population l that equal the kth ranking vector in $\mathcal{S}_m$. The test proceeds as usual, with the statistic having an asymptotic $\chi^2_{(L-1)(m!-1)}$ distribution under H_0. (Throughout this chapter, asymptotics are as $n \to \infty$ such that for each l, $n^{(l)}/n \to \lambda^{(l)}$ for some $\lambda^{(l)} > 0$.) For example, from Table 2.5, we find χ^2 for homogeneity to be 406.39 on 10 degrees of freedom, which tells us what we already know: there are differences due to the judges' place of residence. Typically, there will not be enough data to apply that test, although it is universally consistent in that no matter in what way H_0 fails, the power will approach 1 as $n \to \infty$.

Schucany and Frawley (1973) and Li and Schucany (1975) have a different perspective, looking for agreement between the populations rather than disagreement. The main idea is that if a population has a uniform distribution, then it has no opinion about the objects, and hence it makes no sense to ask whether it agrees or disagrees with another population. Thus we test whether the populations are uniform, where the alternative is that there is some agreement:

$$H_0^C : P^{(l)} = u_{m!}, l = 1, \ldots, L \text{ versus } H_A^C : \text{Agreement.} \quad (4.3)$$

In Section 4.4 we explicitly define "Agreement," which is that the expected distance between two random vectors from different populations is less than the expected distance under uniformity. The "C" means "concordance." It is very possible that both H_0^D and H_0^C will be rejected, i.e., that the populations are different, but agree more than if they were uniform.

In the next section we look at the multivariate analysis of variance tests. Section 4.3 compares populations on their modal rankings and spreads. These sections are aimed at H_0^D. Section 4.4 considers between and within measures of agreement, using both H_0^D and H_0^C.

4.2 The basic statistics: Multivariate analysis of variance tests

As in Chapter 3, there are any number of characteristics of a population of judges that could be of interest when comparing several populations. The vector P of probabilities is the most general, as in the χ^2 test mentioned above. The average ranking is quite popular, although it is possible for the means of two populations to be the same, while the overall populations are quite different. Comparing the Marginals or the Pairs gives more detailed differences, but of course they can be fooled, too.

Distances from modal rankings provide additional possibilities. The question may be whether the populations have the same modal ranking, or whether they are equally clustered about their own, or a common, modal ranking. In Chapter 3, one way to use distances and avoid the search for modal rankings is to use concordance, or inter-ranking distances. This subject is covered in Section 4.4.

Given any $p \times 1$ statistic z as in Chapter 3, let

$$\overline{z}^{(l)} = \frac{1}{n^{(l)}} \sum_{i=1}^{n^{(l)}} z(Y^{(li)}) \tag{4.4}$$

be the average of the statistic for the lth sample. To derive the multivariate analysis of variance statistic, as in, e.g., Mardia, Kent, and Bibby (1979), first define the $m \times L$ matrix

$$W = (\overline{z}^{(1)} - \overline{z}, \ldots, \overline{z}^{(L)} - \overline{z}), \ \overline{z} = \frac{1}{n} \sum_{l=1}^{L} \sum_{i=1}^{n^{(l)}} z^{(li)}. \tag{4.5}$$

Exercise 4.1 shows that, under H_0,

$$E(W) = 0, \ Cov(W) = \Sigma(z) \otimes (\Omega_n^{-1} - \frac{1}{n} J_L), \tag{4.6}$$

where $\Sigma(z)$ is the common covariance matrix of the Z's, $\Omega_n = diag(n^{(1)}, \ldots, n^{(L)})$, and "*diag*" is the diagonal matrix with the given elements on the diagonal. Then

$$Cov(\Sigma^{-1/2}(z) W \Omega_n^{1/2}) = I_p \otimes A_n, \tag{4.7}$$

where $A_n = I_p - (1/n)\Omega_n^{1/2} J_L \Omega_n^{1/2}$. Since $\Sigma(z)$ is unknown, we use the pooled estimate

$$\widehat{\Sigma}(z) = \frac{1}{n-L} \sum_{l=1}^{L} \sum_{i=1}^{n^{(l)}} (z(y^{(li)}) - \overline{z}^{(l)})(z(y^{(li)}) - \overline{z}^{(l)})'. \tag{4.8}$$

Table 4.1 *TV Preferences: the entries are effect (Z for effect).*

Show ↓; T^2 →	Status (G−U) 41.99		Gender (F−M) 81.89		Interaction 7.51	
Sitcom	0.71	(3.25)	−0.27	(−1.25)	−0.12	(−0.53)
Drama	0.15	(0.67)	−1.21	(−5.27)	0.15	(0.66)
News	−1.10	(−5.52)	0.30	(1.52)	0.16	(0.81)
Game	0.37	(1.91)	0.25	(1.31)	0.01	(0.06)
Talk	−0.06	(−0.31)	−0.69	(−3.33)	0.09	(0.42)
Sports	0.20	(0.77)	1.73	(6.85)	−0.09	(−0.35)
Movie	0.10	(0.54)	−0.45	(−2.44)	0.06	(0.30)
Cartoon	−0.38	(−1.55)	0.20	(0.82)	−0.59	(−2.39)
TV Magazine	0.02	(0.08)	0.14	(0.67)	0.33	(1.58)

We will assume that $\Sigma(z)$ is invertible. [To obtain invertible $\Sigma(z)$, we often must take a subvector of the z. For the Probability P and the Mean μ, drop the final element, and for the Marginals M, drop the final row and column. The Pairs matrix will typically have an invertible covariance.]

Exercise 4.2 proves that A_n is idempotent, and $\text{tr}(A_n) = L - 1$. Whence, asymptotically,

$$T^2 = \text{tr}(W'\widehat{\Sigma}^{-1}(z)W\Omega_n) \longrightarrow \chi^2_{p(L-1)} \tag{4.9}$$

under H_0. Exercise 4.3 shows that T^2 can be rewritten as

$$T^2 = \sum_{l=1}^{L} n^{(l)}(\overline{z}^{(l)} - \overline{z})'\widehat{\Sigma}^{-1}(z)(\overline{z}^{(l)} - \overline{z}). \tag{4.10}$$

Example 4.1. TV Preferences: Means Test
Kyungmee Choi asked students to rank nine types of TV show: Situation Comedy, Drama, News, Game Show, Talk Show, Sports, Movie, Cartoon, and TV Magazine. The students are categorized by status, G = Graduate or U = Undergraduate, and gender, F = Female and M = Male. Sample sizes for the groups are GF: 10, GM: 37, UF: 27, UM: 58. We used the test (4.9) with z being the

average. We find $T^2 = 171.32$, which compared to a χ^2_{24}, is very significant. (Here $p = m - 1 = 8$ since we drop the last element of the mean vector.) To see where the differences lie, we look at some effects. Let $\overline{y}^{(l)}$ denote the average rank vector for group l, $l =$ GF, GM, UF, UM. The effects are then Status: $(\overline{y}^{(GF)}+\overline{y}^{(GM)}-\overline{y}^{(UF)}-\overline{y}^{(UM)})/4$; Gender: $(\overline{y}^{(GF)}-\overline{y}^{(GM)}+\overline{y}^{(UF)}-\overline{y}^{(UM)})/4$; and Interaction: $(\overline{y}^{(GF)}-\overline{y}^{(GM)}-\overline{y}^{(UF)}+\overline{y}^{(UM)})/4$. The estimated covariance matrix for each effect, under H_0, is the same, namely $\widehat{\Sigma}_{Effect} = \widehat{\Sigma}(z)\ (1/10+1/37+1/27+1/58)/16$. Table 4.1 contains the effects and the individual components' Z's, where the standard errors are the square roots of the diagonals of the covariance matrix. We also include T^2's, which are $(Effect)'\widehat{\Sigma}^{-1}_{Effect}(Effect)$. From the table, we see significant differences in status, where graduate students are relatively more fond of the News, and undergraduates of Situation Comedies, and in Gender, where Females like Dramas, Talk Shows, and Movies, while Males like to watch Sports. There is not much evidence of interaction, but what there is involves Cartoons. Female Graduate students and Male Undergraduates are more likely to prefer them, although I would guess the two groups enjoy different types of cartoons. □

4.3 Distances from a modal ranking

In Chapter 3, when trying to estimate a modal ranking, we were forced to use resampling methods rather than direct analytical methods. We do the same in this section. Given a distance d, let $\omega^{(l)}$ be the modal ranking for population l with respect to d, that is, $\omega^{(l)}$ minimizes $E_l(d(Y, \omega))$ over $\omega \in \mathcal{S}_m$. The null hypothesis (4.2) implies that

$$\omega^{(1)} = \cdots = \omega^{(L)} \equiv \omega^{(T)}, \tag{4.11}$$

and

$$E_1(d(Y, \omega^{(T)})) = \cdots = E_L(d(Y, \omega^{(T)})) \equiv \delta_T. \tag{4.12}$$

The "T" stands for the total sample. If the $\omega^{(l)}$'s are not unique, we take them to be the set of modes. Testing (4.11) is testing whether the centers of the populations are equal, and (4.12) whether the spreads about $\omega^{(T)}$ are equal.

Start with (4.11). Let $\widehat{\omega}^{(l)}$ be a minimizer over ω of $\sum_{i=1}^{n^{(l)}} d(y^{(li)}, \omega)$ for each group l, and $\widehat{\omega}^{T}$ be a minimizer for the combined sum $\sum_{l=1}^{L} \sum_{i=1}^{n^{(l)}} d(y^{(li)}, \omega)$. If the sample modes $\widehat{\omega}^{(l)}$ are reasonably close to each other, then we should accept (4.11). To assess closeness, we will look at the difference between the distances to the individual modes compared to the distances to the overall mode. Let

$$R^2 = 1 - \frac{\sum_{l=1}^{L} \sum_{i=1}^{n^{(l)}} d(y^{(li)}, \widehat{\omega}^{(l)})}{\sum_{l=1}^{L} \sum_{i=1}^{n^{(l)}} d(y^{(li)}, \widehat{\omega}^{(T)})}. \tag{4.13}$$

(Note that R^2 is well-defined even if some of the sample modes are not unique.) If the sample modes are equal, then $R^2 = 0$, and if there is only one distinct ranking within each sample, but these rankings are not all the same between samples, then $R^2 = 1$. Thus R^2 measures the proportion of the spread explained by the differences in the populations. We wish to reject H_0 if R^2 is too large. The asymptotic sampling properties of this statistic are not very helpful in developing a hypothesis test since, under H_0, $P[R^2 = 0] \to 1$ as $n \to \infty$. A useful alternative is the *randomization p*-value.

The randomization p-value in this case is calculated as follows. (See Feigin and Alvo, 1986.) Suppose H_0 is true, and condition on the collection of observed rankings. Then each partition of the observations into groups of size $n^{(1)}, \ldots, n^{(L)}$ has the same probability, i.e., $1/\binom{n}{n^{(1)},\ldots,n^{(L)}}$. Let x represent the actual data as an $n \times m$ matrix, where the first $n^{(1)}$ columns are from sample 1, the next $n^{(2)}$ from sample 2, ..., and the last $n^{(L)}$ from sample L. Let x^* be a randomization sample from x, i.e., the matrix x with columns randomly permuted, where these permutations are uniformly distributed on $\mathcal{S}_n$. Imagine B independent such permuted matrices $x_1^*, \ldots, x_B^*$. Then the estimated randomization p-value for R^2 is $\widehat{p} = \#\{j | R^2(x_j^*) \geq R^2(x)\}/B$. Under H_0, as $n \to \infty$, $\widehat{p}$ has an approximately $\mathit{Uniform}(0, 1)$ distribution, so that rejecting H_0 when $\widehat{p} \leq \alpha$ gives an approximate level α test.

Example 4.2. Ethnic Preferences: Comparisons
For each ethnic group in Example 2.6, we obtained the sample modal ranking under Spearman's distance. In cases where there is no unique mode, we give the midrank vector, that is, assign to the tied objects the average of the ranks they could have received. See Chapter 11 for more details on midranks. Corresponding to choices in the order (*Asian, Black, Latino, Jewish, Native*

American, White), the modal rankings for the groups are Asian: $(1,5,3,6,4,2)$; Black: $(5,1,2,6,4,3)$; Latino: $(5.5,3,1,5.5,4,2)$; Jewish: $(5,4,6,2,3,1)$; and White: $(3.5,5,6,3.5,2,1)$. As in Examples 2.6 and 2.9, people tend to feel most comfortable within their own ethnic group. To see how the coefficient R^2 behaves, we find the within group sums to be 388, 86, 166, 38, 420, and the overall sum in (4.13) to be 1964, the year of the Civil Rights Act. (The $\widehat{\omega}^{(T)} = (2,3,4,6,5,1)$, reflecting mainly the relative sample sizes.) Thus $R^2 = 1 - 1098/1964 = 0.4409$, i.e., 44% of the spread can be explained by the grouping. To test statistical significance, we calculated 1000 $R^2(x^*)$'s, which ranged between 0.005 and 0.1589 with a median of 0.0530. Thus $\widehat{p} = 0$, and in fact none of the values even came close to the observed R^2, showing that there is indeed a difference between groups. Of more interest is where the differences lie. See Example 9.6 for further analysis. □

The second expression (4.12) is relevant when the populations agree on the modal ranking of the objects, but it is suspected that some populations are more clustered about the mode than others. The situation is basically a one-way analysis of variance, where the observations in group l are $d_{li} \equiv d(y^{(li)}, \widehat{\omega}^{(T)})$, $i = 1, \ldots, n^{(l)}$. The only kicker is that $\widehat{\omega}^{(T)}$ is estimated from the data. Asymptotically there is no problem since under H_0, $P[\widehat{\omega}^{(T)} = 0] \to 1$. Letting $\overline{d}_l$ be the sample average of the distances for population l, and $\overline{d}$ be the average of the entire sample, the F-test is based on

$$F = \frac{n-L}{L} \frac{\sum_{l=1}^{L} n^{(l)} (\overline{d}_l - \overline{d})^2}{\sum_{l=1}^{L} \sum_{i=1}^{n^{(l)}} (d_{li} - \overline{d}_l)^2}. \tag{4.14}$$

Under H_0, $L \times F$ is asymptotically χ^2_{n-L}. Alternatively, one can use the randomization test.

Remark 4.1. Paired Rankings

Paired rankings occur when each judge ranks the objects twice. It may be the rankings are assigned before and after some treatment, as in Example 4.3 below, or that each judge is a husband–wife pair, each person supplying a ranking. In such cases, the random quantity is $(X, Y) \in \mathcal{S}_m \times \mathcal{S}_m$, and data are $(x^{(i)}, y^{(i)})$, $i = 1, \ldots, n$, where the pairs are independent and identically distributed. Note that X and Y are not necessarily independent. The null hypothesis is that X and Y have the same distribution. One approach to testing is to choose a statistic z on $\mathcal{S}_m$ and find the average difference:

$$\overline{w} = \frac{1}{n}\sum_{i=1}^{n} w_i, \; w_i = z(x^{(i)}) - z(y^{(i)}). \tag{4.15}$$

Under the null, $E(\overline{W}) = 0$. The $Cov(z(X) - z(Y))$ can be estimated by

$$\widehat{\Sigma}(w) = \frac{1}{n-1}\sum_{i=1}^{n}(w_i - \overline{w})(w_i - \overline{w})'. \tag{4.16}$$

Assuming $\widehat{\Sigma}$ is invertible, the statistic $T^2 = n\overline{W}'\widehat{\Sigma}^{-1}(w)\overline{W}$ is asymptotically χ_p^2 as in (4.9).

Example 4.3. Literary Criticism: Testing Equality
Critchlow and Verducci (1992) present Rogers' (1991) data wherein 38 ninth graders ranked four styles of literary criticism, once at the beginning of a course in literary criticism, and once at the end. The prime interest is whether the rankings at the end tended to be closer to the teacher's ranking than at the beginning. The four styles are A = Authorial, C = Comparative, P = Personal, and T = Textual. The teacher's ranking is $(3, 2, 1, 4)$, corresponding to the order PCAT. The average rankings for Before and After are, respectively, $(2.00, 2.68, 3.03, 2.29)$ and $(2.32, 1.87, 3.11, 2.71)$. We see that the ranking of Comparative improved by almost a full point, while the others went down a bit.

To test whether the observed difference is significant, we set $z(y) = (y_1, \ldots, y_{m-1})'$, and find that $T^2 = 16.82$. Compared to a χ_3^2, T^2 is indeed large.

To address whether the After rankings are closer to the teacher's, we select $z(y) = d(y, \omega)$, where $\omega = (3, 2, 1, 4)$ is the teacher's. We will use Kendall's distance. The Before and After average distances are 3.7895 and 3.2105, respectively. Thus $\overline{w} = 0.5789$, and since $\widehat{\sigma}^2(w) = 4.0341$, $T^2 = 3.16$. Compared to a χ_1^2, this T^2 is not significant. Thus though there is definitely a movement between before and after, it does not appear to be strongly in the direction of the teacher's ranking. Further analysis appears in Example 9.5. □

4.4 Concordance and diversity

One drawback to the procedure in Section 4.3 is that the modal rankings need to be found. (Of course, they may be of interest in themselves.) As in Section 3.5, we can use the interpoint distances to circumvent the need for the modes. The basic idea stems from Rao (1982), which Feigin and Alvo (1986) apply to rankings. Given a distance d, the population diversity is $\alpha = E(d(X,Y))$, where X and Y are independent observations from the population. With P being the Probability vector, and D the distance matrix as in (3.75), the diversity is the population analog of $\widehat{\alpha}$:

$$\alpha = P'DP. \tag{4.17}$$

As per Remark 3.5, we will assume throughout this section that $H_{m!}DH_{m!}$ is negative semi-definite, so that in particular $\alpha \leq \delta_0$. Now let $P = \lambda^{(1)}P^{(1)} + \ldots + \lambda^{(L)}P^{(L)}$ be the "total" probability, and $\alpha_{\text{Total}} = P'DP$. Rao's apportionment of diversity is given by

$$\alpha_{\text{Total}} = \alpha_{\text{Within}} + \alpha_{\text{Difference}}, \tag{4.18}$$

where

$$\alpha_{\text{Within}} = \sum_{l=1}^{L} \lambda^{(l)}\alpha_l, \quad \alpha_l = P^{(l)\prime}DP^{(l)}, \tag{4.19}$$

and

$$\begin{aligned} \alpha_{\text{Difference}} &= \alpha_{\text{Total}} - \alpha_{\text{Within}} \\ &= -\sum_{1\leq k<l\leq L}\sum \lambda^{(k)}\lambda^{(l)}(P^{(k)} - P^{(l)})'D(P^{(k)} - P^{(l)}) \\ &= \sum_{l=1}^{L} \lambda^{(l)}(P^{(l)} - P)'D(P^{(l)} - P). \end{aligned} \tag{4.20}$$

The α_{Within} measures the average diversity within the populations, and $\alpha_{\text{Difference}}$ measures the difference between the populations. Clearly, $\alpha_{\text{Difference}} = 0$ if the populations are equal. Exercise 4.4 shows that $\alpha_{\text{Difference}} \geq 0$ in any case. The sample analogs of the above quantities are obtained by substituting $\widehat{P}^{(l)}$ for $P^{(l)}$, and $n^{(l)}/n$ for $\lambda^{(l)}$. The diversity formulas simplify for certain distances as in Chapter 3, and the $\widehat{\alpha}_{\text{Difference}}$ is most easily obtained by subtraction.

Feigin and Alvo (1986) discuss coefficients for assessing the degree of similarity between the populations. The ρ^2 below measures the proportion of the spread in the total population that is explained by the differences between populations:

$$\rho^2_{D:T} = \frac{\alpha_{\text{Difference}}}{\alpha_{\text{Total}}}. \tag{4.21}$$

The sample analog is

$$R^2_{D:T} = \frac{\widehat{\alpha}_{\text{Difference}}}{\widehat{\alpha}_{\text{Total}}}. \tag{4.22}$$

Thus if there is no difference between the populations, $R^2_{D:T} = 0$. If the populations are maximally homogeneous within themselves, which means that all the rankings are the same, but there is some difference between the populations, then $R^2_{D:T} = 1$. (Feigin and Alvo, 1986, use $1 - R^2_{D:T}$, which measures how similar the populations are.) Alternatively, one may compare the diversity between the populations relative to the absolute maximum. In that case we would use

$$\rho^2_{D:U} = \frac{\alpha_{\text{Difference}}}{\delta_0}. \tag{4.23}$$

Zero still means the populations are the same, but now for $\rho^2_{D:U}$ to be 1 we need the rankings within each sample to be the same, and those between samples to spread out uniformly. Letting

$$\rho^2_{T:U} = 1 - \frac{\alpha_{\text{Total}}}{\delta_0} \tag{4.24}$$

be the concordance coefficient (3.79) for the total population, we have that

$$\rho^2_{D:U} = \rho^2_{D:T}(1 - \rho^2_{T:U}), \tag{4.25}$$

so that the more stringent assessment of differences between populations can be decomposed into that relative to the total, and the total's relative to the uniform.

Schucany and Frawley (1973) and Li and Schucany (1975) approach the problem from the other side: How much do the populations agree relative to the uniform? That is, the null hypothesis is that the populations are uniform, and the alternative is that they show some agreement, as in (4.3). Populations agree if they tend to be closer to each other than if they were uniform. The measure of disagreement among populations is

$$\begin{aligned}\beta &= E(d(Y^{(K)}, Y^{(K')})) \\ &= \frac{\sum\sum_{1\le k<l\le L} \lambda^{(k)}\lambda^{(l)} E(d(Y^{(k)}, Y^{(l)}))}{\sum\sum_{1\le k<l\le L} \lambda^{(k)}\lambda^{(l)}} \\ &= \frac{\sum\sum_{1\le k<l\le L} \lambda^{(k)}\lambda^{(l)} P^{(k)\prime} D P^{(l)}}{\sum\sum_{1\le k<l\le L} \lambda^{(k)}\lambda^{(l)}}.\end{aligned} \tag{4.26}$$

Here, $Y^{(K)}$ and $Y^{(K')}$ are independent representatives from populations K and K', respectively, and K and K' are chosen with probabilities $P[K = k] = P[K' = k] = \lambda^{(k)}$, conditioned on $K \neq K'$. Small values of β indicate agreement, the maximal agreement being when $\beta = 0$, i.e., all populations are concentrated at the same single ranking. Neither agreement nor disagreement is represented by $P^{(l)} = u_{m!}$ for all l, in which case β equals δ_0. The coefficient is normalized so that 0 means uniformity, and 1 means total agreement. Negative values indicate disagreement. Thus,

$$\gamma^2_{B:U} = 1 - \frac{\beta}{\delta_0}, \tag{4.27}$$

and the alternative hypothesis in (4.3) is

$$H_A^C : \gamma^2_{B:U} > 0, \tag{4.28}$$

which does depend on the distance chosen. The sample coefficient is

$$C^2_{B:U} = 1 - \frac{\widehat{\beta}}{\delta_0}, \tag{4.29}$$

where $\widehat{\beta}$ is the sample average of all distances between rankings from different samples. Note that $\gamma^2_{B:U} = 0$ implies equality of the populations, but not vice versa. In fact, if H_0 of (4.2) holds then $\gamma^2_{B:U}$ can take on any value between 0 and 1. Note that $R^2_{D:T}$ being significant means there are differences in the populations,

Table 4.2 *Ethnic Preferences: diversity coefficients.*

Statistic	Kendall	Spearman	Footrule	Hamming
$\hat{\alpha}_{\text{Total}}$	6.38	28.48	10.14	4.69
$\hat{\alpha}_{\text{Within}}$	4.11	15.83	6.79	3.72
$\hat{\alpha}_{\text{Difference}}$	2.27	12.65	3.36	0.97
$\hat{\beta}$	7.14	32.79	11.28	5.03
δ_0	7.50	35.00	11.67	5.00
$100 \cdot R^2_{D:U}$	30.29	36.15	28.77	19.46
$100 \cdot R^2_{D:T}$	35.62	44.43	33.09	20.74
$100 \cdot (1 - R^2_{T:U})$	85.03	81.37	86.93	93.83
$100 \cdot C^2_{B:U}$	4.78	6.32	3.28	−0.59

and $C^2_{B:U}$ being significant means there are similarities. It is perfectly reasonable to have a situation where both are true, i.e., even though the populations are different, there is more concordance than under uniformity.

Remark 4.2
Feigin and Alvo (1986) discuss a coefficient related to $\gamma^2_{B:U}$:

$$\frac{\delta_0 - \alpha_{\text{Total}}}{\delta_0 - \alpha_{\text{Within}}} = \frac{\rho^2_{T:U}}{\rho^2_{T:U} + \rho^2_{D:U}} \tag{4.30}$$

It is 1 when the populations are the same, but 0 when the combined sample is uniform. □

Example 4.4. Ethnic Preferences and Ballots: Diversity Coefficients
To illustrate the above discussion, Table 4.2 contains all the statistics one would desire for four distances, using the data in Example 4.2. The coefficients are presented as percentages. (The coefficients in the table are based on $\widehat{\alpha}$'s with more significant digits than presented in the table.) The $R^2_{T:U}$'s for Kendall, Spearman, and Footrule indicate that the diversity in the combined sample is about 80–85% of the maximum, and the $R^2_{D:T}$'s show that 35–45% of that diversity is explained by the differences in the groups. Hamming is slightly different: the combined sample is 94% as diverse as the uniform, but the group differences explain only about 21% of

Table 4.3 *Ballots: diversity coefficients.*

Statistic	Kendall	Spearman	Footrule	Hamming
$100 \cdot R^2_{D:U}$	0.79	0.81	0.78	0.74
$100 \cdot R^2_{D:T}$	0.80	0.83	0.79	0.75
$100 \cdot (1 - R^2_{T:U})$	98.28	98.10	98.37	98.64
$100 \cdot C^2_{B:U}$	1.59	1.77	1.09	1.21

the diversity. The difference between Hamming and the rest may be due to its relative lack of sensitivity to strong agreement on a single object, which is the main feature here, the object being the ethnicity of the group.

Either of $R^2_{D:T}$ and $R^2_{D:U}$ can be used to test H_0 in (4.2). Analytic approximations to the null distribution seem difficult, but randomization is straightforward. Since both statistics are monotone in $\widehat{\alpha}_{\text{Within}}$, for the same distance they produce the same randomization p-value. In this case, all four estimated p-values are 0.

The concordance coefficient $C^2_{B:U}$ measures the similarity relative to the uniform. The first three distances show there is some similarity, but far from the maximal. To assess significance, we randomly generated samples from the uniform. Using 1000 samples, we found all to have $C^2_{B:U}$ less than the observed. Hamming shows a slight negative value, meaning the samples are less similar than expected under the uniform. Since 959/1000 of the random samples showed $C^2_{B:U}$ larger than the observed, there is even some evidence this difference is significant.

Another example is the Ballots from Example 2.12. Brook and Upton (1974) separate the rankings into six geographic areas. See Table 4.3. Now the total diversity is over 98% of the maximum, but the difference in areas explains less than 1% of the diversity. Note that all four distances give very similar results. The estimated p-values are 0.092, 0.114, 0.077, and 0.119, respectively, which are not quite significant. On the other hand, even though the concordances in the fourth row are very small, they are significantly different from 0 since their estimated p-values are 0. Thus there is agreement, but it is not very strong. □

Remark 4.3
Feigin and Alvo (1986) show that, under (4.2), the distance measures are in fact based on $X'\widehat{P}$'s (2.50). Write

$$\delta_0 - \alpha_{\text{Total}} = P'(\delta_0 J_{m!} - D)P. \tag{4.31}$$

Since we are assuming that $\delta_0 J_{m!} - D$ is positive semi-definite, and it is symmetric, there exists an $m! \times q$ matrix X such that

$$XX' = \delta_0 J_{m!} - D. \tag{4.32}$$

Thus $\delta_0 - \widehat{\alpha} = ||X'\widehat{P}||^2$, so that by Lemma 3.1, under uniformity $n||X'\widehat{P}||^2$ approaches $\chi^2[X'H_{m!}X]$. Similarly, we have that

$$n\widehat{\alpha}_{\text{Between}} = \sum_{l=1}^{L} n^{(l)}(\widehat{P}^{(l)} - \widehat{P})'XX'(\widehat{P}^{(l)} - \widehat{P}), \tag{4.33}$$

which is the same as T^2 in (4.9) but without the estimated covariance. □

4.5 Exercises

4.1. Prove (4.6). [For the covariance, note that $\overline{z} = \sum_{l=1}^{L} (n_l/n)\, \overline{z}^{(l)}$.]

4.2. Show that A_n in (4.7) is idempotent, and $\text{tr}(A_n) = L - 1$. [Note that $J_L \Omega_n J_L = nJ_L$.]

4.3. Prove (4.10).

4.4. Show that under the condition that $H_{m!}DH_{m!}$ is negative semi-definite, $x'Dx \leq 0$ if $x'1_{m!} = 0$; hence $\widehat{\alpha}_{\text{Difference}} \geq 0$ in (4.20).

CHAPTER 5

Overview of Models

5.1 Introduction

The exploratory techniques described in the previous chapters begin the investigation into the structure of rank data. Formal modeling of the data can provide even deeper understanding. Two different approaches have developed over the years:

I Model the ranking process;
II Model the population of rankers.

The dichotomy is far from strict, but is useful to keep in mind when evaluating and implementing models. Approach I is hard modeling: how to describe mathematically the physical/behavioral/mechanical/neurological processes that a judge undergoes to produce a ranking of the objects. Approach II is more data-analytic: how to describe parametrically the distribution of rankings attached to a population of judges.

Thurstone (1927) pioneered Approach I by defining "discriminal processes." Each object elicits a physical response, measurable on a one-dimensional scale, and the corresponding ranking is based on the order of the responses. Canonical examples are assessing shades of gray, or frequencies of musical tones. There is a true underlying scale, and the measuring instrument (your eye or ear, for example) attempts to place the objects on the scale. Alternatively, ranks can be built from "choice" processes: given a set of objects, choose the best. Luce (1959) carefully studied such experiments, developing an axiom for choice behavior. For example, one might start by picking the best of all the objects, then the best of the remaining, etc., which leads to multistage models as in Plackett (1975), Henery (1981), and Fligner and Verducci (1988). Unfolding and multidimensional scaling models posit relationships between objects and judges, usually represented in some space. The closer the judge is to an object, the more preferred is the object. See Coombs (1964) as the basic tome.

Think about how you go about ranking a set of objects. If the objects are cookies and you have a sweet tooth, you taste each cookie, assess its sweetness, and out pops the ranking Thurstonianly. Suppose you need to rank several candidates for a job. It is likely you will make judgments in a highly multidimensional and subjective context. Perhaps you first pick off the obviously least qualified, then separate the rest into groups, then make paired comparisons to modify and refine the groupings. It is an iterative procedure. For an interesting early discussion of these issues, read Moran (1950), Whitfield (1950), Daniels (1950b), and the ensuing discussion.

When modeling rankings from a sample of judges, Approach I presumes each judge (or at least each judge in some subset of judges) makes basically the same judgment about the items. This assumption might make sense when there is an objective scale, such as sweetness, the judges are trying to reproduce. However, by scanning the examples in the first three chapters, most do not easily fit that profile.

Approach II is less ambitious. Rather than trying to explain how judges perform their rankings, it takes the rankings as the units and attempts a simple description of the population based on the sample. Approach I considerations help in developing models for Approach II, and vice versa, so that it is unnecessary to worry too much about the dichotomy. In any case, modeling has several benefits:

- Theory-based models can be fit and tested;
- Main features of the data can be revealed;
- After fitting the main features, other important features may become clear;
- Formal tests will determine that what one is seeing in the model is actually there (significance testing);
- Formal tests will determine if anything is there beyond what one is seeing (goodness-of-fit testing).

5.2 Probability models — General

We have to be careful of the distinction between rankings and orderings. Recall from Chapter 1 that $\mathcal{T}_m$ represents the set of permutations of the objects in $\mathcal{O}$, while $\mathcal{S}_m$ is the set of permutations of $\{1, 2, \ldots, m\}$. Since $\mathcal{O}$ is often the same set of integers, it is difficult to distinguish between $\mathcal{S}_m$ and $\mathcal{T}_m$. The two sets are in one-to-one correspondence through the inverse function. Thus the ranking $y \in \mathcal{S}_m$ of the objects corresponding to $Q^{(y)} \in \mathcal{Q}_m$ satisfies $Q^{(y)}e_m = y$ (3.11), while the corresponding ordering $w \in \mathcal{T}_m$ is

$$w = Q^{(y)'}e_m. \tag{5.1}$$

Some models are naturally defined on the rankings, some on the orderings. Of course, a model on one immediately induces a model on the other. In addition, it is possible to apply a model on one to the other to create a new, possibly nonsensical, model.

In Chapter 2 we take a distribution P to be a vector in the simplex $\mathcal{P}_{m!}$ of (2.16). A probability model is a family of probability distributions, i.e., a subset of $\mathcal{P}_{m!}$. Typically, this subset will be parametrized by θ in a space Θ: $\{P(\theta) \mid \theta \in \Theta\} \subset \mathcal{P}_{m!}$, where $P(\theta)$ is a function from Θ to $\mathcal{P}_{m!}$. Usually, Θ will be a (low-) dimensional Euclidean space. The distribution on $\mathcal{S}_m$ will be given through its density $f_\theta(y)$ with respect to the *Uniform* distribution (2.18). [Or substitute $(w, \mathcal{T}_m)$ for $(y, \mathcal{S}_m)$.] Thus

$$P_y(\theta) = P[Y = y; \theta] = \frac{1}{m!} f_\theta(y), \quad y \in \mathcal{S}_m, \quad \theta \in \Theta. \tag{5.2}$$

(Throughout the book, densities on finite sample spaces will be given with respect to probability distributions, most often the *Uniform*, rather than counting measure.)

There are two extreme models: the *Uniform* has but one distribution, whose density is $f \equiv 1$, and the *saturated*, which is the whole space $\mathcal{P}_{m!}$. These extreme models are good to use as benchmarks. The *Uniform* is the null model; one often first tests whether to reject it before trying to look for structure in the data as in Chapter 3. It is a submodel of almost all models. On the other hand, every model is contained in the saturated model. Thus one can test for goodness-of-fit of a particular model by comparing it to the saturated model. The hope is to find a model that is meaningful and fits the data well.

Although we generally consider (2.17), i.e., we have a sample from a distribution P, modeling can be useful even without such formal structure. Just as the statistics and graphs from Chapters 2 and 3 help to see the structure in the set of observations at hand, so can fitting a model be a purely descriptive enterprise. Think of it as smoothing the observed $\widehat{P}$ to enhance interpretability. Modeling consists of finding the distribution $P(\theta)$ that is closest to $\widehat{P}$. Throughout this book, we use likelihood to measure closeness. See Section 5.14.

The bulk of this chapter consists of introductions to specific models. Chapters 6 through 10 go into some of these models in more depth. Section 5.13 considers some overall evaluation of models, including Luce's choice axiom and notions of consensus and unidimensionality. Section 5.14 presents the basic likelihood methods that will be used extensively.

5.3 Thurstonian ≡ Order statistic models

Let W represent the random ordering of the objects. Each object has a continuous but unobserved random variable associated with it. Call these variables $Z_1, Z_2, \ldots, Z_m$, where Z_i is that associated with Object i. They are meant to represent a (continuous) response the judge has to the objects, e.g., assessment of the cookies' sweetness. A joint model is assumed for the vector Z of the Z_i's. The observed ordering of the objects is then given by the order of the Z_i's, that is,

$$W = (i_1, i_2, \ldots, i_m) \text{ if and only if } Z_{i_1} < Z_{i_2} < \cdots < Z_{i_m}. \quad (5.3)$$

For example, if $\mathcal{O} = \{Red,\ Yellow,\ Green\}$, then $W = (Yellow,\ Green,\ Red)$ if and only if $Z_{Yellow} < Z_{Green} < Z_{Red}$. The probability of any ranking of the objects is the probability of the corresponding order of the Z_i's.

The parameters for the model are those needed to describe the distribution of the vector Z. Thurstone (1927), who invented the model, proposed using the normal distribution; hence the parameters included the m means, m variances, and $\binom{m}{2}$ correlations. He also suggested simplifications such as equating the correlations, equating the variances, and/or setting the correlations to 0 so that the Z_i's are independent.

Daniels (1950a) continued by looking at cases in which the Z_i's are independent and from a location-family model with possibly different location parameters. That is, for some continuous density g, the density of Z is

$$\prod_{i=1}^{m} g(z_i - \eta_i) \quad (5.4)$$

with respect to Lebesgue measure on $\Re^m$. Luce (1959) uses the Gumbel distribution [see Yellott (1977)], and Henery (1983) and Stern (1990) use Gamma distributions. The class of Thurstonian models is much richer than suggested by (5.4). Any multivariate model contemplated on means and variances can be extended to rank data through the Z's in (5.3). See Böckenholt (1993) for more complex models.

5.4 Distance-based models

Often, one assumes there is a "modal" ranking $\omega \in \mathcal{S}_m$ of the objects, and judges are expected to have rankings more or less close to ω. Appropriate models would give higher probability to rankings closer to the modal ranking; the question is how to measure the distance. Section 2.5.1 presents a number of choices. Models based on these distances are usually exponential family models, as presented later in Section 5.7. For given mode ω, distance function d, and real parameter θ, the density with respect to the $Uniform(\mathcal{S}_m)$ is

$$f_\theta(y;\omega) = \exp(\theta d(\omega, y) - \psi(\theta)), \tag{5.5}$$

where ψ is the normalizing constant. Generally, θ is to be estimated; ω may or may not be known.

A famous model, Mallows' (1957) ϕ model, is (5.5) with Kendall's distance (2.29). Mann (1945), Schulman (1979), and Feigin and Cohen (1978) have also developed this model. Mallows also suggests Spearman's distance, which is called Mallows' θ model. See Section 5.5.2. These and other such models can be found in Fligner and Verducci (1986), including the models based on Cayley and Hamming distances. Chapter 6 delves further into distance models.

5.5 Paired comparison models – Babington Smith

An alternative to asking someone to rank the m objects is to have the ranker choose which of each pair of objects is preferred. Thus instead of giving the ordering (*Yellow, Red, Green*), one would say "Yellow is preferred to Red," "Red is preferred to Green," and "Yellow is preferred to Green." For m objects there are $\binom{m}{2}$ such comparisons to make. It should be clear that given a ranking, one can easily determine what the pairwise preferences are. However, it is not always the case that a set of pairwise preferences corresponds to a ranking. For example, one might prefer Yellow to Red, Red to Green, and Green to Yellow. There are a number of papers that look at paired comparison data in which such nontransitivities are allowed. Kendall and Babington Smith (1940) is perhaps the seminal one. David (1988) is the most comprehensive volume on paired comparisons. Also of interest is Pendergrass and Bradley (1960), which analyzes sets of rankings of three objects out of m. In a stochastic setting, there is nothing surprising about nontransitivities. For example, in any sport it is not overly unusual that A beats B, B beats C, and C beats A. Exercise 5.1 gives a game-theoretic example. We are taking the view that people construct a legitimate ranking by starting with paired comparisons, but report their preferences only after having a consistent set of comparisons, i.e., a set that yields an unambiguous ranking. The resulting ranking model is referred to as the "Babington Smith model" (Smith, 1950).

We look at orderings, so that w_i is the index of the object ranked i^{th}. The general model is based on the $\binom{m}{2} \times 1$ parameter vector p whose indices ij, $i < j$, are ordered as in (2.58). The p_{ij} is interpreted as the probability Object i would be preferred to Object j if only that comparison were to be made. Ties are not allowed, so that $p_{ji} = 1 - p_{ij}$. A ranking is obtained by making independently all the pairwise comparisons using those probabilities. If the comparisons yield a consistent ordering, that ordering is the W. If not, start over with the pairwise comparisons. One continues until the comparisons are consistent.

The probability that the paired comparisons are consistent is

$$c(p) = \sum_{w \in \mathcal{T}_m} \prod_{i<j} p_{w_i w_j}. \tag{5.6}$$

For $m = 3$ and $\mathcal{O} = \{A, B, C\}$, $c(p) = p_{AB}p_{AC}p_{BC} + p_{AC}p_{AB}p_{CB} + p_{BA}p_{BC}p_{AC} + p_{BC}p_{BA}p_{CA} + p_{CA}p_{CB}p_{AB} + p_{CB}p_{CA}p_{BA}$. The probability of an ordering w given that the pairwise comparisons are consistent is the probability that the comparisons yield w divided by the probability they are consistent. Thus the Babington Smith model has density

$$f_p(w) = \frac{m!}{c(p)} \prod_{i<j} p_{w_i w_j} \tag{5.7}$$

with respect to *Uniform*$(\mathcal{S}_m)$. Note that it is not necessary that the p_{ij}'s be consistent. That is, it may be odd but it is legitimate to have $p_{12} = 0.9$ and $p_{23} = 0.95$ but $p_{13} = 0.1$.

This model can be unwieldy if m is at all large since there are many parameters, and the constant $c(p)$ is not very compact: it consists of a sum of $m!$ products of $\binom{m}{2}$ terms. The next two subsections consider some simplifications.

Warning
Recall that p_{ij} is the probability of Object i being preferred to Object j if that comparison is the whole experiment. However, the probability that i is preferred to j after the entire ranking experiment has been performed is $P[Y_i < Y_j]$, which is different from p_{ij}. □

5.5.1 *Bradley–Terry/Mallows*

Here, the parameters p_{ij} are given special forms. Bradley and Terry (1952) proposed positive constants $v = (v_1, \ldots, v_m)'$, where v_i is associated with Object i, setting

$$p_{ij} = \frac{v_i}{v_i + v_j}. \tag{5.8}$$

The idea is that the larger v_i, the more preferred is Object i. Bradley and Terry were thinking of just paired comparisons, but Mallows (1957) suggested substituting the Bradley–Terry form of the p_{ij}'s into the Babington Smith model for ranks. Then for an ordering $w \in \mathcal{T}_m$, inserting (5.8) into (5.7) yields

$$f_v(w) = \frac{1}{c^*(v)} \prod_{i=1}^{m} v_{w_i}^{m-i} \tag{5.9}$$

since "v_{w_i}" appears in the numerator the same number of times as Object w_i is preferred to the other objects. [Here, $c^*(v) = c(v)\prod_{i<j}(v_i + v_j)$.] This model has only m parameters; in fact, since the probabilities are invariant to multiplying the v_i's by a positive constant, there are only $m - 1$ free parameters.

5.5.2 Mallows' models

Mallows (1957) simplified the Babington Smith model (5.7) by assuming other special forms for the p_{ij}'s. His two-parameter model sets

$$p_{ij} = \frac{1 + \tanh((j-i) \cdot \log(\eta) + \log(\phi))}{2}. \tag{5.10}$$

We will just look at the two one-parameter subfamilies obtained by setting either η or ϕ to 1. Mallows' ϕ model sets $\eta = 1$. The paired-comparison probability of ranking Object i before Object j then depends only on whether $i < j$ or $j > i$. Thus the p_{ij}'s for $i < j$ are all equal. It is usual to parametrize this probability as

$$p_{ij} = \frac{\exp(\theta \cdot I[i > j])}{1 + \exp(\theta)}. \tag{5.11}$$

Placing these (5.11) in the Babington Smith model (5.7) yields the ϕ model. Exercise 5.2 shows that the result is the distance model (5.5) based on d_{Ken} of (2.29).

Mallows' "θ" model sets $\phi = 1$. It is a Bradley–Terry/Mallows model with the special form for the v_i's being $v_i(\gamma) = \exp(\gamma \cdot i)$; hence $p_{ij} = \exp(\gamma \cdot i)/(\exp(\gamma \cdot i) + \exp(\gamma \cdot j))$. Inserting these into the Babington Smith model (5.7) gives

$$g_\gamma(w) = \frac{1}{c^{**}(\gamma)} \exp(-\gamma \cdot \sum_{i=1}^{m} i w_i) \tag{5.12}$$

for an appropriate constant c^{**}. Since $d_{\mathrm{Spear}}(e_m, w) = (const) - 2\sum i \cdot w_i$, we see that (5.12) is (5.5) where d_{Spear} is Spearman's distance, (2.23) with $p = 2$. (Set $\gamma = 2\theta$.) The difference between Mallows' ϕ and θ models is that in the latter p_{ij} depends on the difference $i - j$, while in the former the p_{ij} depends only on the sign of $i - j$.

5.6 Multistage models

5.6.1 Plackett–Luce

Plackett (1975) presents a model motivated by an m-horse race. See also Luce (1959). Each horse has a positive number v_i assigned to it as in the Bradley–Terry/Mallows model, but v_i denotes the probability Horse i wins when all m horses race. If Sea Biscuit, Citation, Mr. Ed, and Man-o-War are the $m = 4$ horses, then $v_{Citation}$ is the probability that Citation wins when all four horses race. The probability Mr. Ed comes in second given that Citation is first is assumed to be the same as the probability he comes in first when just he, Sea Biscuit, and Man-o-War race, which probability is given to be $v_{Mr.\ Ed}/(v_{Mr.\ Ed} + v_{Sea\ Biscuit} + v_{Man\text{-}o\text{-}War})$. The general model is

$$P[W = w; v] = \prod_{i=1}^{m} \frac{v_{w_i}}{v_{w_i} + v_{w_{i+1}} + \cdots + v_{w_m}}. \tag{5.13}$$

Note that the last term in the product is 1 since it is the probability that the last horse wins when it is running alone. This model turns out to be the same as Thurstone's model with the Gumbel density. See Section 5.13.1.

Silverberg (1980, 1984) calls this model a vase model. The idea is that one has a vase with an infinite number of balls, a proportion of v_{Red} are Red, v_{Blue} are Blue, etc. The ordering W is generated by first randomly drawing a ball, and assigning its color to W_1. Another ball is drawn. If it is a different color, that color is assigned to W_2. If it is the same as W_1, it is thrown away and another ball is drawn. This process continues until a new color is drawn for W_2. More draws are made until a color distinct from W_1 and W_2 is found, etc.

Example 5.1: Feed Flavors

Nombekela, Murphy, Gonyou, and Marden (1993) report on several experiments used to determine cows' preferences of various flavors added to their food. The experiments proceed stepwise. The first day, the cow is presented five food bins, each having a different flavoring. At the end of the day, the favorite flavoring, the one for which the largest amount is consumed, is given rank 1. The next day, the cow has only four bins of food, the favorite from the previous day having been eliminated. The favorite of these is given rank 2. The process continues, with the cow being presented each day with the previously unchosen flavors. (Bins were randomized, and extra empty bins were placed at the extremes to counter edge effects.) In Example 8.1 we fit the Plackett–Luce model to the data. □

Table 5.1 *Stages in the component model.*

e_m	**Stage 1**	**Stage 2**	**Stage 3**	**Stage 4** $\equiv W$
A	**B**	B	B	B
B	A	**D**	D	D
C	C	A	**E**	E
D	D	C	A	**C**
E	E	E	C	A
	$U_1 = 1$	$U_2 = 2$	$U_3 = 2$	$U_4 = 1$

5.6.2 *Free and ϕ-component models*

Fligner and Verducci (1986, 1988) define a different multistage model. It has the same stages as the Plackett–Luce model, but the probabilities at each step are different. The ordering W is obtained by operating iteratively on the vector e_m as follows. First, choose which object will be ranked #1, assign it to W_1, and move it to the first slot in the vector, keeping the other objects in the same relative order. Let U_1 be the number of places the object moved. The second ranked object is chosen from the remaining. Assign it to W_2, move it to the second slot, and keep the other $m - 2$ objects in the same order. Now U_2 is the number of places W_2 moved. Continue until there are two objects left. Decide which is preferred, and place it in the $(m-1)^{st}$ slot. Then U_{m-1} is the number of places it moved; hence is either 0 or 1. Thus,

$$U_i = \sum_{j=i+1}^{m} I[w_i > w_j]. \tag{5.14}$$

It is easier to see an example. Suppose $\mathcal{O} = \{A, B, C, D, E\}$, and the resulting ordering is $W = (B, D, E, C, A)$. Table 5.1 exhibits the stages in transforming ABCDE to BDECA.

Of course, these U_i's can be defined for any model. However, the crux of the Fligner and Verducci model is that the distribution for W is that induced from one on $U \equiv (U_1, U_2, \ldots, U_{m-1})$ in which the U_i's are independent. The range of U_i is $\{0, \ldots, m-i\}$, so that the space of U is

$$\mathcal{U}_m \equiv \{0,\ldots,m-1\} \times \{0,\ldots,m-2\} \times \cdots \times \{0,1\}. \qquad (5.15)$$

It can be shown that there is a one-to-one correspondence between $\mathcal{U}_m$ and $\mathcal{T}_m$, so that knowing the U_i's is enough to know w. (See Lemma 7.5.) The models for U are of the form $h(u) = h_1(u_1)\cdots h_{m-1}(u_{m-1})$, $u \in \mathcal{U}_m$, where h_i is a density on $\{0, \ldots, m-i\}$ with respect to *Uniform*$(\{0,\ldots,m-i\})$. The *free* model allows the individual h_i's to be unrestricted. Other models restrict the h_i's to some parametric form. A special case is the ϕ-component model in which U_i is assumed to have a truncated geometric distribution with parameter θ_i, i.e., the density with respect to *Uniform*$(\{0,\ldots,m-i\})$ is

$$h_{i\theta_i}(u_i) = \exp(\theta_i \cdot u_i - \phi(\theta_i; m-i+1)),\ u_i = 0,\ldots,m-i, \quad (5.16)$$

where for positive integer q,

$$\phi(\gamma; q) = \log\left(\frac{1-\exp(\gamma q)}{q(1-\exp(\gamma))}\right). \qquad (5.17)$$

Then the density of $U = (U_1,\ldots,U_{m-1})$ is

$$h_\theta(u) = \exp\left(\sum_{i=1}^{m-1} \theta_i \cdot u_i - \sum_{i=1}^{m-1} \phi(\theta_i; m-i+1)\right) \qquad (5.18)$$

with respect to *Uniform*$(\mathcal{U}_m)$. This model (5.18) is called the ϕ-component model because when the θ_i's are all equal, the density is that of Mallows' ϕ model (5.5) and (2.29) since $d_{\text{Ken}}(e_m, w) = \sum u_i$ of (5.14).

[When $\theta_i = 0$ for all i, then the U_i's have the same distribution as in Lemma 3.5. The only difference between the U_i's here and those in (3.44) is that here they are defined on the orderings, while there they are defined on the rankings. Under uniformity, the distributions are the same.]

Each stage above involves choosing the one favorite object among those left. More general stages could choose sets of objects, so that, for example, if $\mathcal{O} = \{A, B, C, D, E\}$, at the first stage one might divide the objects into two groups $\{A, C, D\}$ and $\{B, E\}$, the interpretation being that A, C, and D are all preferred to B and E. The second stage may choose A as the favorite among $\{A, C, D\}$, the third stage choose C above D, and the fourth choose E above B. The resulting ordering is then ACDEB. A model analogous to the Fligner and Verducci ones would then declare the choices at each stage independent. See Section 7.9.

5.7 Sufficient statistic models

The models up until now have had a hard-modeling flavor. Each model assumes that ranking is performed by building upon smaller units, whether paired comparisons or stages or distances. Once the probability density is determined, it is possible to determine whether there are any sufficient statistics, that is, whether the information in the data about the parameters can be obtained from some functions of the data, which is especially useful when there is a large number of judges.

Formally, suppose A is a random vector or matrix with space $\mathcal{A}$. We will treat only the case that $\mathcal{A}$ is finite. We assume we have a set of densities $\{g_\theta(a) | \theta \in \Theta\}$ with respect to $Uniform(\mathcal{A})$. Suppose $S(a)$ is a function on the space $\mathcal{A}$.

Definition 5.1: Sufficiency
S is sufficient for the model if the conditional distribution of A given S does not depend on θ. That is, $P_\theta[A = a | S(A) = s]$ is independent of θ. □

Fisher's Factorization Theorem (Hogg and Craig, 1978) states that S is sufficient if and only if there exist functions $c(a)$ and $b_\theta(s)$ such that

$$g_\theta(a) = c(a) \cdot b_\theta(S(a)). \tag{5.19}$$

Thus the parameter θ "touches" the data a only through the function $S(a)$.

For many models, the sufficient statistic can give a great reduction of the data. For example, if one has a sample (2.17) following a distance model (5.5), then the sufficient statistic is $\sum_{i=1}^{n} d(y^{(i)}, \omega)$. Thus one number summarizes the data. Other models show other reductions. The Thurstonian models typically do not allow much reduction. For the Babington Smith paired comparison models the $\binom{m}{2}$-dimensional sufficient statistic consists of the unique elements of the Pairs matrix, $\widehat{K}^*$ of (2.60), i.e., the number of people who prefer Object i to Object j, $i < j$. The Bradley–Terry/Mallows model has sufficient statistic $\overline{y}$. See Exercises 5.4 and 5.5. Such reductions when possible can be very helpful.

A softer, more data-analytic approach to modeling does not start with a mechanism and then try to find a density and hence sufficient statistic; rather, one starts with some numerical summaries of the data one thinks capture the main features of the data, and then creates a model that has those summaries for the sufficient statistics. For example, the data $y^{(1)}, \ldots, y^{(n)}$ might be summarized by the Marginals $\widehat{M}$ of (2.10). The corresponding model would be as in (5.19) with $S = \widehat{M}$. One still must choose c, b, and θ. Section 9E of Diaconis (1988), and some references therein, motivate the *exponential family* distributions: if $S = (S_1, S_2, \ldots, S_p)$, then

$$f_\theta(y^{(1)}, \ldots, y^{(n)}) = \exp\left(\sum_{j=1}^{p} \theta_j \cdot S_j - n\psi(\theta)\right), \tag{5.20}$$

where θ is $p \times 1$. This is a very flexible and rich set of models. The next two subsections exhibit some particular examples of exponential family models.

5.8 Loglinear models

A loglinear model posits a linear model on the logarithms of the probabilities. Assume that $P_y > 0$ for all y, and let $\log(P)$ be the vector of logarithms, where $\log(P)_y = \log(P_y)$. As in Section 2.6, X is an $m! \times v$ matrix whose columns represent interesting functions on $\Re^{m!}$. We still require $X'1_{m!} = 0$. The corresponding loglinear model is then given by

$$\log(P(\theta)) = X\theta - (\psi(\theta) + \log(m!))1_{m!} \tag{5.21}$$

for $\theta \in \Theta = \Re^v$, where $\psi(\theta)$ is the normalizing constant, i.e., $\log(E(\exp(\theta' X' \zeta^Y)))$, ζ is the indicator vector for y in (2.47), and the expected value is over $Y \sim \mathit{Uniform}(\mathcal{S}_m)$. Such models are special cases of the generalized linear models treated in McCullagh and Nelder (1989). Given a sample, the joint density of the data under $P(\theta)$ of (5.21) is shown in Exercise 5.5 to be

$$f_\theta(y^{(1)}, \ldots, y^{(n)}) = \exp(\theta' X' N - n\psi(\theta)). \tag{5.22}$$

Thus $X'N$, or equivalently, $X'\widehat{P}$, is a sufficient statistic.

Equation (5.22) turns any set of statistics into a model. Paul Holland [see Diaconis (1988), p. 172] and Verducci (1982) suggest using the Marginals $\widehat{M}$. There are potentially $(m-1)^2$ statistics, hence parameters, but simpler distributions can be found by making some restrictions. For example, let S_i be the number of people who rank Object i #1. Another model might let S_{ki} be the number of people who rank Object i k^{th}, but only $k = 1$ and 2 are used. More complicated models can be built with the extended Marginals (2.72). See Silverberg (1980) and Plackett (1975) in addition to Diaconis (1988, 1989). Using $\widehat{K}$, we obtain the Babington Smith model (5.7), and using the higher-order comparisons (2.83), we obtain the inversion models of McCullagh (1993a).

5.9 ANOVA-like models

In contingency table analysis, an important set of hierarchical models consists of those that mimic the usual models in analysis of variance. Given T factors, models defined by setting various interactions to zero (respecting a certain hierarchy) include models for the independence of the factors, conditional independence, two-way interaction, three-way interaction, etc. These models have proven very useful. See Bishop, Fienberg, and Holland (1975); Agresti (1990); and Santner and Duffy (1989).

Rank data can be placed in this framework, where there are many structural zeroes, i.e., many cells are *a priori* known to be empty. Specifically, consider the m-way contingency table where each factor has m levels. The factors are the rankings, and the levels within each factor are the objects. Each ordering corresponds to a unique cell, but there are many cells that correspond to no ordering. Table 5.2 is an example for $m = 3$ with $\mathcal{O} = \{A, B, C\}$. Thus the ordering CAB has factor 1 being C, factor 2 being A, and factor 3 being B. Note that the upper-left cell (and many others) is empty since all three factors cannot be A simultaneously, i.e., A cannot be ranked 1, 2, and 3.

The usual independence, conditional independence and other models do not make sense since the rankings are inherently nonin-

Table 5.2 *The Orderings for* $m = 3$ *arranged in a* 3×3 *times* 3 *contingency table: "–" indicates an unobservable cell.*

Rank 1 → Rank 2 ↓	Rank 3 ↓	A	B	C
A	A	–	–	–
	B	–	–	CAB
	C	–	BAC	–
B	A	–	–	CBA
	B	–	–	–
	C	ABC	–	–
C	A	–	BCA	–
	B	ACB	–	–
	C	–	–	–

dependent. However, quasi-independence, quasi-conditional independence, etc., models can be defined using the same parametrization as for tables without empty cells. One selling point for this approach is that it easily extends (or contracts) to cases where judges rank only their top q, $q < m$: just drop factors $q+1, \ldots, m$ from consideration. Silverberg (1980, 1984) proposed such models. The spectral analysis of Diaconis (1988) resolves degrees-of-freedom problems in the higher-order models. Croon (1989a) develops these models further, especially as applied to latent structure, and gives natural constraints on the parameters to achieve identifiability (see Condition R1 in Section 5.14) in some cases.

For ordering $w \in \mathcal{T}_m$, the quasi-independence model sets

$$log(P_w) = \mu + \alpha^1_{w_1} + \alpha^2_{w_2} + \cdots + \alpha^m_{w_m}, \tag{5.23}$$

where α^i_j is the effect of Object j having rank i. The μ is the normalizing constant. In a full contingency table, model (5.23) is the regular independence model. The idea for ranking is that there is an effect for each object at each rank that acts on the probability multiplicatively, but that there is no interaction between objects other than that imposed by the fact that the data consist of rankings. Two-way interactions are modeled by adding to (5.23) parameters of the form α_{kl}^{ij} that represent the effect of Objects k and l being ranked i and j, respectively. Three- and higher-way interactions are similarly defined.

The roles of Objects and Ranks can be interchanged, so that the factors are the objects and the levels are the ranks. As long as for each k, either all or none of the k-factor interactions are present, the two models are the same. However, quasi-conditional independence models will differ, so that having Ranks 1 and 2 quasi-conditionally independent given Rank 3 is different than having Objects A and B quasi-conditionally independent given Object C.

It turns out that the quasi-independence model (5.23) is the same as the Marginals model in the previous section. Similarly, the model with all two-way interactions is equivalent to that based on all counts of the form $\#\{y \mid y_k = i \text{ and } y_l = j\}$.

Tallis and Dansie (1983) propose models in which the raw P_y's are linear in the parameters. These models need extra linear constraints on the parameters in order for P to remain in $\mathcal{P}_{m!}$.

5.10 Nested orthogonal contrast models

The previous section showed how to turn a set of interesting statistics into a model. Can the same be accomplished with the clustering diagram in Figure 2.10? There we saw the objects form a hierarchical grouping based on the judges' preferences. The largest grouping compared the Team sports to the others. The smallest compared Football to Basketball, and Cycling to Swimming.

The approach taken in Chung and Marden (1991, 1993) and Marden (1992) is to define "contrasts" and "nested orthogonal contrasts" for objects. A contrast compares two or more groups of objects. Two contrasts are orthogonal if the comparisons they represent are not confounded. For example, the Team versus Non-team contrast is not confounded with the Football versus Baseball contrast since the overall preference of Team sports to Non-team sports does not necessarily imply anything about the relative preferences within the Team sports, specifically Football versus Baseball. These ideas are formalized next.

Definition 5.2
A *contrast* is an element $C \in \lambda(\mathcal{O})$ of (2.69) for some ordered partition λ (Definition 2.4.b) with $\lambda_+ \leq m$. □

Note that a given contrast

$$C = (\mathcal{O}_1, \mathcal{O}_2, \ldots, \mathcal{O}_K) \tag{5.24}$$

carries along the set $\mathcal{O}^C \equiv \cup_{k=1}^K \mathcal{O}_k$ and the vector $\lambda^C = (\#\mathcal{O}_1, \ldots, \#\mathcal{O}_K)$.

The idea is that contrast C represents a comparison between the sets of objects $\mathcal{O}_k$. For example, the contrasts for the Sports data represented in Figure 2.10 are

$$\begin{aligned} C_1 &= (\{\textit{Baseball, Football, Basketball}\},\\ &\qquad \{\textit{Tennis, Cycling, Swimming, Jogging}\});\\ C_2 &= (\{\textit{Baseball}\}, \{\textit{Football, Basketball}\});\\ C_3 &= (\{\textit{Football}\}, \{\textit{Basketball}\});\\ C_4 &= (\{\textit{Jogging}\}, \{\textit{Tennis, Cycling, Swimming}\});\\ C_5 &= (\{\textit{Tennis}\}, \{\textit{Cycling, Swimming}\}) \text{ and}\\ C_6 &= (\{\textit{Cycling}\}, \{\textit{Swimming}\}). \end{aligned} \tag{5.25}$$

The above all contain just $K = 2$ groups of objects. One with three groups is

$$\begin{aligned} C_7 =&(\{\textit{Baseball, Football, Basketball}\}, \{\textit{Tennis}\},\\ &\{\textit{Cycling, Swimming, Jogging}\}), \end{aligned} \tag{5.26}$$

which compares Team to Tennis to Individual sports.

The contrasts in (5.25) are orthogonal; C_1 and C_7 are *not* orthogonal since knowing the Team versus Non-team preferences will certainly reveal much about the Team versus Tennis versus Individual preferences. The formal definition of orthogonality follows.

Definition 5.3
Two contrasts, C as in (5.24) and

$$D = (\mathcal{O}_1^*, \ldots, \mathcal{O}_L^*) \in \lambda^D(\mathcal{O}), \tag{5.27}$$

are *orthogonal* if either (i) $\mathcal{O}^C \cap \mathcal{O}^D = \emptyset$; (ii) $\mathcal{O}^C \subseteq \mathcal{O}_l^*$ for some l; or (iii) $\mathcal{O}^D \subseteq \mathcal{O}_k$ for some k. □

Part (i) means the two contrasts compare totally different sets of objects (disjoint); (ii) and (iii) state that all the objects in one of the contrasts are contained in the same group of the other contrast (nested). Thus C_1 and C_2 are orthogonal by virtue of nesting (iii), while C_2 and C_5 are disjoint (i). Contrast C_7 is orthogonal to C_2, C_3, and C_6, but not to C_1, C_4, or C_5.

Any set of orthogonal contrasts can be represented by a tree diagram as in Figure 2.10, and any tree diagram represents a set of orthogonal contrasts. With m objects, a set of orthogonal contrasts can have at most $m - 1$ contrasts, and then only if each contrast contains exactly two groups.

In order to build a model, we have to relate contrasts to rankings. The value a particular judge has for a particular contrast C is defined to be the set of ranks of the objects in the subsets $\mathcal{O}_k$ relative to the objects in $\mathcal{O}^C$. The relative ranks of the objects *within* the groups are irrelevant. Consider contrast C_1 above. A judge who likes the three Team sports better than any of the others would have a value of $(\{1,2,3\},\{4,5,6,7\})$ for the contrast. This value is not meant to give any information about which of the Team sports is preferred, i.e., $(\{2,3,1\},\{6,5,7,4\})$ is the same value. For C_2, a value $(\{1\},\{2,3\})$ means Baseball is the favorite of the three Team sports; $(\{2\},\{1,3\})$ means Baseball is second favorite, but we do not know which is first and which is third.

The precise meaning of "value" of a contrast is given next.

Definition 5.4
For $y \in \mathcal{S}_m$ and C a contrast (5.24), the *value* of the contrast at the ranking is

$$C(y) = (y^*[\mathcal{O}_1], \ldots, y^*[\mathcal{O}_K]) \in \mathcal{V}(\lambda) \equiv \lambda(\{1, \ldots, \lambda_+\}), \quad (5.28)$$

where y_i^* is the rank of y_i relative to the ranks $y[\mathcal{O}^C]$. See (2.71). □

Note that this definition is the same as (2.70) if $\lambda_+ = m$. If $\lambda_+ < m$, then the relevant values in y are reranked to provide an element of $\mathcal{S}_{\lambda_+}$.

Continuing the Sports example, suppose $y = (3,7,5,6,2,4,1)$. The values of the seven contrasts in (5.25) and (5.26) are $C_1(y) = (\{3,5,7\},\{1,2,4,6\})$, $C_2(y) = (\{1\},\{2,3\})$, $C_3(y) = (\{2\},\{1\})$, $C_4(y) = (\{1\},\{2,3,4\})$, $C_5(y) = (\{3\},\{1,2\})$, $C_6(y) = (\{1\},\{2\})$, and $C_7(y) = (\{3,5,7\},\{6\},\{1,2,4\})$. The value for C_2 is found by first finding the ranks for the Team sports: $(3,7,5)$. Then these are reranked relative to each other to obtain the y_k^*'s: $(1,3,2)$. Since $\mathcal{O}_1 = \{1\}$, the first group is assigned Rank 1, and the second group Ranks 2 and 3. The value for C_3 is $(\{2\},\{1\})$ since, of Football and Basketball, the latter is preferred.

Knowing the value of a contrast gives only partial information about the entire ranking y. However, if there are enough contrasts it is possible to reconstruct the entire ranking. In fact, y itself is the value of the contrast $(\{1\},\ldots,\{m\})$. Section 7.6.1 gives conditions under which a set of orthogonal contrasts is enough.

The following lemma is a partial justification for the designation "orthogonal."

Lemma 5.1
If $(C_1, C_2, \ldots, C_T)$ is a set of orthogonal contrasts and $Y \sim$ Uniform $(\mathcal{S}_m)$, then $C_1(Y), C_2(Y), \ldots C_T(Y)$ are independent, and $C_t(Y) \sim$ Uniform $(\mathcal{V}(\lambda^{C_t}))$ for each t. □

See Theorem I of Terpstra (1952), and Exercise 5.6. The crux is that the relative rankings within a group are independent of the rankings between groups. Below we give special cases of orthogonal contrast models. These models are nested: $\phi \subset$ free $\subset$ contingency table. In each, we are presupposing a set $(C_1,\ldots,C_T)$ of nested orthogonal contrasts.

5.10.1 The free model

Start with densities $h_1,\ldots,h_T$, where h_t is a density for contrast C_t. The free model states that the contrasts are independent but their distributions are otherwise unrestricted. Thus $f(y) = \prod_{t=1}^T h_t(C_t(y))$. Compare this to the free model in Section 5.6.2. Since each $C_t(y)$ can take on $\binom{\lambda_+}{\lambda_1,\ldots,\lambda_K}$ values, the number of parameters in the free model can be quite large. The next model reduces the number of parameters significantly.

5.10.2 The ϕ model

Analogous to (5.18), the ϕ model declares a particular parametric form for each of the h_t's. First we need a single number for each value of a contrast to measure the preferences in the contrast. The statistic we choose is the Jonckheere–Terpstra statistic, which measures trend in the groups given in the order they are listed in the contrast. That is, define

$$d_{\text{Ken}}(C(y)) = \sum\sum_{1\leq k<l\leq K} \sum_{i\in\mathcal{O}_k} \sum_{j\in\mathcal{O}_l} I[y_i > y_j]. \tag{5.29}$$

The lower the number, the more the earlier groups are preferred to the later ones. The "Kendall" designation refers to the fact that if $C = Brack(\omega)$ for some $\omega \in \mathcal{S}_m$, where for any $l \times 1$ vector a,

$$Brack(a) = (\{a_1\}, \ldots, \{a_l\}), \tag{5.30}$$

then $d_{\text{Ken}}(C(y)) = d_{\text{Ken}}(\omega, y)$ from (2.29). (The Brack function just places brackets around the elements.) When there are only two groups in the contrast, those familiar with nonparametrics will recognize (5.29) as the Mann–Whitney/Wilcoxon statistic for testing the equality of two populations.

Now the ϕ model has a $T \times 1$ parameter vector θ and density

$$f_\theta(y) = \exp\left(\sum_{t=1}^{T} \theta_t \cdot d(C_t(y)) - \psi(\theta)\right). \tag{5.31}$$

If there is just the one contrast $C = Brack(\omega)$ as above, then model (5.31) is Mallows ϕ model of (5.5) and (2.29). One of the more interesting properties is that under f_θ, the distributions of the $d(C_t(Y))$'s are independent, each depending only on its parameter θ_t. See Lemma 7.3. In (7.11) and (7.12), we see that (5.31) is a submodel of Babington Smith (5.7).

Barton, David, and Mallows (1958) present the model with one contrast with two groups, and Critchlow (1985) extends this model to one contrast with several groups.

5.10.3 Contingency table models

The ϕ model, as the free model, assumes that the contrasts are independent. Such a requirement is not always tenable. A contingency table model generalizes the free model by considering the contrasts as the factors in a T-way contingency table. Any well-known contingency table model can be contemplated. The free model is just the independence model. One could also look at the model with all second-order interactions, or all third-order interactions, or one in which the first two contrasts are conditionally independent given the others, or the saturated model.

Remark 5.1
Orthogonal contrasts can be thought of as representing another style of multistage models. Whereas in Section 5.6, one first chose the object to be #1, then one of the remaining to be #2, etc., one could first decide upon the rank of Object 1, then of Object 2, choosing among the remaining ranks, then of Object 3, etc. The corresponding contrasts are $(\{1\}, \{2, \ldots, m\})$, $(\{2\}, \{3, \ldots, m\})$, ..., $(\{m-1\}, \{m\})$. This interpretation assumes each judge evaluates the objects in the same order, which may be valid if the experiment is so set up. However, the interpretation appears less tenable than that in Section 5.6 if judges are left on their own. Thus I prefer to think of the orthogonal contrast models as being good descriptive models, rather than behavioralist. Another interpretation, at least of the ϕ models, arises from its Babington Smith-ness, as in Chapter 7. □

Remark 5.2
Basing models on sets of non-orthogonal contrasts is another possibility. An example is the Babington Smith model, where the contrasts are all the $(\{i\}, \{j\})$'s for $i < j$. The advantages of having the contrasts orthogonal include easier interpretation, easier calculation, and meaningful independence models. Models with non-orthogonal contrasts may still be preferable in particular situations. □

Remark 5.3
Orthogonal contrasts can also be used to develop orthogonal subspaces as in Section 2.6. For contrast C, define the subspace

$$S^C = Span\{x^{(C,\Gamma)} \mid \Gamma \in \mathcal{V}(\lambda)\}. \tag{5.32}$$

See (2.74). Note that the differences between S^C and U^λ in (2.75) are that S^C uses a single contrast C, while U^λ requires $\lambda_+ = m$. Now for a set of orthogonal contrasts, let $S^{(t)}$ be the orthogonal complement of S_0 within S^C for $C = C^{(t)}$. Then $S^{(1)}, \ldots, S^{(T)}$ are orthogonal subspaces. Each measures the main effect of its contrast. (These subspaces are those for the independence model in the contingency table orthogonal contrast model.) Interaction subspaces are obtained by multiplying, e.g., the subspace measuring two-way interactions between contrasts 1 and 2 is the span of component-wise products of the generators of $S^{(1)}$ and $S^{(2)}$. □

5.11 Unfolding models

Whitfield (1950) motivated unfolding models using Goldilocks and the Three Bears. The Papa Bear's porridge was too hot, the Mama Bear's was too cold, but the Baby Bear's was just right. Presumably, Papa's ordering is (*Hot, Medium, Cool*); Mama's is (*Cool, Medium, Hot*); and Goldilocks and Baby like Medium best. Whitfield added his own spin: Goldilocks likes Hot second best; Baby likes Cool second best. The porridges are placed along the table in order of hotness, and the judges will sit so that their favorite is closest, second favorite is next closest, and least favorite is farthest:

Compare this arrangement to that in Example 2.16, where empirically the data mostly fell on the City–Suburb–Country continuum. Such ideas can be extended to any number of objects and any number of judges.

Coombs (1964) has a general theory for preference data, which includes placing everything in higher-dimensional space. The word "unfolding" arises since one can obtain the ordering of a particular judge by folding the scale at that judge's point. The objects are then in the correct order. In typical data sets, it is impossible to find an arrangement such that all orderings can be obtained. For example, we cannot find a place for one who likes (*Hot, Cool, Medium*). (Think of borscht.) Thus some way to incorporate slight deviations from the scale must be incorporated. One approach is to allow people to make errors with certain probabilities. For a given placement of the objects on a scale, there is a set of orderings $\Omega \subset \mathcal{O}$ that contains all the orderings that are exactly obtainable from the scale. Then any particular ranking y is either in Ω, or a certain distance from some element of Ω. Van Blokland-Vogelesang (1989, 1993) uses Kendall's distance to create the following model that allows variance in the rankings. Assume that for each $\omega \in \Omega$, there is a proportion π_ω of people in the population with ω as their "true" ranking. The distribution of the ranking Y for those people then follows Mallows' ϕ model. Thus the density is a mixture,

$$f_\theta(y) = \sum_{\omega \in \Omega} \pi_\omega \exp(\theta \cdot d(\omega, y) - \psi(\theta)). \tag{5.33}$$

5.12 Generalizing the models

The models discussed so far are very useful and informative in their own right, but they also have value as building blocks in more complicated models. They can be applied to reduced parts of data, e.g., on a subset of the objects or subset of the judges; multisample testing proceeds by fitting one model to the entire data and comparing the fit to that with a separate model for each sample; estimated parameters from a model can be fed into another model. Some of the models have direct generalizations into hierarchies of models. Babington Smith (5.7) is based on paired comparisons; an extended model could be based on rankings of triples of objects. McCullagh's (1993a) models based on inversions give a systematic formulation of these and more complicated models. See Section 7.11. The extended Marginals models appear in Section 9.3. Plackett (1975) suggests a "logistic" generalization (Chapter 8) of the Plackett–Luce model (5.13) where parameters are indexed by pairs, triples, etc., of objects. Silverberg (1980) presents another generalization of (5.13) based on the vase model interpretation in Section 5.6.1. There, draws were made from the same vase at each stage. The generalized model allows each stage to have a different vase. The contingency table version of the orthogonal contrast model in Section 5.10.3 fits immediately into the usual loglinear hierarchy.

Recall the clustering method in Section 2.5.5. There, the data was grouped according to closeness to particular rankings. Rank data often reveal such clustering of judges. The unfolding model in Section 5.11 is another example, where the clusters' centers belong to a set of unfolding rankings. Any model can similarly be extended by assuming a mixture model, that is, that there are L groups in the population, the groups following the same model but with possibly different parameters. See Croon and Luijkx (1993). Thus if π_l is the proportion of the population in group l, and the parameter for that group is $\theta^{(l)}$, then the mixture of f_θ densities is the density

$$f(y;\theta^{(1)},\ldots,\theta^{(L)};\pi)=\sum_{l=1}^{L}\pi_l f(y;\theta^{(l)}). \tag{5.34}$$

Such models can reveal detailed structure in the data not evident in the simpler models or exploratory statistics.

5.13 Some axiomatics

The chapter up until now has been an overview of specific models for rank data. In this section, we touch on some approaches to assessing general properties of models, especially notions of unidimensionality/concordance/consensus and of independence/non-interaction. The former is an attempt to take a distribution or sample on $\mathcal{S}_m$ and identify one ranking that best summarizes the population or sample. We have seen this problem as one of estimation in Chapter 2, but different distances give different results. Is there a more fundamental definition? Arrow (1963) says, "No." See Marley (1993) for more recent theory. In regular multivariate analysis, it is easy to define independence of variables. With rank data it is more problematic since the ranks assigned to objects are inherently dependent. One attempt to define an analogy of independence is Luce's choice axiom, discussed in Section 5.13.1. It turns out to lead to the Plackett–Luce model. From this model an overall consensus ranking is immediate, but we do not want to restrict ourselves to one model. It would make this book quite a bit shorter.

5.13.1 Luce's choice axiom

Given a set of objects $\mathcal{O}$, a set of *choice probabilities* consists of probabilities $p_{\mathcal{A}}(\cdot)$ for all subsets $\mathcal{A} \subset \mathcal{O}$, where $p_{\mathcal{A}}(\cdot)$ is a probability measure on $\mathcal{A}$, and for $a \in \mathcal{A}$, $p_{\mathcal{A}}(a)$ is interpreted as the probability that a is chosen as the favorite among the elements of $\mathcal{A}$. Luce (1959) wished to place some restrictions on these sets, since unrestrained they could include unusual values. For example, if $\mathcal{O}$ = {*Coke, Pepsi, 7-Up, Sprite*}, one could have that the probability Coke is the favorite among all four is 90%, while the probability that Coke is the favorite among {*Coke, Pepsi*} is only

10%. A set of choice probabilities is said to satisfy Luce's *choice axiom* if all subsets $\mathcal{A} \subset \mathcal{O}$ with at least two elements satisfy

L1 If $p_{\{a,b\}}(a) > 0$ for all $a, b \in \mathcal{A}$, then for any $c \in \mathcal{B} \subset \mathcal{A}$, $p_{\mathcal{A}}(c) = p_{\mathcal{B}}(c) p_{\mathcal{A}}(\mathcal{B})$;

L2 If $p_{\{a,b\}}(a) = 0$ for some $a, b \in \mathcal{A}$, then if $b \in \mathcal{A}$ and $b \neq a$, $p_{\mathcal{A}}(b) = p_{\mathcal{A}-\{a\}}(b)$.

The essential idea is that the probability that a is the favorite in $\mathcal{A}$ is the same as the product of the probability that the favorite in $\mathcal{A}$ is contained in the set $\mathcal{B}$ and the probability that the favorite in $\mathcal{B}$ is element a. Thus the probability Coke is the favorite overall equals the probability that a cola ($\mathcal{B} = \{Coke,\ Pepsi\}$) is the favorite, times the probability that Coke is the favorite cola. (This notion looks like conditional probability, but it is not.)

A key result is that, under the condition that $p_{\{a,b\}}(a) > 0$ for all $a, b \in \mathcal{O}$, the axiom is satisfied if and only if there are positive constants, $v_a, a \in \mathcal{A}$, such that $p_{\mathcal{A}}(a) = v_a / \sum_{b \in \mathcal{A}} v_b$. These constants are similar to those in the Plackett–Luce model. In fact, if one uses these choice probabilities to obtain a multistage ranking model, where the stages involve first choosing the favorite, then the second favorite out of the remaining, etc., then the model is the Plackett–Luce model. The result goes the other way, also. That is, for given distribution P on $\mathcal{S}_m$, define the corresponding choice probabilities $p_{\mathcal{A}}(a)$ by summing P_y over all $y \in \mathcal{S}_m$ such that Object a is ranked ahead of all other objects in $\mathcal{A}$. If, again, $p_{\{a,b\}}(a) > 0$ for all a and b, then only the Plackett–Luce model has P that leads to choice probabilities that satisfy the axiom. This result, and many others, follow from Yellott (1977). He shows that a Thurstonian model of the form (5.4) satisfies the axiom only if the $-Z_a$'s are Gumbel, where the Gumbel has distribution function $G(z) = \exp(-\exp(-z))$ for z real, which model is Plackett–Luce.

McCullagh (1993a) makes an even more interesting point. Imagine the reverse set of choice probabilities, that is, the probabilities of choosing the least favorite among each subset of objects. Suppose both the "choosing favorite" and "choosing least favorite" sets of choice probabilities satisfy the axiom simultaneously, and the Plackett–Luce ranking probabilities using these sets are the same. (For the second set, the model starts by choosing the least favorite, then the second least favorite out of the remaining, etc.) If $p_{\{a,b\}}(a) > 0$ for all a and b, then we must have the uniform model. This fact may call into question whether we should accept the axiom, or at least force us to explain why we wish it in one direction but not the other.

Remark 5.4. L-Decomposability
Critchlow, Fligner, and Verducci (1991) define L-Decomposability for ranking models, which is weaker than satisfying the choice axiom. They require axiom L1 to apply only to "one at a time" sequences:

$$f(w) = p_{\mathcal{O}}(w_1)p_{\{w_2,\ldots,w_m\}}(w_2)\cdots p_{\{w_{m-1},w_m\}}(w_{m-1}). \quad (5.35)$$

Their necessary and sufficient condition is that for $k \geq 3$, $P[W_k = a \mid W_1 = b_1, \ldots, W_{k-1} = b_{k-1}]$ is invariant under permuting the $b_1, \ldots, b_{k-1}$. Clearly Plackett–Luce is L-Decomposable, but so are the Babington Smith model and multistage models of Fligner and Verducci.

5.13.2 Unidimensionality, unimodality, and consensus

The idea that there is an overall ranking for the objects is very important in ranking situations. For example, in competitions, several judges' rankings must be combined into one overall ranking; in sociology, the problem is that of arriving at a social consensus. The question can also be asked of an individual, that is, whether after repeated observations of the person's rankings, one can ascertain an overall ranking. Basically, one is asking for a placing of the objects on a one-dimensional scale, where objects on the left are more likely to be preferred.

There are a number of ways to define an overall ranking given a set of rankings or a distribution on $\mathcal{S}_m$ that yield a result no matter how inconsistent the individual rankings may be. Thus one actually desires more than just a composite ranking; one wishes to be assured that the ranking tells the whole story. Suggested requirements that a probability density f on the ranks in $\mathcal{S}_m$ should meet in order for there to be some sense of agreement on an overall ranking follow.

The density f is *unimodal* if there exists one ranking ω that has higher probability than any other, i.e., $f(\mu) > f(y)$ for all $y \in \mathcal{S}_m - \{\omega\}$. The ranking ω is the *modal* ranking. *Strong unimodality* with respect to a given partial ordering $<_*$ obtains if the density is unimodal, and whenever $\omega <_* y <_* z$, $f(y) > f(z)$.

That is, the density decreases as the rankings become farther from the mode. *Order in expectation* of the objects holds if $E(Y_1) < E(Y_2) < \cdots < E(Y_m)$. The objects are *marginally stochastically ordered* if the marginal distributions of the Y_i's are distinct, and for any number k, $P[Y_1 \leq k] \geq P[Y_2 \leq k] \geq \cdots \geq P[Y_m \leq k]$.

The ordering w is a *consensus ordering* if $P[Y_{w_i} < Y_{w_j}] > 1/2$ for all $i < j$. Note that, empirically, the ordering in Table 2.2 is a consensus ordering. There is *complete consensus* about the ordering if for all $i < j$

$$P[Y_i < Y_j \mid Y_k = r_k, \text{ for } all\ k \neq i, j] > 1/2 \qquad (5.36)$$

for all sets of ranks r_k. See Critchlow, Fligner, and Verducci (1991); Henery (1981); and Fligner and Verducci (1988). In this section, we will refer to these three articles as CFV, H, and FV, respectively. The definition of consensus here differs from that in FV.

Unimodality and order in expectation are very weak properties. They are satisfied as long as there are enough differences among the probabilities of the rankings. They do not require any real cohesion among the rankings. Marginal stochastic ordering and consensus are stronger conditions, but do not eliminate the possibility of interactions among the objects. For example, one might prefer A to B 95% of the time if C is ranked third, but prefer B to A 55% of the time when C is ranked first. Strong unimodality (with nontrivial $<_*$) and complete consensus do restrict the possible interactions, at least qualitatively.

Different partial orderings $<_*$ yield different strong unimodalities, of course. We will give three possibilities, all based on the modal ranking being $\omega = e_m$. (If ω is not e_m relabel the objects so that it is.) CFV suggest $y <_{CFV} z$ if for some $i < j$,

$$y_j = z_i = y_i + 1 = z_j + 1 \text{ and } y_k = z_k \ for\ k \neq i, j. \qquad (5.37)$$

H's ordering (3.33) says $y <_H z$ if

$$y_i = z_j < y_j = z_i \text{ and } y_k = z_k \text{ for } k \neq i, j. \qquad (5.38)$$

Kendall's ordering says $y <_{Ken} z$ if $d_{\text{Ken}}(e_m, y) < d_{\text{Ken}}(e_m, z)$, where d_{Ken} is Kendall's distance as in (2.29). Other orderings immediately arise from other distances. The partial orderings are themselves ordered: $[y <_{CFV} z] \Rightarrow [y <_H z] \Rightarrow [y <_{Ken} z]$. The idea is that for CFV, if two rankings are identical except for two objects with adjacent ranks, then the one with the two objects in the wrong order (with respect to e_m) is farther from e_m than the other. The H ordering is similar, except that the requirement that the ranks have to be adjacent is dropped. The Kendall ordering just looks at the number of discordances between ω and each of the rankings. Block, Chhetry, Fang, and Sampson (1990, 1991, 1992) discuss even more possibilities.

Some implications:

$$\begin{aligned} &\textit{Complete consensus} \Leftrightarrow \textit{Strong unimodality } [<_H] \\ &\qquad \Rightarrow \begin{cases} \textit{Unimodality} \\ \textit{Marginal stochastic ordering} \\ \quad \Rightarrow \textit{ Order in expectation} \\ \textit{Consensus} \end{cases} . \end{aligned} \tag{5.39}$$

Unidimensionality is not precisely defined, but the idea is that there is a real number attached to each object, and the lower the number the more likely the object will be preferred. CFV call these numbers *item parameters.* See also Critchlow and Fligner (1993). For example, the Thurstone models (5.4) have the η_a associated with Object a, and the Bradley–Terry/Mallows (5.9) and Plackett–Luce (5.13) have ν_a. The modal rankings, ω_a, in the distance models (5.5) can also be considered item parameters.

Now we shall see which models satisfy which of the unimodality and consensus conditions. Thurstonian models of the form (5.4) have complete consensus if $\eta_1 < \cdots < \eta_m$, and the density g has monotone likelihood ratio, i.e.,

$$\frac{g(z-\eta_a)}{g(z-\eta_b)} \text{ is decreasing in } z \text{ for } a < b. \tag{5.40}$$

See H. Even without the monotone likelihood ratio property, any Thurstonian model with the η_a's strictly increasing will have marginal stochastic ordering, hence order in expectation. See CFV.

Babington Smith models do not have item parameters in general. However, there are notions of qualitative ordering. A Babington

Smith model has *weak stochastic transitivity* if for i, j,

$$p_{ij} \geq \frac{1}{2} \text{ and } p_{jk} \geq \frac{1}{2} \Rightarrow p_{ik} \geq \frac{1}{2}, \tag{5.41}$$

and has *strong stochastic transitivity* if

$$p_{ij} \geq \frac{1}{2} \text{ and } p_{jk} \geq \frac{1}{2} \Rightarrow p_{ik} \geq \max\{p_{ij}, p_{jk}\}. \tag{5.42}$$

See David (1988) for these definitions. CFV show that a Babington Smith model is strongly unimodal [$<_{CFV}$] if and only if it has weak stochastic transitivity, and if it has strong stochastic transitivity, it has complete consensus.

Bradley–Terry/Mallows and Plackett–Luce models show complete consensus whenever $v_1 > v_2 > \cdots > v_m$. See H and CFV.

Distance-based models show complete consensus whenever $f(y)$ decreases as $d(e_m, y)$ increases, and $y <_H z$ implies that $d(e_m, y) < d(e_m, z)$. It has already been noted above that the Kendall distance has this property; hence Mallows' ϕ model, (5.5) with $d = d_{\text{Ken}}$ of (2.29), has complete consensus if the parameter $\theta < 0$. Also, if $\theta > 0$, the model has complete consensus but with the modal ranking being $e_m^c = (m, \ldots, 1)$.

The ϕ-component model (5.18) has complete consensus if $\theta_i / \theta_{i+1} \geq ((m-1)-i)/(m-i)$ for $i = 1, \ldots, m-2$. See FV, Theorem 2.3.

Orthogonal contrast models in general do not lend themselves to such unidimensionality questions. In fact, part of their usefulness lies in being able to capture interesting deviations from consensus and unimodality. Unfolding models (Section 5.11) fit in to this framework by considering the mixture model (5.33). The idea is that for a given unfolding scale, there are up to $\binom{m}{2} + 1$ possible rankings, and for each of these rankings there is a population of judges and corresponding distribution of rankings. Consensus, unimodality, and the other properties are said to hold for the model (5.33) if they hold for each of the subpopulations, where the modal ranking is that from the unfolding scale connected with that population.

Remark 5.5. Arrow's Impossibility Theorem
The above aimed at conditions for defining an overall ranking based on a sample, or population, of rankings. In voting situations, the goal is to find one overall ranking no matter what the configuration of the data. The candidates may be people running for the School Board, or they may be social choices, e.g., a list of priorities on which the board is to concentrate. There are many schemes for translating a set of ballots into an overall ranking of the candidates (or a partial ranking, if that is all that is needed), but they all seem to be susceptible to counterintuitive behavior. See Fishburn and Brams (1983) for an entertaining example. Arrow (1963) approached the general problem. He developed a set of criteria a voting scheme should possess, and showed that no scheme could possess all of them. Perhaps this situation is not too surprising. Note how different distances yield different rankings in the Sports example 2.11. See Stern (1993) for a concise summary of various schemes, and applications to rank data.

5.14 Likelihood methods and exponential families

In this section we set out some of the basic notation and techniques for fitting models. We will concentrate on likelihood methods such as maximum likelihood estimation (MLE) and likelihood ratio testing (LRT). There are many examples in which likelihood methods fail, or can be much improved, or are extremely difficult to implement computationally. However, for most of the ranking models we are considering, they work remarkably well.

The generic experiment will have observations taking values in the finite set $\mathcal{A}$ with K elements. For a typical ranking model, $\mathcal{A}$ is the set of rankings $\mathcal{S}_m$ or orderings $\mathcal{T}_m$, and $K = m!$. There is a corresponding vector P of probabilities, where P_a is the probability the observation is element a. The space of P is the K-dimensional simplex $\mathcal{P}_K$ from (2.16). For a sample of independent and identically distributed random variables, $A_1, \ldots, A_n$, we have

$$N \sim \mathit{Multinomial}_K(n, P), \text{ where } N_a = \#\{A_i = a\}. \tag{5.43}$$

The results in this chapter are standard. More details and proofs can be found in Rao (1973), Serfling (1980), and Lehmann (1983). A model for the multinomial distribution is given by specifying a subset of $\mathcal{P}_K$ to which p is restricted. We will be mainly concerned with *regular* models, which will be defined in this section. Start with a parametrization of the model, that is, assume there exists a v-dimensional vector θ with space Θ and a function $P(\theta)$ such that $P_a(\theta) = P[A = a; \theta]$. See (5.2). The following regularity conditions are required of a regular model.

R1 Identifiability. If $\theta^{(1)} \neq \theta^{(2)}$, then $P(\theta^{(1)}) \neq P(\theta^{(2)})$ for $\theta^{(1)}, \theta^{(2)} \in \Theta$.

R2 Θ is an open subset of $\Re^v$.

R3 $P_a(\theta) > 0$ for all a and θ, and as functions of θ, the $P_a(\theta)$'s have that all first, second, and third (mixed) partial derivatives are continuous in θ.

(In most treatments, there are additional boundedness conditions on the derivatives. In our case, since $\mathcal{A}$ is finite and the derivatives are continuous in θ, the boundedness conditions can be shown to hold.)

These regularity conditions are automatically satisfied by regular exponential families as in Section 5.7, and in fact by almost all the models except the mixture models for unfolding scales. Thurstonian models may also violate the assumptions if the distributions of the underlying variables are irregular.

5.14.1 The likelihood function and Fisher information

The likelihood function L is a function of θ for a fixed value of the data $a_1, \ldots, a_n$, or equivalently, N. It is proportional to the probability, i.e.,

$$L(\theta; N) = \prod_{i=1}^{n} f_\theta(a_i) = \prod_{a \in \mathcal{A}} f_\theta(a)^{N_a}, \tag{5.44}$$

where $f_\theta(a) = (1/K)P_a(\theta)$ is the density with respect to *Uniform* ($\mathcal{A}$). The loglikelihood is $l(\theta; N) = \log(L(\theta; N))$. The values of L are supposed to measure how "likely" various values of θ are in light of the particular data N of (5.43). The *maximum likelihood estimate* (MLE) of θ based on x is the value of $\theta \in \Theta$ that maximizes the likelihood. Denote it $\widehat{\theta}$. It may be that there is no maximum, or the maximum is not unique, or the maximum lies outside of Θ. In such cases, we say that the MLE does not exist.

If the MLE does exist, then for a regular model, the derivatives of l are 0 at $\widehat{\theta}$, that is,

$$\nabla l(\widehat{\theta}; x) = 0; \quad \nabla l(\widehat{\theta}; x)_j = \frac{\partial}{\partial \theta_j} l(\theta; x) \text{ for } 1 \leq j \leq v. \tag{5.45}$$

The first equation in (5.45) is termed the *likelihood equation.*

Once the $\widehat{\theta}$ is obtained, it is also of interest to ascertain how confident one should be in it as an estimate. The likelihood approach is to look at the likelihood for values of θ near $\widehat{\theta}$. If the likelihood tends to drop off precipitously, then one is quite sure of the estimate. If the likelihood is relatively flat around the MLE, then there are many other estimates that are nearly as likely as the MLE; hence not as much reliance can be put on the estimate. One way to measure how quickly the likelihood falls off is to look at the second derivative matrix. If it is large negative, then the likelihood has relatively much information. If it is near 0, then the likelihood has little information. Thus the *observed Fisher information* is defined to be

$$\widehat{I}(\widehat{\theta}) = -\left\{ \frac{\partial^2}{\partial \theta_i \partial \theta_j} l(\theta; x) \mid_{\theta = \widehat{\theta}} \right\}_{i,j=1}^{v}. \tag{5.46}$$

(It is the matrix of partial second derivatives.)

The $\widehat{I}$ is a data-dependent measure of information. The corresponding population quantity is the *expected* Fisher information, which is the information about θ one expects from a sample. It is usually called simply the Fisher information. The Fisher information for one observation is defined to be

$$I_1(\theta) = -\left\{ E_\theta [\frac{\partial^2}{\partial \theta_i \partial \theta_j} l(\theta; A_1)] \right\}_{i,j=1}^{v}, \tag{5.47}$$

and the information for the sample of n independent observations is $I_n(\theta) = nI_1(\theta)$.

5.14.2 Maximum likelihood estimation

In this section, we present asymptotic results for the MLE, asymptotic as $n \to \infty$. In order to state the result, we will assume the following additional regularity condition.

R4 For any θ, the Fisher information $I_1(\theta)$ is invertible.

Theorem 5.1
Suppose the regularity conditions R1 through R4 hold. Then as $n \to \infty$, for any $\theta \in \Theta$, $P[\text{The MLE exists}; \theta] \to 1$ as $n \to \infty$, and $\sqrt{n}(\widehat{\theta}_n - \theta) \to N(0, I_1^{-1}(\theta))$, where $\widehat{\theta}_n$ is the MLE based on the first n observations when it exists, and any arbitrary value when it does not. Also, $(1/n)\widehat{I}_n(\widehat{\theta}_n) \to I_1(\theta)$ in probability. □

The value of the theorem lies in the fact that for any regular model, the distribution of the MLE can be approximated by a normal with covariance matrix being the inverse of the observed Fisher information, i.e., $\widehat{\theta}_n \approx N(\theta, \widehat{I}_n^{-1}(\widehat{\theta}_n))$. If you think about this result, it is amazing.

5.14.3 Likelihood ratio tests

It often (always?) happens that one has more than one model in mind, and wishes to find which is best. One model may be the same as another except for a restriction on the parameter, or it may be that the models have different parametrizations. Here, we present a result for testing two nested models, nested in the sense that the set of allowable P's for one is strictly contained in that for the other. Specifically, we assume that we have two regular models, both satisfying conditions R1 through R4. Let the smaller one have parametrization $\phi \in \Phi$, where ϕ is $v_0 \times 1$, and corresponding probabilities $R(\phi)$, and the larger one be as in (5.2), where $v_0 < 0$. The containment is $\{R(\phi) \mid \phi \in \Phi\} \subset \{P(\theta) \mid \theta \in \Theta\}$.

We wish to test

$$H_0 : \text{The } \phi\text{-model holds} \quad \text{versus} \quad H_A : \text{The } \theta\text{-model holds.} \tag{5.48}$$

Let l_0 and l_A be the respective loglikelihoods (5.44) for the two hypothesized models. The likelihood ratio test compares the best likelihood under the null model to the best under the alternative model. The latter will always be larger since there is a larger family of distributions. The question is how much larger must it be for us to reject the null model in favor of the alternative? It is convenient to consider the *likelihood ratio statistic*

$$LRS \equiv 2 \cdot [l_A(\widehat{\theta}_n; N) - l_0(\widehat{\phi}_n; N)]. \tag{5.49}$$

The MLE's are calculated relative to the respective models.

Theorem 5.2
Under the above assumptions, if H_0 in (5.48) holds, then as $n \rightarrow \infty$, $LRS \rightarrow \chi^2_{v-v_0}$. □

Even if one has only one model to consider, there are two particular hypothesis tests one is almost obligated to perform. Suppose the model has the θ-parametrization. The first is to see whether there is "anything going on," that is, to see if it is possible that the uniform model holds. Then the H_0 model in (5.48) is the uniform, $R_a = 1/K$ for all a, whence $v_0 = 0$. If the LRS is not sufficiently large ($> \chi^2_{p,\alpha}$) then one cannot conclude that the model is any better than the uniform. The other important test is the goodness-of-fit test. That is, is there anything going on that the model does not detect? Now the θ-model is the H_0 model, and the H_A model is the saturated model. The saturated model has dimension $K-1$ since it can be parametrized by $P_1, \ldots, P_{K-1}$. It is not hard to show that the MLE of P_a under the saturated model is N_a/n. Thus the goodness-of-fit test statistic is asymptotically $\chi^2_{K-1-v_0}$ under H_0, and can be written

$$GOF = 2 \cdot \sum N_a \cdot \log\left(\frac{N_a}{nP_a(\widehat{\theta})}\right). \tag{5.50}$$

An alternative statistic, which has the same asymptotic null distribution, is the Pearson chi-squared, $\sum (N_a - nP_a(\widehat{\theta}))^2 / nP_a(\widehat{\theta})$, which may be more familiar. They are equally fine. Santner and Duffy (1989) discusses these and others in a unified manner.

In linear regression, the multiple correlation coefficient R^2 is often used to measure the percentage of variation in the data that is explained by the regression model. An analogous quantity for us is the percentage of nonuniformity in the data that is explained by the model. Take LRS to be the measure of nonuniformity. Then we define

$$R^2 = 1 - \frac{GOF}{Denom}, \quad Denom \equiv 2\sum N_a \log(\frac{N_a}{n/K}). \qquad (5.51)$$

Thus if the model exactly fits the data, $R^2 = 1$, and if it performs no better than the uniform, $R^2 = 0$.

5.14.4 Exponential families

Section 5.7 presents exponential family models. A ***regular*** exponential family has the form (5.20) with the additional requirements that

E1 The parameter space Θ is $\Re^v$;

E2 The statistic has full dimension, i.e., there is no $v \times 1$ vector $b \neq 0$ and constant c such that $P[b'S = c; \theta] = 1$ for any θ.

Requirement E2 just makes sure that there are no superfluous dimensions in θ. It guarantees identifiability and the invertibility of I_1. Thus a regular exponential family model is a regular model as at the beginning of this section. See Exercise 5.7. Requirement E1 can be replaced with the requirement that Θ be open with a nonempty interior.

Many of the models in this chapter are regular exponential family models. The ones in Sections 5.7 and 5.8 are such because they are constructed to be. For example, Babington Smith (5.7) uses the pairwise comparisons $I[y_i < y_j]$ as sufficient statistics, Bradley–Terry/Mallows (5.9) uses the elements of y itself, and the Marginals model uses the $I[y_i = j]$'s. About the only ones that are not this type are the Thurstonian models and some of the L-Decomposable models such as Plackett–Luce. The statistics for a sample consist of the sums of the statistics for the individual observations.

A v-dimensional exponential family will have sufficient statistic $t(a_i)$ for each observation, so that the density with respect to $Uniform(\mathcal{A})$ is

$$f_\theta(a_i) = \exp(\theta' t(a_i) - \psi(\theta)), \qquad (5.52)$$

where $\theta' t = \sum \theta_i t_i$, and

$$\psi(\theta) = \log(E_0(\exp(\theta' t(A)))), \qquad (5.53)$$

E_0 being expectation with respect to the uniform. Differentiating ψ shows that

$$\beta(\theta) \equiv E_\theta(t(A_i)) = \nabla\psi(\theta) \tag{5.54}$$

and

$$\Sigma(\theta) \equiv Cov_\theta(t(A_i)) = \left\{ \frac{\partial^2\psi(\theta)}{\partial\theta_j\partial\theta_k} \right\}_{j,k=1}^{D}. \tag{5.55}$$

Thus the mean and the covariance matrix of the sufficient statistic vector can be found by differentiating rather than integrating.

Now from (5.44) and (5.52), the loglikelihood for the data is

$$\begin{aligned} l(\theta; N) &= \sum_{i=1}^{n} \theta' t(a_i) - n \cdot \psi(\theta) \\ &= S'\theta - n \cdot \psi(\theta), \end{aligned} \tag{5.56}$$

where S is the vector of sufficient statistics for the sample, $S(N) = \sum t(a_i)$. Using the loglinear notation from Section 5.8, we have the $K \times v$ matrix X such that $X'N = S$; hence (5.56) can also be written

$$l(\theta; N) = N'X\theta - n\psi(\theta). \tag{5.57}$$

The likelihood equation (5.45) is thus $0 = \nabla l(\widehat{\theta}, N) = S - n\beta(\widehat{\theta})$, or

$$\beta(\widehat{\theta}) = \frac{S}{n} \equiv X'\widehat{P}. \tag{5.58}$$

That is, the MLE of θ is that value for which the population mean of the sufficient statistic equals the sample mean.

Next, to find the observed and expected Fisher information, use (5.46) and (5.47) on (5.56). The first term drops out since it is linear in θ, and what are left are second derivatives of $-\psi$, so that from (5.46) and (5.55),

$$\widehat{I}(\widehat{\theta}) = n \cdot \Sigma(\widehat{\theta}) \text{ and } I_1(\theta) = \Sigma(\theta). \tag{5.59}$$

The dual central limit theorems for exponential families are

$$\sqrt{n}(\frac{S}{n} - \beta(\theta)) \to N(0, \Sigma(\theta)) \text{ and } \sqrt{n}(\widehat{\theta} - \theta) \to N(0, \Sigma^{-1}(\theta)). \tag{5.60}$$

Those familiar with the Δ-method (see Exercise 11.16) will realize these are the same theorem.

Remark 5.6. Finding the MLE
It is typical that the likelihood equation (5.45) cannot be solved in closed form, so that some numerical methods must be employed. A very appropriate method for current purposes is the Newton–Raphson method. It is derived by expanding the likelihood equation in a two-term Taylor Series around $\theta = \widehat{\theta}$, the MLE. One starts with a guess of θ, say $\theta^{(0)}$, then iteratively finds $\theta^{(i)}$ from $\theta^{(i-1)}$ via

$$\theta^{(i)} = \theta^{(i-1)} + \widehat{I}_n^{-1}(x)\nabla l(\theta^{(i-1)}, x). \tag{5.61}$$

In the exponential family case, this simplifies to

$$\theta^{(i)} = \theta^{(i-1)} + \Sigma^{-1}(\theta^{(i-1)}) \cdot [\frac{S}{n} - \beta(\theta^{(i-1)})]. \tag{5.62}$$

A possible drawback to the Newton–Raphson method occurs when v is large, so that $\widehat{I}_n$ may be difficult to invert. There are many modifications and other possible methods. Other general methods for maximizing functions can be found in Press, Teukolsky, Vetterling, and Flannery (1992). See also the EM algorithm, which we discuss in Section 10.3.

Iterative proportion fitting is a method to find the MLE's of $P(\theta)$ rather than the θ directly. It is a special case of cyclic ascent, and avoids any matrix inversions. It works in exponential family models for which the sufficient statistics are categorical in nature. See Darroch and Ratcliff (1972) for the general theory, and Silverberg (1984) and Croon (1989a) for applications to ranking data, especially the loglinear models of Section 5.8. It is also useful in the extended Marginal model, Babington Smith model, and the inversion models. □

5.15 Exercises

5.1. A voter in the 1992 U.S. Presidential elections had a choice of Bush, Clinton, and Perot. The voter rated the candidates on three criteria: Change, Stability, and Budget Deficit. The higher the rating, the better the candidate is perceived to be on the issue. The results: Bush $(0, 15, 2)$; Clinton $(10, 8, 5)$; Perot $(15, 0, 10)$. In choosing the favorite of any two candidates, the voter picked the candidate whose worst issue rating relative to the other candidate is smallest. Thus, between Bush and Clinton, Clinton is 10 points better on Change, 7 points worse on Stability, and 3 points better on the Deficit. Since Clinton's worst relative to Bush is 7, and Bush's worst relative to Clinton is 10, Clinton is preferred to Bush. Show that the same reasoning yields that Bush is preferred to Perot, and Perot is preferred to Clinton.

5.2. Define the p_{ij}'s as in (5.11). Show that $p_{ij} + p_{ji} = 1$. Show that the Babington Smith model (5.7) with those p_{ij}'s equals the distance model (5.5) with the distance being $d_{\text{Ken}}(e_m, y)$ from (2.29), where y is the ranking associated with the ordering w in (5.7).

5.3. Show that the Babington Smith model (5.7) for a sample has sufficient statistic $\widehat{K}$.

5.4. Show that the Bradley–Terry/Mallows model (5.9) for a sample has sufficient statistic $\overline{y}$. [Note that $v_{w_i}^{m-i} = v_j^{m-y_j}$, where y is the ranking associated with the ordering w.]

5.5. Prove (5.22). [Recall that $N = \sum_{i=1}^{n} \zeta^{y^{(i)}}$.]

5.6. (a) For $m = 4$, take the orthogonal contrasts $C_1 = (\{1,2\}, \{3,4\})$, $C_2 = (\{1\},\{2\})$, and $C_3 = (\{3,\},\{4\})$. Write out the values of the contrasts for all 24 rank vectors. Find the spaces and distributions of the $C_t(Y)$'s under uniformity, and show they are independent. (b) Do the same for the contrasts $D_1 = (\{1,2,3\},\{4\})$, $D_2 = (\{1,2\},\{3\})$, and $D_3 = (\{1\},\{2\})$.

5.7. Show that if conditions E1 and E2 hold for an exponential family (5.20), then conditions R1, R2, and R3 must hold. (We are assuming a finite sample space.) [Conditions R2 and R3 are fairly immediate. For R1, note that if $\theta^{(1)} \neq \theta^{(2)}$, then the set where $\{x | f_{\theta^{(1)}}(x) = f_{\theta^{(2)}}(x)\}$ is of the form $\{x | b'S = c\}$. By E2, the probability of this set is strictly less than 1; hence R1 must hold.]

CHAPTER 6

Distance-Based Models

6.1 Introduction

The distance-based models are often convenient first steps in modeling a set of rank data. They capture the main feature of the data with only one parameter, and provide a basis for further investigation. As discussed in Chapters 2 and 3, one has to decide on a distance and whether the modal ranking is known or to be estimated. Some consideration has to be given to the ease of implementation, especially for moderate or large numbers of objects.

In Section 6.2 we fit the models assuming the mode is known and in Remark 6.1 present a "universal" coefficient to compete with the t's of (2.21). Sections 6.3 and 6.4 discuss estimating the mode, the latter presenting a Bayesian approach by Fligner and Verducci (1990). In Section 6.5 we look at the models as the number of objects becomes large. It is typically unrealistic to expect a one-parameter model to fully explain a set of data. In Section 6.6 we present ways to assess goodness-of-fit. One approach is to test the distance model versus a larger one, e.g., the saturated model. The other is to compare observed and expected values of some interesting statistics, such as those in Chapter 2.

Distance models have been used in paired-sample situations, e.g., Example 4.3. Although we will not pursue the models here, the reader is urged to read the articles by Critchlow and Verducci (1992), Feigin (1993), and Tubbs (1993).

6.2 Fitting the models: Known mode

In principle, there is no problem in fitting the models (5.5). Difficulties do occur when the sufficient statistic $S \equiv \sum d(\omega, y^{(i)})$, where ω is the modal ranking, equals its minimum 0 or its maximum $n \cdot Max(d)$. As $n \to \infty$, the probability such happens goes quickly to 0, but in small samples it is possible. In either case, the maximum likelihood estimate as defined in Section 5.14 does not exist, but it is reasonable to adopt the convention that

$$S = 0 \Rightarrow \widehat{\theta} = -\infty; \quad S = n \cdot Max(d) \Rightarrow \widehat{\theta} = +\infty. \tag{6.1}$$

See Exercise 6.1.

When $0 < S < n \cdot Max(d)$, the main goal is to find ψ and its first two derivatives. Rather than adding over all $m!$ elements in $\mathcal{S}_m$, it is more efficient to first find the probabilities for d under the *Uniform*, $c(k) = \#\{y \in \mathcal{S}_m \mid d(\omega, y) = k\}/m!$, for k in the range Λ of $d(\omega, y)$. Label invariance (Definition 3.3) implies that the c does not depend on ω. Section 3.4.2 discusses particular cases of c. Then $P_\theta[d(\omega, Y) = k] = c(k)\exp(\theta k - \psi(\theta))$, where $\exp(\psi(\theta)) = \sum_{k\in\Lambda} c(k)\exp(\theta k)$. The mean and variance of d, $\delta(\theta)$ and $I_1(\theta)$, are easily found as the first and second derivatives of ψ, respectively. Newton–Raphson is generally easy to apply:

$$\theta^{(i)} = \theta^{(i-1)} + \frac{\overline{d} - \delta(\theta^{(i-1)})}{I_1(\theta^{(i-1)})}. \tag{6.2}$$

See (5.62). It is wise to bracket the sequence of $\theta^{(i)}$'s. That is, first search for $\theta^{(l)}$ and $\theta^{(u)}$ that are sure to surround $\widehat{\theta}$. Any values for which $\delta(\theta^{(l)}) < \overline{d} < \delta(\theta^{(u)})$ will suffice. While iterating (6.2), if $\theta^{(i)}$ is found outside the brackets, then instead set $\theta^{(i)} = (\theta^{(l)} + \theta^{(u)})/2$. At each step the bracketing values are updated, i.e., narrowed. See Press, Teukolsky, Vetterling, and Flannery (1992) for further discussion. Iteration continues until the loglikelihood ratio, $\theta^{(i)} \cdot S - n \cdot \psi(\theta^{(i)})$, is not changing appreciably. Usually the procedure takes fewer than ten iterations, and often around four or five.

The estimated standard error for $\widehat{\theta}$ is $(I_1(\widehat{\theta})n)^{-1/2}$, and that for $\widehat{\tau} = \tau(\widehat{\theta}) = t$ is estimated by $(I_1(\widehat{\theta})/n)^{1/2}/\delta_0$. See Theorem 5.1. Confidence intervals and hypothesis tests for θ and τ easily follow.

Three of the distances have clever expressions for ψ that have only m terms and do not require finding c. Not surprisingly, the three are Kendall, Cayley, and Hamming, the ones that have simple null distributions. Fligner and Verducci (1986) present the results for these distances. That for Kendall also appears in Mann (1945), Feigin and Cohen (1978), and Schulman (1979).

6.2.1 *Kendall*

Using (3.44) and Lemma 3.5, we have that

$$\begin{aligned}\psi_{\text{Ken}}(\theta) &= \log(E_0(\exp(\theta d_{\text{Ken}}(e_m, Y)))) \\ &= \log(E_0(\exp(\theta \sum U_i))) \\ &= \sum_{i=1}^{m-1} \log(E_0(\exp(\theta U_i))) \\ &= \sum_{i=1}^{m} \phi(\theta; i) \equiv \psi(\theta; m)\end{aligned} \tag{6.3}$$

where $\phi(\theta; q)$ is given in (5.17). By continuity, set $\phi(0; i) = 0$. Differentiating yields

$$\begin{aligned}\delta_{\text{Ken}}(\theta) &= \frac{m \cdot \exp(\theta)}{1 - \exp(\theta)} - \sum_{i=1}^{m} \frac{i \cdot \exp(i\theta)}{1 - \exp(i\theta)} \quad \text{and} \\ I_{\text{Ken},1}(\theta) &= \frac{m \cdot \exp(\theta)}{(1 - \exp(\theta))^2} - \sum_{i=1}^{m} \frac{i^2 \cdot \exp(i\theta)}{(1 - \exp(i\theta))^2}.\end{aligned} \tag{6.4}$$

The moments in (6.4) are the null moments as in Table 3.3 if we let $\theta \to 0$. For small θ some asymptotic expansions are more accurate. See Exercise 6.2.

6.2.2 *Cayley*

The process here is the same as for Kendall, but the U_i's are Bernoulli as in (3.51). Details are in Exercise 6.3:

$$\begin{aligned}\psi_{\text{Cay}}(\theta) &= \sum_{i=1}^{m} \log((i-1)\exp(\theta) + 1) - \log(m!), \\ \delta_{\text{Cay}}(\theta) &= \sum_{i=1}^{m} \frac{(i-1)\exp(\theta)}{1 + (i-1)\exp(\theta)}, \quad \text{and} \\ I_{Cay,1}(\theta) &= \sum_{i=1}^{m} \frac{(i-1)\exp(\theta)}{(1 + (i-1)\exp(\theta))^2}.\end{aligned} \tag{6.5}$$

6.2.3 Hamming

Let $X = m - d_{\text{Ham}}(\omega, Y)$. To find ψ_{Ham}, write $\exp(\psi_{\text{Ham}}(\theta)) = \exp(m\theta)E_0(\exp(-\theta X))$. Set $\lambda = \exp(-\theta)$ so that

$$E_0(\exp(-\theta X)) = E_0(\lambda^X) \equiv g_m(\lambda). \tag{6.6}$$

Now g_m is the factorial moment generating function of X, that is, $g_m^{(k)}(1)$ (the kth derivative at $\lambda = 1$) equals the kth factorial moment of X: $E_0(X(X-1)\cdots(X-k+1))$. It is well known that the kth factorial moment of a $Poisson(\eta)$ is η^k. Diaconis (1988, page 135) points out that under the *Uniform*, X has its first m factorial moments equaling those of $Poisson(1)$, that is, they are all 1. The moments for $k > m$ are 0 since $X = m - d_{\text{Ham}} \leq m$. See Exercise 6.4. Thus by the Taylor Series for g_m,

$$g_m(\lambda) = \sum_{k=0}^{\infty} \frac{1}{k!} g_m^{(k)}(1)(\lambda - 1)^k = \sum_{k=0}^{m} \frac{1}{k!}(\lambda - 1)^k; \tag{6.7}$$

hence

$$\begin{aligned} \psi_{\text{Ham}}(\theta) &= m\theta + \log(g_m(\lambda)), \\ \delta_{\text{Ham}}(\theta) &= m - \frac{\lambda g_{m-1}(\lambda)}{g_m(\lambda)} \text{ and} \\ I_{Ham,1}(\theta) &= \frac{\lambda g_{m-1}(\lambda)}{g_m(\lambda)} + \frac{\lambda^2 g_{m-2}(\lambda)}{g_m(\lambda)} - \frac{\lambda^2 g_{m-1}^2(\lambda)}{g_m^2(\lambda)}. \end{aligned} \tag{6.8}$$

Remark 6.1. A Universal Coefficient?
In Section 2.5 the idea of normalizing d to obtain a more interpretable t (2.21) was discussed. We also mentioned that the t's for different distances are not necessarily comparable, since they are all on different scales. The goal in general is to find a transformation of d so that "0" means uniformity and "1" means all the data collect at the mode. Is there a transformation that can be applied to all distances to put them on the same scale? One suggestion is to use the R^2 proposed in (5.51), where for each distance the model is the exponential family model corresponding to the distance. This is clearly an *ad hoc* suggestion, but then it is

Table 6.1 *Ballots: distance models.*

Distance	S	t	(se)	θ	(se)	LRS	$100 \cdot R^2$
Spearman	3338	0.120	(0.023)	−0.060	(0.012)	27.28	53.67
Footrule	2244	0.112	(0.019)	−0.128	(0.021)	36.93	72.66
Kendall	1263	0.112	(0.021)	−0.184	(0.034)	29.16	57.38
Hamming	1697	0.105	(0.025)	−0.192	(0.030)	39.25	77.24
Cayley	981	0.113	(0.019)	−0.275	(0.047)	34.50	67.89
Ulam	857	0.096	(0.019)	−0.288	(0.056)	26.21	51.56
Maximum	1122	0.112	(0.019)	−0.256	(0.042)	36.93	72.66

also arbitrary to use d instead of d^2 or $\sqrt{d}$ in the t coefficients. The R^2 measures how much of the nonuniformity in the data can be explained by the particular distance model. In all cases, the denominator is the same, being the total amount of nonuniformity in the data, *Denom* of (5.51). One nice feature is that for all the distances, one can test uniformity using $(1 - R^2) \cdot Denom$, which is asymptotically χ_1^2 under the null. □

Example 6.1. Ballots: Distance Models
Recall Example 2.12. The modal ranking was given as $(1, 2, 3)$. The total nonuniformity is $Denom = 50.8214$. Table 6.1 contains the statistics for the seven distances. Although the t's are similar, the R^2's differ widely. Hamming has the highest, explaining 77% of the nonuniformity. Spearman and Ulam do worse, explaining not much more than half. □

Example 6.2. Draft Lottery: Distance Models
Without some tricks, it is fairly obvious that it is impossible to fit Spearman, Footrule, or Ulam to the data in Examples 3.4 amd 3.5. Maximum is possible since the chance $d_{\text{Max}} < m/2$ is negligible. Kendall, Hamming, and Cayley are quite easy. In fact, for Hamming we may as well assume that $m - d_{\text{Ham}} \sim Poisson(\exp(-\theta))$. Thus for 1969, $m - d_{\text{Ham}} = 2$, and $\widehat{\theta} = -\log(2)$. See Example 3.5. Table 6.2 contains the results. They are not surprising in light of Example 3.5. The total nonuniformity in the sample of $n = 1$ is $Denom = 2\log(m!) = 3596.47$. The goodness-of-fit test does not make sense since there is only one observation and 366! cells. Note

Table 6.2 *Draft Lottery: distance models.*

Distance	Year	Mode	$d(y,\omega)$	θ	se	LRS	$100 \cdot R^2$
Kendall	1969	e_m	38369	0.0037	0.0009	18.24	0.51
	1970	e_m	32883	−0.0002	0.0009	0.08	0.00
Hamming	1969	e_m	364	−0.6931	0.7071	0.77	0.02
	1969	e_m^c	364	−0.6931	0.7071	0.77	0.02
	1970	e_m	363	−0.6931	0.7071	0.77	0.02
	1970	e_m^c	361	−1.3863	0.5000	5.09	0.14
Cayley	1969	e_m	359	−0.1028	0.4360	0.05	0.00
	1969	e_m^c	360	0.1037	0.4744	0.05	0.00
	1970	e_m	359	0.1032	0.4750	0.05	0.00
	1970	e_m^c	357	−0.2802	0.4060	0.44	0.01
Maximum	1969	e_m	362	0.4808	0.3760	3.80	0.11
	1969	e_m^c	335	−0.1449	0.0917	2.25	0.06
	1970	e_m	356	0.1546	0.1797	1.01	0.03
	1970	e_m^c	354	0.0995	0.1538	0.51	0.01

how small the R^2's are, which is not surprising in light of Figure 3.1. The highest, for Kendall, is only half a percent. However, the likelihood ratio statistic of 18.24 is highly significant as a χ_1^2, showing that the data are not uniform. None of the other statistics show significance, except Hamming in 1970. We will ascribe that to random variation. □

6.3 Unknown mode – Likelihood

The density is now a function of θ and $\omega \in \mathcal{S}_m$,

$$f_\theta(y;\omega) = \exp(\theta d(y,\omega) - \psi(\theta)), \tag{6.9}$$

and the loglikelihood is

$$l(\theta,\omega;N) = \theta S(\omega) - n\psi(\theta); \quad S(\omega) = \sum_{i=1}^{n} d(y^{(i)},\omega). \tag{6.10}$$

For fixed $\theta > 0$ in (6.10), the maximum over ω occurs where $S(\omega)$ is maximized. If $\theta < 0$, then the maximum over ω occurs where $S(\omega)$ is minimized. To find the maximum likelihood estimate of (θ, ω), first find $\omega^{(1)}$ and $\omega^{(2)}$ that minimize and maximize, respectively, $S(\omega)$. For each $\omega^{(j)}$, find the maximum over θ in (6.10). Call the maximizers $\theta^{(1)}$ and $\theta^{(2)}$, respectively. Finally, set $(\widehat{\theta}, \widehat{\omega}) = (\theta^{(j)}, \omega^{(j)})$ for the j that corresponds to the maximum of $l(\theta^{(j)}, \omega^{(j)}; N)$ over $j = 1, 2$.

It may be that $\widehat{\theta} = 0$ ($\Rightarrow$ *Uniform*), in which case there is no unique $\widehat{\omega}$; in fact all ω's give the same likelihood since it does not depend on S. For Spearman and Kendall, $d(y, \omega) + d(y, \omega^c) = Max(d)$ (see 2.45); hence $(\omega^{(1)})^c = \omega^{(2)}$, $\theta^{(1)} = -\theta^{(2)}$, and $l(\theta^{(1)}, \omega^{(1)}; N) = l(\theta^{(2)}, \omega^{(2)}; N)$. Thus it is enough find just $\omega^{(1)}$ and $\theta^{(1)}$. See Exercise 6.5. There is an indeterminacy in the MLE, but it is conventional to take the $\omega^{(i)}$ that minimizes S.

It should be clear that the model (6.10) is not a regular model since the parameter space $\Re \times \mathcal{S}_m$ has a discrete component. However, Critchlow (1985) has shown that asymptotically the likelihood can be regarded as the one with ω known. In this case, we have that Theorem 5.1 holds except that we must restrict $\theta \neq 0$. Theorem 5.2 applies to the goodness-of-fit test (5.50), again if $\theta \neq 0$. The test of uniformity, i.e., LRS of (5.49) for (5.48) when $H_0 : \theta = 0$ is uniformity, is not asymptotically χ_1^2. One could simulate the distribution, as in the next example.

Example 6.3. Sports: Effect of Estimating the Mode
Fix the mode ω and consider the Spearman model under uniformity, $\theta = 0$. The expected value of $\overline{d}(\omega)$ is 56, and the information I_1 is 522.67. To find out the effect of estimating ω, we simulated the experiment 1000 times, each trial consisting of 130 vectors from $Uniform(\mathcal{S}_7)$. For each, we find the optimal $\widehat{\omega}$, $\widehat{\theta}$, $\overline{d}_{Spear}(\widehat{\omega})$, $\widehat{I}_1(\widehat{\omega})$, and the likelihood ratio statistic. The average $\widehat{\theta}$ is -0.009 (with sampling standard error 0.001), the average $\overline{d}_{Spear}(\widehat{\omega})$ is 51.31 (se = 0.04), and the average $\widehat{I}_1$ is 514.74 (se = 0.07). The average LRS is 5.98 (se = 0.11), high for a putative χ_1^2. In fact, 34% of the LRS's are significant at the 1% level, and 68% at the 5% level. Thus estimating the mode does have an important effect at the uniform. We note that the maximum of the 1000 LRS's is 25.45, which is smaller than the observed value 27.08 based on the data; hence in this case the uniform is still not accepted. □

6.4 Unknown mode – Bayesian

In Remark 3.3, we suggest a method for obtaining confidence sets for the center. An alternative for assessing the reasonableness of any ranking as the mode is to use Bayesian methods. Many mathematical statistics books introduce Bayesian methods, e.g., Hogg and Craig (1978). For a more detailed introduction, see Box and Tiao (1973). The basic idea is that the unknown parameter and data enjoy a joint distribution. The marginal distribution of the parameter is called the *prior* π since it reflects the researcher's belief about the parameter prior to the collection of the data. The conventional data density, $f_\theta(x)$, is now the conditional density of the data x given the parameter θ. Once the data are observed, one uses Bayes' Theorem to find the *posterior* distribution of the parameter given the data. Inferences about the parameter are then based on this posterior distribution.

For a distance model, ω and θ are the parameters. Starting with a prior on (ω, θ), upon seeing the data $y^{(1)}, \ldots, y^{(n)}$, one can find the posterior probability of each possible modal ranking. Fligner and Verducci (1990) consider the ϕ model and the ϕ component model. The presentation here will concentrate on Spearman's model. It should be clear how to treat other distances.

Our prior distribution will assume θ and ω are independent, with $\omega \sim$ *Uniform* $(\mathcal{S}_m)$. We will place the *conjugate* prior on θ, i.e., given parameters d_0 and n_0, θ has density

$$\pi(\theta; d_0, n_0) = \exp(\theta n_0 d_0 - n_0 \psi(\theta)). \tag{6.11}$$

As written, π does not necessarily integrate to 1, nor need it have a finite integral at all. If $n_0 > 0$ and $0 < d_0 < Max(d) = m(m^2 - 1)/3$, then it will have a finite integral. See Exercise 6.6. The interpretation of the parameters is that the prior distribution for θ is as if one had n_0 prior observations whose distance from ω averaged d_0. If n is large, the choice of d_0 and n_0 is not crucial, and in fact can be taken to be 0. Upon observing the data, the joint posterior distribution for θ and ω is

$$\pi(\theta, \omega; y^{(1)}, \ldots, y^{(n)}) = \frac{\exp(\theta(S(\omega) + n_0 d_0) - (n + n_0)\psi(\theta))}{\sum_{\eta \in \mathcal{S}_m} \nu(\eta)}, \tag{6.12}$$

where

$$\nu(\eta) = \int \exp(\alpha(S(\eta) + n_0 d_0) - (n + n_0)\psi(\alpha)) \, d\alpha. \qquad (6.13)$$

See Exercise 6.7. Since we are mainly interested in ω, the posterior probability for each ω is obtained by integrating (6.12) over θ, which yields $\nu(\omega)/\sum_{\eta \in \mathcal{S}_m} \nu(\eta)$. We have purposely omitted the limits in the integral. If the integral is over the entire real line, then the symmetry in Spearman's distance will mean that ω and ω^c have the same posterior probability. By restricting to $\theta \leq 0$, ω's that are close to the data have higher probability. The integral (6.13) can be performed numerically, or using an approximation. Fligner and Verducci (1990) use an approximation specially tailored to the ϕ models. We will use an alternative approximation that can also be used for other distances, at least if n is reasonably large.

Let $\widehat{\theta}(\omega)$ be the maximizer of $\theta(S(\omega) + n_0 d_0) - (n + n_0)\psi(\theta)$ over $\theta \in \Re$. Thus it is the maximum likelihood estimator of θ with ω fixed based on $n + n_0$ observations with sufficient statistic $S(\omega) + n_0 d_0$. If $n_0 = 0$, then it is the regular maximum likelihood estimate. Now expand the exponent in (6.13) about $\theta = \widehat{\theta}(\omega)$, to obtain

$$\begin{aligned} &\theta(S(\omega) + n_0 d_0) - (n + n_0)\psi(\theta) \\ &\qquad \approx \frac{LRS(\omega)}{2} - (n + n_0) I_1(\widehat{\theta}(\omega)) \frac{(\theta - \widehat{\theta}(\omega))^2}{2}, \end{aligned} \qquad (6.14)$$

since $S(\omega) + n_0 d_0 = (n + n_0)\psi'(\widehat{\theta}(\omega))$, where

$$LRS(\omega) = 2(\widehat{\theta}(\omega)(S(\omega) + n_0 d_0) - (n + n_0)\psi(\widehat{\theta}(\omega))). \qquad (6.15)$$

Thus we have that

$$\begin{aligned}\nu(\omega) &\approx \exp(LRS(\omega)/2))\\ &\quad \int_{-\infty}^{0} \exp(-(n+n_0)I_1(\widehat{\theta}(\omega))(\theta-\widehat{\theta}(\omega))^2/2)\, d\theta\\ &= \frac{\sqrt{2\pi}\exp(LRS(\omega)/2)\Phi(-\sqrt{(n+n_0)I_1(\widehat{\theta}(\omega))}\widehat{\theta}(\omega))}{\sqrt{(n+n_0)I_1(\widehat{\theta}(\omega))}},\end{aligned} \tag{6.16}$$

where Φ is the standard normal distribution function. If n is large, then $\nu(\omega)$ will be practically 0 when $\widehat{\theta}(\omega) > 0$. If there is a large number of objects, then the calculations become tedious since (6.16) is needed for each $\omega \in \mathcal{S}_m$. If there is symmetry, as for Kendall and Spearman, then the $LRS(\omega)$ is needed only for half the rankings. See Exercise 6.8.

Example 6.4. Voting and Sports: Mode
The confidence region for ω in Example 3.2, the Voting data, when ω is defined via Spearman's distance, has four elements. The best ranking is 15243. Using the Bayes method with $d_0 = n_0 = 0$, $\nu(15243) \approx 100\%$. The next three highest probabilities belonged to the other three rankings in the confidence set, but their probabilities are only 0.01%, 0.001% and 0.001% for 14253, 15342, and 25143.

For the Sports data, using the same $d_0 = n_0 = 0$, the probabilities are much more spread out, as expected from Example 3.3. There are 29 rankings whose posterior probabilities are over 1% for a total of 63.26%. The highest is $\nu(4625137) = 5.07\%$. All 29 have Jogging ranked 7 and Football ranked 6. Choosing the smallest set that has total probability at least 95%, we find 163 rankings. All have Jogging either 6 or 7, and Football either 4, 5, 6, or 7. All other sports are represented at each ranking from 1 to 6. It is possible to obtain other interesting quantities, such as the posterior marginals of the ranks. For example, $P[\omega_{Jogging} = 7] = 95.96\%$, and $P[\omega_{Cycling} = 1] = 60.64\%$. □

6.5 Asymptotics as $m \to \infty$

In Section 3.4.3 we present the asymptotic distribution under the uniform of the distances as the number of objects increases. In this section we do the same but assuming the non-null distance models. The ones we can handle are Kendall, Cayley, Hamming, and Maximum. We will start with Kendall. In addition to finding the asymptotic distribution, we find approximations to the likelihood calculations. These approximations work quite well for the Draft Lottery example for Kendall, Hamming, and Cayley, but not as well for Maximum.

6.5.1 Kendall

If $\theta \neq 0$ is fixed and $m \to \infty$, then $\tau(\theta) \to \pm 1$, the sign being the opposite of θ's. See Exercise 6.9. Instead we set $\theta = \alpha/m$, which will ensure that τ has a nontrivial limit. The asymptotic mean and variance of d_{Ken} under $Kendall_m(\alpha/m)$ can be found using Equation (6.4) for $\alpha \neq 0$ as

$$\begin{aligned}\mu_{\text{Ken}}(\alpha) &\equiv \lim_{m\to\infty} \frac{1}{m^2}\delta_{\text{Ken}}(\alpha/m)\\ &= \lim_{m\to\infty}\Big(\frac{\exp(\alpha/m)}{m(1-\exp(\alpha/m))}\\ &\quad - \frac{1}{m}\sum_{i=1}^{m}\frac{(i/m)\exp(i\alpha/m)}{1-\exp(i\alpha/m)}\Big)\\ &= -\frac{1}{\alpha} - \int_0^1 \frac{z\exp(z\alpha)}{1-\exp(z\alpha)}\,dz,\end{aligned} \tag{6.17}$$

and

$$\begin{aligned}\sigma^2_{\text{Ken}}(\alpha) &\equiv \lim_{m\to\infty} \frac{1}{m^3}I_{\text{Ken},1}(\alpha/m)\\ &= \lim_{m\to\infty}\Big(\frac{\exp(\alpha/m)}{m^2(1-\exp(\alpha/m))^2}\\ &\quad - \frac{1}{m}\sum_{i=1}^{m}\frac{(i/m)^2\exp(i\alpha/m)}{(1-\exp(i\alpha/m))^2}\Big)\\ &= \frac{1}{\alpha^2} - \int_0^1 \frac{z^2\exp(z\alpha)}{(1-\exp(z\alpha))^2}\,dz.\end{aligned} \tag{6.18}$$

We also obtain an approximation for ψ in (6.3):

$$\begin{aligned}\rho_{\mathrm{Ken}}(\alpha) &\equiv \lim_{m\to\infty} \frac{1}{m}\psi_{\mathrm{Ken}}(\alpha/m) \\ &= \frac{1}{m}\sum_{i=1}^{m} \log\left(\frac{1-\exp(i\alpha/m)}{i(1-\exp(\alpha/m))}\right) \\ &= \int_0^1 \log\left(\frac{1-\exp(z\alpha)}{-z\alpha}\right)\, dz.\end{aligned} \tag{6.19}$$

Note that μ_{Ken} and σ^2_{Ken} are the first and second derivatives of ρ_{Ken}. Further asymptotic moments can be found by repeatedly differentiating. It is interesting that ρ is related to the *dilog* function. See Section 27.7 of Abramowitz and Stegun (1972), and Exercise 6.10. The asymptotic normality of d_{Ken} is given next.

Theorem 6.1
Let $\theta = \alpha/m$ for $\alpha \neq 0$. Then as $m \to \infty$, $\sqrt{m}((d_{\mathrm{Ken}}/m^2) - \mu_{\mathrm{Ken}}(\alpha)) \to N(0, \sigma^2_{\mathrm{Ken}}(\alpha))$. □

The proof is very similar to that for the uniform case, below Theorem 3.4. See Exercise 6.11. We can use (6.17), (6.18), and (6.19) to perform approximate likelihood calculations, as in the next example.

Example 6.5. Draft Lottery: Distance Models
In Table 6.2 we have the results of the likelihood analysis of the Draft Lottery data. Consider the Kendall model. To use the approximation, the estimate of α is that found by solving $\mu_{\mathrm{Ken}}(\widehat{\alpha}) = d^2_{\mathrm{Ken}}/m^2$. Then $\widehat{\theta} \approx \widehat{\alpha}/m$; $I_{\mathrm{Ken},1}(\widehat{\alpha}) \approx m^3\sigma^2_{\mathrm{Ken}}(\widehat{\alpha})$, and $LRS \approx 2(\widehat{\alpha} d_{\mathrm{Ken}}/m - m\rho_{\mathrm{Ken}}(\widehat{\alpha}))$. For this example, the approximations are amazingly accurate, giving $\widehat{\theta} = 0.003702$ with se $= 0.0008785$ and $LRS = 18.24$ for 1969; $\widehat{\theta} = -0.0002448$ with se $= 0.0008588$ and $LRS = 0.08127$ for 1970. (The exact LRS for 1970 to 4 significant digits is 0.08127, too.) That is, only the first $\widehat{\theta}$ is off by even one digit in the fourth place. The approximation is best for large m and small θ, the situation here. □

6.5.2 Cayley

The approximations for Cayley are quite simple. The quantities in (6.5) are approximated by substituting an integral for the sum of all but the first $k-1$ terms in the summations, the integral being for θ from $k - 1/2$ to $m + 1/2$ with the integrand being the summand. The larger k is, the better the approximation.

Applying the approximation to the Draft Lottery data, using $k = 10$, we obtain the same results as in Table 6.2, with the approximations usually off only in the last decimal place.

Cayley's distance is asymptotically normal under the non-null model. The proof is as for the null model, noting that the variance in Table 3.3 grows at the rate of $\log(m)$.

Theorem 6.2
As $m \to \infty$, $\sqrt{m}(d_{\text{Cay}} - \mu_{\text{Cay}}(\theta)) \to N(0, \sigma^2_{\text{Cay}}(\theta))$. □

6.5.3 Hamming

We have practically taken care of Hamming already. If m is large, then $g_m(\lambda) \approx \exp(\lambda - 1)$ in (6.7). Set $X = m - d_{\text{Ham}}(\omega, Y)$ as in Section 6.2.3. From (3.66), for fixed k,

$$\begin{aligned} P_\theta[X = k] &= \exp(\theta(m-k) - \psi_{\text{Ham}}(\theta))P_0[X = k] \\ &\to \exp(-\theta k - (\exp(-\theta) - 1))\exp(-1)/k! \qquad (6.20) \\ &= P[\mathit{Poisson}(\exp(-\theta)) = k)]. \end{aligned}$$

It is then especially easy to fit the approximate model: $\widehat{\lambda} = m - d_{\text{Ham}}$, $I_1 = \lambda$, and $LRS = 2((m - d_{\text{Ham}})\log(\lambda) - \lambda + 1)$. The results in Table 6.2 for Hamming are exactly duplicated using the approximations.

6.5.4 Maximum

Under uniformity, we have the limit in (3.67) for $(m - d_{\text{Max}})/\sqrt{m}$. The asymptotic non-null distribution is found by setting $\theta = -\alpha/\sqrt{m}$. The ψ function is approximated as follows:

$$\begin{aligned} \psi(\theta) &= \log(E_0(\exp(\theta d_{\text{Max}}))) \\ &= m\theta + \log(E_0[\exp(-\sqrt{m}\theta((m - d_{\text{Max}})/\sqrt{m}))]); \end{aligned} \qquad (6.21)$$

hence

$$\lim_{m \to \infty} (\psi(\theta) - m\theta) = \log(E_0[\exp(\alpha W)]) \equiv \rho_{\text{Max}}(\alpha), \qquad (6.22)$$

where W has the distribution in (3.67). Thus, asymptotically under α, $W = (m - d_{\text{Max}})/\sqrt{m}$ has density

$$f_\alpha(w) = 2w\exp(\alpha w - w^2 - \rho_{\text{Max}}(\alpha)). \tag{6.23}$$

Integration by parts (Exercise 6.12) shows that

$$\rho_{\text{Max}}(\alpha) = \log(1 + \sqrt{\pi}\alpha\exp(\alpha^2/4)\Phi(\alpha/\sqrt{2})). \tag{6.24}$$

The mean and variance of the asymptotic W are found by taking the first two derivatives of ρ_{Max}. The approximate likelihood calculations proceed by finding the $\widehat{\alpha}$ to satisfy $(m - \overline{d})/\sqrt{m} = \mu_{\text{Max}}(\widehat{\alpha})$, from which we find $\widehat{\theta} \approx -\sqrt{m}\alpha$, $I_1(\widehat{\theta}) \approx m\sigma^2_{\text{Max}}(\widehat{\alpha})$, and $LRS \approx 2n(\widehat{\alpha}(m - d_{\text{Max}})/\sqrt{m} - \rho_{\text{Max}}(\widehat{\alpha})))$. The estimates of $\widehat{\theta}$ for the four cases in Table 6.2 are 0.4679, -0.1292, 0.1536, and 0.0999. These estimates are close, but especially for 1969, are not nearly as good as the previous approximations. □

6.6 Assessing fit

For most data sets, the one-parameter distance models are not rich enough to capture the full structure in the data. One hopes that they can give a good first approximation, and that by contrasting the fitted model to the data one can find more insight into the data, and possibly ideas for further models to apply. The two approaches to assessing fit in this section are (1) formally embedding in a larger model, and (2) comparing expected statistics to the observed. The test of the model versus the saturated model is an example of the first approach. It is very effective when there are enough data (e.g., $n \geq 3 \cdot m!$). Otherwise, testing against a model smaller than the saturated, but larger than the distance model, is indicated. In any case, comparing the basic statistics to what is expected under the model can be helpful.

In order to use approach (1), one must have convenient larger models. The main ones we will discuss here are those based on the basic statistics, i.e., Bradley–Terry/Mallows, Babington Smith, and Marginals. Later chapters will suggest other possibilities. The next equation exhibits some of the inclusions:

$$\begin{aligned} Hoeffding &\subset Marginals; \\ Spearman &\subset Bradley\text{–}Terry/Mallows \\ &\subset (Marginals \cap Babington\ Smith); \\ Kendall &\subset Orthogonal\ contrast\ \phi \subset Babington\ Smith; \\ Kendall &\subset \phi - component; \\ Cayley &\subset Cayley\ component. \end{aligned} \tag{6.25}$$

It is straightforward to test a distance model versus one that includes it using the likelihood ratio test. By comparing parameters within the models, one can determine where fit is lacking if the distance model is not adequate. Examples will follow.

6.6.1 Kendall

The proofs of Theorems 6.3 and 6.5 will come in Chapter 7 since we need a certain amount of machinery. We start with the Pairs. The result is due to Mallows (1957).

Theorem 6.3: Mallows
Suppose Y has Mallows' ϕ distribution with mode e_m and parameter θ. Then for $i < j$,

$$K_{ij} = P_\theta[Y_i < Y_j] = 1 - h(j - i + 1; \theta) + h(j - i; \theta), \tag{6.26}$$

where for positive integer k, $h(k; \theta) = k/(1 - \exp(-k\theta))$. □

For $i > j$, $K_{ij} = 1 - K_{ji}$. Note that (6.26) depends only on the difference of the indices $j - i$. Thus the matrix K is constant on parallels to the diagonal. The next result, for the mean, is found in Chung (1989).

Theorem 6.4
Suppose Y has Mallows' ϕ distribution with mode e_m and parameter θ. Then with $\mu = E_\theta(Y)$, $\mu_k = h(k; -\theta) + h(m - k + 1; \theta)$. □

Proof
Since for any rank vector, $y_k = 1 + \sum_{i \neq k} I[y_k > y_i]$,

$$\begin{aligned} E_\theta(Y_k) &= 1 + \sum_{i \neq k} P_\theta[Y_k > Y_i] \\ &= 1 + \sum_{i<k} (1 - h(k-i+1;\theta) + h(k-i;\theta)) \\ &\quad + \sum_{i>k} (h(i-k+1;\theta) - h(i-k;\theta)). \end{aligned} \tag{6.27}$$

Note that in the final two summations, the summands collapse. Further canceling the $h(1;\theta)$'s and combining k and $h(k;\theta)$ proves the result. □

The Marginals, i.e., $M = E_\theta(Q^{(Y)})$ of (3.11), takes a bit more notation. For integers i, k, define

$$g_{i,k} = h_{i\theta}(k)/(m-i+1), \; k = 0, \ldots, m-i, \tag{6.28}$$

from (5.16), and

$$G_{i,k} = \sum_{l=0}^{k} g_{i,l} = \frac{1 - \exp(\theta(k+1))}{1 - \exp(\theta(m-i+1))}. \tag{6.29}$$

Now for $1 \leq i \leq m-1$, define the $(m-i+1) \times (m-i+1)$ matrix

$$A_\theta^{(i)} = \begin{pmatrix} g_{i,0} & g_{i,1} & g_{i,2} & \cdots & g_{i,m-i-1} & g_{i,m-i} \\ 1-G_{i,0} & G_{i,0} & 0 & \cdots & 0 & 0 \\ 0 & 1-G_{i,1} & G_{i,1} & \cdots & 0 & 0 \\ \vdots & \vdots & \vdots & \ddots & \vdots & \vdots \\ 0 & 0 & 0 & \cdots & 1-G_{i,m-i} & G_{i,m-i} \end{pmatrix}. \tag{6.30}$$

Embed these matrices in $m \times m$ matrices as follows:

$$B_\theta^{(i)} = \begin{pmatrix} I_{i-1} & 0 \\ 0 & A_\theta^{(i)} \end{pmatrix}. \tag{6.31}$$

We are ready for the result.

Theorem 6.5
Suppose Y has Mallows' ϕ distribution with mode e_m and parameter θ. Then $M = E_\theta(Q^{(Y)}) = B_\theta^{(m-1)} \times B_\theta^{(m-2)} \times \cdots \times B_\theta^{(1)}$. □

The proof is in Section 7.6.4. See Chung and Marden (1990) for an efficient algorithm.

6.6.2 *Hamming and Cayley*

Hamming and Cayley are both bi-invariant distances. Much of what we have to say is true for all bi-invariant distances. The following shows that M has equal diagonals, and equal off-diagonals, and μ is linear in the mode.

Theorem 6.6
If d is bi-invariant and the mode is e_m, then under the distance model there exists a constant A such that $M_{ii} = A$ for all i, $M_{ij} = (1-A)/(m-1)$ if $i \neq j$, and

$$\mu = \frac{Am-1}{m-1} e_m + \frac{(m+1)m}{2} \frac{1-A}{m-1} 1_m. \qquad \square \quad (6.32)$$

Proof
From Section 3.3, any bi-invariant function of y is invariant under conjugation, $Q^{(y)} \to QQ^{(y)}Q'$ for $Q \in \mathcal{Q}_m$. Since $d(e_m, y)$ is given to be bi-invariant, the density is bi-invariant; hence $Q^{(Y)}$ and $QQ^{(Y)}Q'$ have the same distribution for any $Q \in \mathcal{Q}_m$. By taking expected values, $M = QMQ'$. Average over Q: $M = (1/m!) \sum_{Q \in \mathcal{Q}_m} QMQ'$. From Exercise 3.10, the right-hand side has the form $aI + bJ$. Thus the result for M follows since the rows and columns of M sum to 1. Then $\mu = Me_m$. □

For any particular model, all one needs to do is find M_{ii} for one i. For Cayley, $M_{11} = 1 - E(U_m) = 1/(1 - (m-1)\exp(\theta))$, where U_m from (3.51) is *Bernoulli* $((m-1)/m)$). For Hamming, $E(d_{\text{Ham}}(e_m, y)) = m - \sum M_{ii} = m(1-A)$; hence $M_{11} = 1 - \delta_{\text{Ham}}/m$. When using the maximum likelihood estimator for θ, we know that it is chosen to satisfy $\delta_{\text{Ham}}(\widehat{\theta}) = \overline{d}_{\text{Ham}}$; hence $\widehat{M}_{11} = 1 - \overline{d}_{\text{Ham}}/m$.

Table 6.3 *Political Goals: Bradley–Terry/Mallows parameters.*

		Order	Say	Prices	Freedom
v:					
	Spearman	**27.48**	**20.36**	**37.08**	**15.08**
	BTM	**32.35**	**17.97**	**34.40**	**15.38**
μ:					
	Spearman	**2.270**	**2.730**	**1.843**	**3.157**
	BTM ($\bar{y}$)	**2.027**	**2.905**	**1.947**	**3.121**

Example 6.6. Political Goals: Nested Models
We fit a set of nested models to the data from Example 2.3. The models and their LRS_n's and corresponding degrees of freedom are Spearman: 1397.18 on 1 df; Bradley–Terry/Mallows: 1553.96 on 3 df; Babington Smith: 1560.33 on 6 df; and Saturated: 1838.47 on 23 df. It is clear that none of these models fit relative to the saturated model, Bradley–Terry/Mallows fits much better than Spearman, but Babington Smith does not significantly improve upon Bradley–Terry/Mallows. The Bradley–Terry/Mallows parameters are v_i. Spearman's model with parameter θ is a submodel where $v_i = c.\exp(-2\omega_i\theta)$. The estimate of θ under Spearman is -0.1499. Table 6.3 exhibits the estimates of v_i under Spearman and Bradley–Terry/Mallows. They are normalized to sum to 100. The latter model fits the observed means of the ranks exactly. The table also contains those means, and the expected values under Spearman. Recall that the higher v_i, the more preferred the object. Although Spearman's estimates are in the right direction, they underestimate the gap between Order and Say, and overestimate the gaps between Order and Prices, and between Say and Speech. □

Example 6.7. Occupations: Assessing Fit of Mallows' Model
Cohen and Mallows (1983) compared the observed $\widehat{K}$ (2.12) to the expected under the ϕ model for the data in Example 2.4, and found the fit to be inadequate. We will look at the average. Fitting the ϕ model with mode e_m produces $\widehat{\theta} = -0.6523$. Table 6.4 shows the observed $\overline{y}$ and expected values of $\mu = E_\theta(Y)$. The Z

Table 6.4 *Occupations: fit of the ϕ model to the mean.*

	Occupation									
	Fac	**Own**	**Sci**	**OR**	**IE**	**Mgr**	**ME**	**Sup**	**Tech**	**For**
$\bar{y}_i$	**2.43**	**2.87**	**3.15**	**4.71**	**4.73**	**5.71**	**6.10**	**6.66**	**9.10**	**9.53**
μ_i	**2.07**	**2.72**	**3.45**	**4.24**	**5.08**	**5.92**	**6.76**	**7.55**	**8.28**	**8.93**
Z	**2.91**	**1.07**	**−1.95**	**2.91**	**−2.11**	**−1.26**	**−4.08**	**−5.83**	**5.95**	**4.96**

is the difference of the observed and expected means, divided by the standard error of the observed mean under the model. The standard error is found by obtaining the variance of each Y_i from the $E_\theta(Q^{(Y)})$.

The μ is symmetric about $(m+1)/2$ $(\mu_i + \mu_{m-i} = m+1)$ and is fairly linear. The observed means are not as regularly spaced: the gaps between Scientist and OR, and between Supervisor and Technician, are much larger than expected. Thus the ϕ model gives a reasonable first impression of the rankings, but does not capture the grouping of the occupations. □

Example 6.8. Ballots: Assessing Fit of Hamming's Model
Consider the Ballots data, Example 2.12. Hamming's model fits fairly well. Its $LRS = 39.25$, and the goodness-of-fit statistic is $50.82 - 39.25 = 11.57$ on 4 degrees of freedom, which yields a p-value of 0.02. (The $\widehat{\theta} = -0.1918$ with standard error 0.0298.) The next equation gives the sample marginals $100 \times \widehat{M}$, the fitted marginals $100 \times M(\widehat{\theta})$, and Z. Since $\overline{d}_{\mathrm{Ham}} = 1697/948 = 1.7901$, $A = 0.4033$.

$$\widehat{M} = \begin{pmatrix} 38.82 & 30.38 & 30.80 \\ 34.28 & 39.35 & 26.37 \\ 26.90 & 30.27 & 42.83 \end{pmatrix},$$
$$M(\widehat{\theta}) = \begin{pmatrix} 40.33 & 29.83 & 29.83 \\ 29.83 & 40.33 & 29.83 \\ 29.83 & 29.83 & 40.33 \end{pmatrix}, \qquad (6.33)$$
$$Z = \begin{pmatrix} -0.95 & 0.37 & 0.65 \\ 2.99 & -0.62 & -2.33 \\ -1.98 & 0.30 & 1.57 \end{pmatrix}.$$

The sample $\widehat{M}$ looks reasonably close to the form expected by a bi-invariant model. It is only the second row where there are significant differences, with second place on the ballot ranked higher relative to third place than expected by the model. It is interesting that the Marginals model fits almost perfectly, with $LRS = 50.01$, which is significantly better than Hamming, but not overwhelmingly so. □

6.7 Exercises

6.1. Consider the model (5.5). Show that if for a sample, $S \equiv \sum d(\omega, y^{(i)}) = 0$, then the likelihood function is strictly decreasing in θ, and if $S = n \cdot Max(d)$, then the likelihood function is strictly increasing.

6.2. Show that the one-term Taylor expansion of $\delta_{\text{Ken}}(\theta)$ in (6.4) about $\theta = 0$ is

$$\frac{1}{2}\binom{m}{2} + \frac{\theta}{72} m(2m+5)(m-1) + o(\theta),$$

and the two-term expansion of $I_{\text{Ken},1}$ is

$$\frac{m(m-1)(2m+5)}{72} - \frac{\theta^2}{7200} m(m-1)(6m^3+21m^2+31m+31) + o(\theta^2).$$

6.3. Verify (6.5). [For ψ_{Cay}, write $\exp(\psi_{\text{Cay}}(\theta)) = E(\exp(\theta d_{\text{Cay}}(e_m, Y))) = \prod_{k=2}^{m} E(\exp(\theta U_k))$, where the U_k's are Bernoulli from (3.51).]

6.4. Let $X = m - d_{\text{Ham}}(e_m, Y)$ for $Y \sim Uniform(\mathcal{S}_m)$. Show that for $1 \leq k \leq m$, $E(X(X-1)\cdots(X-k+1)) = 1$. [Note that since (3.56) is a probability distribution, where E_{m-k} is given in (3.53), for any $l > 0$, $\sum_{i=0}^{l} E_{l-i}/i! = 1$. Write

$$E(X(X-1)\cdots(X-k+1)) = \sum_{i=k}^{m} i(i-1)\cdots(i-k+1)\frac{E_{m-i}}{i!},$$

then change to summation over $j \equiv i - k$.]

6.5. For the distance model using either Kendall or Spearman distance, show that $f_\theta(y;\omega) = f_{-\theta}(y;\omega^c)$.

6.6. Show that for Spearman's distance, if $0 < d_0 < Max(d)$, then π in (6.11) has finite integral over $-\infty < \theta < \infty$. [Show that $\exp(\psi(\theta)) > \exp(Max(d)\theta)/m!$, so that for large θ, $\pi(\theta; d_0, n_0) < \exp(\theta n_0(d_0 - Max(d))$, which has finite integral at $+\infty$. Similarly, $\exp(\theta) > 1/m!$ can be used for the case $\theta \to -\infty$.]

6.7. Prove (6.13).

6.8. Show that Exercise 6.5 implies that $LRS(\omega) = LRS(\omega^c)$ for the Kendall and Spearman models.

6.9. Show that for Mallows' ϕ model (the distance model using Kendall), $\tau(\theta) \to +(-)1$ as $m \to \infty$ if $\theta < (>)0$, θ fixed. [For $\theta > 0$, show that $(\delta_{\text{Ken}}(\theta) - \sum_{i=1}^{m} i)/m^2 \to 0$; hence $\delta_{\text{Ken}}(\theta)/m^2 \to 1/2$, and $\tau(\theta) = 1 - \delta_{\text{Ken}}(\theta)/(m(m-1)/4) \to -1$. Similarly for $\theta < 0$.]

6.10. The *dilog* function is defined for $x > 0$ to be $dilog(x) = \int_1^x log(t)\,(1-t)^{-1}dt$. Show that $\rho_{\text{Ken}}(\alpha)$ from (6.19) can be written as $1 + \log((\exp(\alpha) - 1)/\alpha) + dilog(\exp(\alpha))/\alpha$. [Make the change-of-variables $u = 1 - \exp(z\alpha)$, and show using integration-by-parts that $\int_0^{1-x} \log(t)(1-t)^{-1}\, dt = -\log(x)\log(1-x) - dilog(x)$.] Show that $\mu_{\text{Ken}} = -1/\alpha - dilog(\exp(\alpha))/\alpha^2$.

6.11. Prove Theorem 6.1. [Use Theorem 3.4. Set $X_i = U_i/m^2$, so that the numerator in (3.65) is of order $m^{1-\nu}$, and the denominator is of order $m^{-\nu/2}$.]

6.12. Use integration by parts to obtain (6.24) from (6.22).

CHAPTER 7

Babington Smith, Phi-Models and Inversions

7.1 Introduction

This chapter ties together the Babington Smith paired comparison model (Section 5.5), Bradley–Terry/Mallows (BTM) (Section 5.5.1), Mallows' ϕ model (Section 5.4), the Fligner and Verducci ϕ-component model (Section 5.6.2), the orthogonal contrast ϕ models (Section 5.10.2), and the inversion models of McCullagh (1993a). All but the last are exponential family models based on statistics that are sums of paired comparisons $I[y_i > y_j]$'s or $I[w_i > w_j]$'s. The inversion model extends the paired comparisons to triples, quadruples, etc.

The ϕ-component models are fundamentally different than the others. The ϕ-component models are based on stages defined by the *ranks*: which object is ranked first, which second, ..., and which last. The other models are built on pre-set comparisons of *objects*. Thus, in Babington Smith, the building blocks are the paired comparisons, and in the orthogonal contrast ϕ models, they are comparisons of groups of objects. Compare the ϕ-component model (5.14) and (5.18) to the orthogonal contrast ϕ model (5.31) based on the contrasts $C^{(1)} = (\{A\}, \{B, C\})$ and $C^{(2)} = (\{B\}, \{C\})$. The sufficient statistics in the two models are

$$\begin{aligned} &\phi\text{-component: } (I[w_1 > w_2] + I[w_1 > w_3], I[w_2 > w_3]) \text{ and} \\ &\text{orthogonal contrast } \phi\text{: } (I[y_A > y_B] + I[y_A > y_C], I[y_B > y_C]), \end{aligned} \tag{7.1}$$

where $w \in \mathcal{T}_m$ is the ordering and $y \in \mathcal{S}_m$ is the corresponding ranking. The behavioralist interpretation of the ϕ-component model is to first decide which object to rank #1, then which of the remaining two to rank #2. That for the orthogonal contrast ϕ model is to first decide on the rank of object A, then decide which of B and C is preferred. Both seem reasonable, and both have the same mathematics. (They have the same normalizing constants, for instance.) However, they have distinctly different interpretations.

Section 7.2 presents details of the full Babington Smith model. The other sections look at various sub- and super-models. In (6.25) we mentioned some sequences of models. One is *Spearman* $\subset$ *BTM* $\subset$ *Babington Smith.* The BTM model fits the Means $\overline{y}$ exactly. Spearman's model places linear restrictions on the BTM parameters, $\log(\nu_i)$'s, in fact, placing them on a straight line. In Section 7.4.2 we look at other linear submodels that are simpler than BTM. The full Babington Smith model is based on the Pairs $\widehat{K}$ of (2.12), which has $\overline{m}(= \binom{m}{2})$ rather than the $m-1$ parameters of BTM. What is between these two models? The BTM model places a strict transitivity on the Babington Smith parameters. For the objects A, B, C, this transitivity in terms of odds is

$$\frac{p_{AB}}{1-p_{AB}} \frac{p_{BC}}{1-p_{BC}} = \frac{p_{AC}}{1-p_{AC}}. \tag{7.2}$$

Thus the differences in the two models are captured by the extent to which (7.2) fails for various triples of objects, i.e., by the nontransitivities. See McCullagh (1993a) and Section 7.4.3.

Another sequence of models is *Mallows'* $\phi \subset$ *orthogonal contrast* $\phi \subset$ *Babington Smith.* In Sections 7.5, 7.6, and 7.7, the orthogonal contrast ϕ models and their relationship with Mallows' ϕ and other models is explored in depth. In Section 7.8, we look at some models between the orthogonal contrasts and Babington Smith models. These involve some orthogonal contrasts mixed with small Babington Smith models within components of the contrasts. See Joe and Verducci (1993).

Since the above models are all subsets of Babington Smith, they all assume that the Pairs contain all the information in the sample. In Section 7.11 we go beyond the Pairs to the inversion models of McCullagh (1993a), which are based on higher-order comparisons. Section 7.10 grows models from the orthogonal contrast models, relaxing first the ϕ-ness, then the independence, to obtain contingency table models. Both the inversion and contingency table models allow expansion to the saturated model, that is, they are flexible enough to expand as far as the data will bear.

In Section 7.9, we interchange the roles of objects and ranks, which leads to some multistage models and models for tied rankings.

7.2 Babington Smith

We will use the exponential family notation (5.20), where the sufficient statistics consist of the $\overline{m}$ comparisons $I[y_i > y_j]$ for $i < j$. Thus, given a sample of size n, the sufficient statistics are the $n\widehat{K}_{ji}$, $i < j$, from (2.12). The density is

$$f_\theta(y) = \exp(\theta'(1_{\overline{m}} - \kappa(y)) - \psi(\theta)), \; y \in \mathcal{S}_m, \tag{7.3}$$

where $\theta = (\theta_{12}, \theta_{13}, \ldots, \theta_{m-1,m})$ is indeed as in (2.61), κ is given in (2.59), and ψ is the normalizing constant. Exercise 7.1 shows that the parameters p in (5.7) are related to the θ's through

$$p_{ij} = \frac{\exp(-\theta_{ij})}{1 + \exp(-\theta_{ij})}, \; i, j \in \mathcal{O}. \tag{7.4}$$

Likelihood analysis is fairly straightforward. The only problem is that ψ can be difficult to calculate if m is large, say $m \geq 10$, since one must sum over all $m!$ elements of $\mathcal{S}_m$. It may be that an EM-like algorithm exists, but I have been unable to find one. Thus we are left with either a regular Newton–Raphson algorithm, or iterative proportional fitting as mentioned in Remark 5.6. For the latter approach, the categorizations are based on the pairs of objects, where for each pair we categorize rankings into those that prefer one and those that prefer the other. Consider Newton–Raphson. Given a sample of n, the maximum likelihood estimator exists if and only if $0 < \widehat{K}_{ij} < 1$ for all $i < j$. If that restriction were violated for some pairs, a "quick-and-dirty fix" would be to add ϵ to each $n\widehat{K}_{ij}$ and 2ϵ to n for some small ϵ. A purist would reduce the problem as follows:

If $\widehat{K}_{ij} = 0 (= 1)$, set $\widehat{\theta}_{ij} = +\infty(-\infty)$. Let $G \equiv \{(i,j) | i < j, 0 < \widehat{K}_{ij} < 1\}$ be the set of pairs remaining, and $\mathcal{S}^*_m$ be the subset of rankings that conform to the sample, i.e.,

$$\mathcal{S}^*_m = \{y \in \mathcal{S}_m \mid y_i > (<) y_j \; if \; \widehat{K}_{ij} = 0 (= 1)\}. \tag{7.5}$$

Replace the loglikelihood in (5.56) by

$$l^*(\theta, \widehat{K}) = \sum\sum_{(i,j)\in G} n\theta_{ij}\widehat{K}_{ji} - n\psi^*(\theta), \tag{7.6}$$

where

$$\exp(\psi^*(\theta)) = \frac{1}{m!}\sum_{y\in\mathcal{S}^*_m} \exp(\sum\sum_{(i,j)\in G}\theta_{ij}I[y_i > y_j]). \tag{7.7}$$

The usual Newton–Raphson can then be applied to l^* to find the estimates of the θ_{ij} for $(i,j) \in G$.

A reasonable starting value for θ is 0. If G contains all pairs, i.e., $\mathcal{S}_m = \mathcal{S}^*_m$ in (7.5), then we know that in (5.62), using (3.23) and thereabove, $\beta(0) = (1/2)1_{\overline{m}}$ and $\Sigma(0) = (I_{\overline{m}} + \Gamma_m\Gamma'_m)/12$. Then, using Lemma 3.4, the first iteration in (5.62) can be shown to yield $\theta^{(1)}_{ij} = 12(\widehat{K}_{ji} - 1/2 - (\overline{y}_j - \overline{y}_i)/(m+1))$.

Example 7.1. Leisure Time: Babington Smith

We fit Babington Smith models to the two groups in Example 2.1, and to the total, to see how well the models fit individually and to compare the groups. Newton–Raphson works well for the Blacks and the Total since all $\widehat{K}_{ij}$'s are strictly between 0 and 1. The Whites, however, always prefer Females to Males. Thus for them, we set $\theta_{MF} = +\infty$. There are only three rankings in $\mathcal{S}^*_m$ of (7.5): $\mathcal{S}^*_m = \{(2,1,3),(3,1,2),(3,2,1)\}$. The reduced likelihood (7.6) is $l^*(\theta,\widehat{K}) = 13\cdot\theta_{MB} + 6\cdot\theta_{FB} - 14\cdot\log(1+\exp(\theta_{MB}) + \exp(\theta_{MB}+\theta_{FB})) + 14\cdot\log(3!)$. It is easy to maximize l^* since there are two statistics and two parameters. In fact, the model fits exactly. Exercise 7.2 shows that $\widehat{\theta}_{MB} = \log(7)$ and $\widehat{\theta}_{FB} = \log(6/7)$, $LRS = 2\cdot l^*(\widehat{\theta},\widehat{K}) = 25.0195$, and $R^2 = 1$ in (5.51). Table 7.1 contains the results of fitting Babington Smith to each group of women, and the combined sample.

Comparing the $\widehat{p}_{ij}$'s, we see that the two groups differ strongly on *Male versus Female* and *Female versus Both*: the Whites strongly prefer Females to Males, while the Black Women are split, and the Blacks much prefer *Both* to *Female*, while the Whites are split. This difference is evident in Figure 2.3. The *GOF* statistics are the goodness-of-fit statistics (5.50), while the *LRS* statistics test the Babington Smith versus the *Uniform* in (5.49). The *LRS*'s

Table 7.1 *Leisure Time: Babington Smith model.*

Pair of Objects	Black Women $\widehat{\theta}_{ij}$ se	$\widehat{p}_{ij}$ se	White Women $\widehat{\theta}_{ij}$ se	$\widehat{p}_{ij}$ se	Combined $\widehat{\theta}_{ij}$ se	$\widehat{p}_{ij}$ se
$(Male, Female)$	0	0.5000	$+\infty$	0	0.7757	0.3152
	0.6095	0.1524	–	–	0.4897	0.1057
$(Male, Both)$	1.1670	0.2374	1.9459	0.1250	1.6495	0.1612
	0.8399	0.1520	1.0690	0.1169	0.6699	0.0906
$(Female, Both)$	2.1310	0.1061	−0.1542	0.5385	0.5030	0.3768
	1.1018	0.1045	0.5564	0.1383	0.4562	0.1071
LRS (3 d.f.)		13.79		25.02		21.42
GOF (2 d.f.)		3.70		0		0.33
$100 \cdot R^2$		78.84%		100%		98.49%

are quite significant, but the *GOF*'s do not show anything wrong with the models. (The sample sizes may be too small to place too much faith in the χ^2 approximations.) To test whether the same Babington Smith model holds for both groups, versus the alternative that they have their own Babington Smith models, one adds the *LRS*'s for the individual models, and subtracts that for the combined model: $25.0195 + 13.7913 - 21.4242 = 17.3866$. Since each individual model has three parameters, as does the combined model, the degrees of freedom for this statistic is 3. Thus, there does seem to be a statistically significant difference between the groups.

Recall the warning in Section 5.5: The $\widehat{p}_{ij}$'s are different than the observed $\widehat{K}_{ji}$'s. For example, in the combined sample 11.11% prefer Male to Both, while $\widehat{p}_{MB} = 16.12\%$. In most cases, the observed paired comparisons derived from the rankings are more extreme than the p_{ij}'s, although it can happen that the p_{ij}'s are even on the other side of 0.5. See Example 7.5. □

Example 7.2. Political Goals, Voting, and Sports: Babington Smith

Table 7.2 gives the results of fitting the Babington Smith model to three familiar data sets. For Voting and Sports, the Babington Smith parameters are statistically significant, but the models

Table 7.2 *Babington Smith models.*

Data Set→	**Political Goals**		**Voting**		**Sports**	
LRS (df)	1560.33	(6)	189.60	(10)	62.82	(21)
GOF (df)	278.14	(17)	1527.92	(109)	900.30	(5018)
Total (df)	1838.47	(23)	1717.51	(119)	963.12	(5039)
$100 \cdot R^2$	84.87%		11.04%		6.52%	

Table 7.3 *Political Goals: Babington Smith parameters.*

Pair (i,j) →	(O,S)	(O,P)	(O,F)	(S,P)	(S,F)	(P,F)
$100 \cdot \widehat{p}_{ij}$	66.90	45.95	67.83	36.34	54.61	68.20
se	1.33	1.30	1.36	1.40	1.31	1.38

explain only a small proportion of the data. For Political Goals, Babington Smith explains 85%. (As in Example 6.6, the Bradley–Terry/Mallows model fits nearly as well.) There is a significant amount left unexplained. The estimates for this data set are given in Table 7.3. There are no surprises given the analysis in Example 6.6. Order and Prices are preferred strongly to Say and Freedom, with $\widehat{p}$'s around 2/3 in favor of the former. Prices are slightly preferred to Order, and Say to Speech. □

7.3 Contrast ϕ models

Contrasts were introduced in Section 5.10 as comparisons of groups of objects. See Definition 5.2. In the examples in the previous section, some contrasts arise naturally from inspecting the data. In Example 7.1, for the White Women and Black Women, respectively, the main contrasts are

$$\begin{aligned} C^{(1)} &= (\{Male\}, \{Female,\ Both\}) \text{ and} \\ C^{(2)} &= (\{Male,\ Female\}, \{Both\}). \end{aligned} \tag{7.8}$$

In Example 7.2, for the Political Goals, the main contrast is

$$D^{(1)} = (\{\mathit{Order},\ \mathit{Prices}\}, \{\mathit{Say},\ \mathit{Freedom}\}). \tag{7.9}$$

Recall also the contrasts for the Sports Data in (5.25).

A *contrast* ϕ model is the exponential family model constructed from a set of T contrasts, $C^{(1)}, \ldots, C^{(T)}$, using the Jonckheere–Terpstra statistics (5.29) as the sufficient statistics. Thus, as in (5.31), the density is

$$f_\theta(y; C^{(1)}, \ldots, C^{(T)}) \\ = \exp(\sum_{t=1}^{T} \theta_t \cdot d(C^{(t)}(y)) - \psi(\theta; C^{(1)}, \ldots, C^{(T)})). \tag{7.10}$$

[In (5.31), the contrasts are orthogonal. Those in (7.10) may or may not be orthogonal.] The interesting fact is that model (7.10) is a submodel of the Babington Smith model, where the Babington Smith parameters are linear functions of θ.

The relation is easiest to see in a one-contrast model. Take the contrast $C = (\mathcal{O}_1, \ldots, \mathcal{O}_K)$. Letting θ be the scalar parameter for the contrast model, and γ_{ij}'s be the Babington Smith exponential family parameters,

$$\begin{aligned} \theta \cdot d(C(y)) &= \sum\sum\nolimits_{1 \le k < l \le m} \sum\nolimits_{a \in \mathcal{O}_k} \sum\nolimits_{b \in \mathcal{O}_l} \theta I[y_a > y_b] \\ &= \sum\sum\nolimits_{1 \le i < j \le m} \gamma_{ij} I[y_i > y_j] + \eta\theta, \end{aligned} \tag{7.11}$$

where η is a constant independent of y and θ. Sorting out the terms, we have that

$$\gamma_{ij} = \begin{cases} \theta & \text{if } i \in \mathcal{O}_k \text{ and } j \in \mathcal{O}_l, k < l; \\ -\theta & \text{if } i \in \mathcal{O}_k \text{ and } j \in \mathcal{O}_l, l < k; \\ 0 & \text{otherwise.} \end{cases} \tag{7.12}$$

Equivalently, we have that for $p = \exp(-\theta)/(1 + \exp(-\theta))$, the Babington Smith p_{ij}'s in (7.4) satisfy

$$p_{ij} = \begin{cases} p & \text{if } i \in \mathcal{O}_k \text{ and } j \in \mathcal{O}_l, l < k; \\ 1 - p & \text{if } i \in \mathcal{O}_k \text{ and } j \in \mathcal{O}_l, k < l; \\ \frac{1}{2} & \text{otherwise.} \end{cases} \tag{7.13}$$

Table 7.4 *One-contrast models.*

Data Set	**Contrast**	$\widehat{\theta}$	**se**	$\widehat{p}$	*LRS*	$100 \cdot R^2$
White Women	$C^{(1)}$	**2.6961**	**0.9771**	**0.0632**	**23.42**	**93.61%**
Black Women	$C^{(2)}$	**1.5889**	**0.5610**	**0.1695**	**13.32**	**76.12%**
Political Goals	$D^{(1)}$	**−0.6910**	**0.0194**	**0.6662**	**1510.72**	**82.17%**

Thus the contrast model equates certain Babington Smith p's, making the model easy to interpret: they depend only upon group membership. In the Political Goals example, the p's for pairs of objects with one from {*Order, Prices*} and one from {*Say, Freedom*} are the same, i.e., each Materialist goal is equally preferred to each Post-materialist goal. Table 7.4 contains the results of fitting one-contrast models to the data in Examples 7.1 and 7.2, using the contrasts in (7.8) and (7.9).

Note first that these models all have R^2's almost as high as those for the full Babington Smith models in Tables 7.1 and 7.2. The statistics for testing the one-contrast model versus the Babington Smith model are 1.60 (2 df), 0.48 (2 df), and 49.61 (5 df), so that only for the Political Goals is the Babington Smith model significantly better. (Keep in mind the single contrasts were chosen by first looking at the data.) Comparing the p's in (7.13), for the White Women, the Babington Smith p's were 0 and 0.1250 versus 0.0632 for the pairs in the contrast, and 0.5383 vs 0.5 for p_{BF}; for the Black Women they were 0.2374 and 0.1061 versus 0.1695 for the contrast, and 0.5 versus 0.5 for p_{MF}. For the Political Goals they were 0.6690, 0.6783, 0.6366, and 0.6820 versus 0.6662 for the contrast , and 0.4596 and 0.5461 versus 0.5 for within the groups. Thus the contrast model performs a reasonable smoothing of the Babington Smith parameters.

As noted below (5.30), Mallows' ϕ model, which is the distance model using Kendall's distance (Section 6.2.1), is a one-contrast model. Supposing the modal ranking is e_m,

$$d_{\text{Ken}}(e_m, y) = \sum\sum\nolimits_{1 \le i < j \le m} I[y_i > y_j] = d(C(y)), \qquad (7.14)$$

where $C = Brack(e_m) = (\{1\}, \ldots, \{m\})$ from (5.30). From (7.12) we see that $\gamma_{ij} = \theta$, and $p_{ij} = p$, for all $i < j$.

For general contrast models (7.10), the equations comparable to (7.12) are more complicated. Specifically, γ_{ij} equals the sum of T terms, where the tth term is either θ_t, $-\theta_t$, or 0 as given by (7.12) applied to contrast $C^{(t)}$. If the contrasts are orthogonal as in Definition 5.3, then each paired comparison appears in at most one contrast; hence each Babington Smith γ_{ij} equals $\pm\theta_t$ for that contrast, or 0 if it appears in none of the contrasts.

Example 7.3. Political Goals: Orthogonal Contrast Model
Let $D^{(1)}$ be the contrast in (7.9), also seen in Table 7.4, and take $D^{(2)} = (\{Order\}, \{Prices\})$, and $D^{(3)} = (\{Say\}, \{Freedom\})$. Then $D^{(1)}, D^{(2)}, D^{(3)}$ constitute a set of orthogonal contrasts. Fitting the model yields $\widehat{\theta} = (-0.6910, 0.1275, -0.2470)$ with standard errors $(0.0194, 0.0421, 0.0424)$. These translate to $\widehat{p}_1 = 0.6662$, $\widehat{p}_2 = 0.4682$, and $\widehat{p}_3 = 0.5615$. The quantities for the first contrast are as in Table 7.4. Compare the other two p_t's to those for Babington Smith: $\widehat{p}_{Order,Prices} = 0.4596$ and $\widehat{p}_{Say,Freedom} = 0.5461$. The LRS for the orthogonal contrast model is 1554.15 (3 df), $100 \times R^2 = 84.54\%$. Comparing these to the Babington Smith model, we see that the orthogonal contrast model is essentially as good. □

This section showed how to relate Babington Smith models to contrast models. In Section 7.4, we investigate the Bradley–Terry/ Mallows model, and some linear submodels. Sections 7.5 through 7.7 develop the mathematics for orthogonal contrast models, from which we see that fitting such models is computationally efficient. Section 7.8 explains a mixed orthogonal contrast/Babington Smith model proposed by Joe and Verducci (1993).

7.4 Bradley–Terry/Mallows and Spearman's distance models

In Example 6.6, Spearman's distance model, the Bradley–Terry/ Mallows model, and Babington Smith were seen to be a nested set of models. Applied to the Political Goals data, Bradley–Terry/ Mallows fits much better than Spearman's, and almost as well as Babington Smith, but even Babington Smith shows lack-of-fit versus the saturated model. Spearman and Bradley–Terry/Mallows are in fact contrast models, with nonorthogonal contrasts.

7.4.1 *Bradley–Terry/Mallows*

Recall that the Bradley–Terry/Mallows model is the Babington Smith model with the probability that Object i is preferred to Object j, p_{ij}, being $\nu_i/(\nu_i + \nu_j)$ for positive constants ν_i, where the preference in the probability is based on the experiment when only the two objects are compared. See (5.8). If γ_{ij}, $i < j$ are the parameters for Babington Smith (7.3), then by (7.4), $p_{ij} = \exp(-\gamma_{ij})/(1 + \exp(-\gamma_{ij})) = \nu_i/(\nu_i + \nu_j)$; hence, $\gamma_{ij} = \theta_i - \theta_j$, where $\theta_k \equiv -\log(\nu_k)$. To see the contrast model, insert the γ_{ij}'s into the exponent in Babington Smith and rearrange terms to obtain

$$\begin{aligned}\sum_{i<j} \gamma_{ij} I[y_i > y_j] &= \sum_{i<j} (\theta_i - \theta_j) I[y_i > y_j] \\ &= \sum_{i=1}^{m} \theta_i \sum_{j \neq i} I[y_i > y_j] + \eta(\theta),\end{aligned} \tag{7.15}$$

where $\eta(\theta)$ is a linear function of θ independent of y. For each i define the contrast

$$C^{(i)} = (\{i\}, \{1, \ldots, m\} - \{i\}), \tag{7.16}$$

so that $d(C^{(i)}(y)) = \sum_{j \neq i} I[y_i > y_j] = y_i - 1$. Thus the final equation in (7.15) equals

$$\sum_{i=1}^{m} \theta_i \cdot d(C^{(i)}(y)) + \eta(\theta) = \sum_{i=1}^{m} \theta_i y_i + \eta^*(\theta), \tag{7.17}$$

where $\eta^*(\theta)$ is another linear function, which yields the following result.

Lemma 7.1
The Bradley–Terry/Mallows model with parameters ν_i, $i = 1, \ldots, m$ is equivalent to the contrast model (7.10) with contrasts $C^{(1)}, \ldots, C^{(m)}$ in (7.16) and $\theta_i = \log(-\nu_i) + (const)$, and also equivalent to the exponential family model with sufficient statistic y and parameter θ. □

7.4.2 Submodels of Bradley–Terry/Mallows, including Spearman

Since Spearman's distance model with modal ranking ω has sufficient statistic $d_{\text{Spear}}(\omega, y) = ||y_i - \omega_i||^2 = (const) - 2 \cdot \omega' y$, the parameter θ_{Spear} in (5.5) relates to the θ in (7.17) via $\theta_i = -2\omega_i \theta_{\text{Spear}}$ Using (7.17), the Babington Smith parameters for Spearman are then $\gamma_{ij} = 2(\omega_j - \omega_i)\theta_{\text{Spear}}$.

Since Bradley–Terry/Mallows is a submodel of Babington Smith, it cannot fit better than Babington Smith. There may be submodels of the BTM that fit almost as well as BTM. Critchlow and Fligner (1993) propose linear models on the parameters θ_i, that is, set $\theta = X\beta$ where X is a fixed $m \times q$ design matrix, and β is the $q \times 1$ parameter. Spearman uses $X = -2\omega$ and $\beta = \theta_{\text{Spear}}$. The resulting model is an exponential family model with sufficient statistic $y'X\beta$, so the usual techniques given in Section 5.14 work well.

Example 7.4. Political Goals: Linear Submodel of BTM
Proceed with the Political Goals Examples 6.6 and 7.3. Instead of using the orthogonal contrast model, we can take orthogonal contrasts of the θ. As in Example 7.3, we wish to compare *Order* and *Prices* to *Say* and *Freedom*, *Order* to *Prices*, and *Say* to *Freedom*. Thus, set

$$X = \begin{pmatrix} 1 & 1 & 0 \\ -1 & 0 & 1 \\ 1 & -1 & 0 \\ -1 & 0 & -1 \end{pmatrix}. \tag{7.18}$$

The model is the same as the BTM model since θ is identifiable only up to an additive constant, and the columns of X span the space of contrasts. We obtain $\widehat{\beta} = (-0.3474, 0.0292, -0.0780)$ with standard errors $(0.0097, 0.0127, 0.0127)$ and $Z's$ $(-35.64, 2.31, -6.13)$. As expected, the first contrast is by far the most significant, with the Materialist goals being preferred. There is almost no difference within the Materialist goals, but *Say* is significantly preferred to *Freedom*. □

7.4.3 Between Babington Smith and Bradley–Terry/Mallows

The Bradley–Terry/Mallows model is an easily interpreted submodel of the Babington Smith model. It implies that there is no interaction between objects, i.e., given an object's overall popularity, the paired comparisons reveal no new information. If the BTM parameters do not explain the paired comparisons adequately, it is of interest to assess the deviation. In terms of the subspaces in Section 2.6, the BTM model uses the Means subspace, and the Babington Smith the Pairs subspace; hence of interest is the orthogonal complement of the Means within the Pairs. An interesting way to measure the effects of this difference is to look at the "nontransitivities" in the estimates. See McCullagh (1993a). Note that if the BTM model holds, then

$$\gamma_{ij} + \gamma_{jk} = \gamma_{ik}, \tag{7.19}$$

where $\gamma_{ij} = -\gamma_{ji}$, which in terms of odds is (7.2). Equation (7.19) suggests that to measure the deviations from the BTM, we should use parameters

$$\tau_{ijk} = \gamma_{ij} + \gamma_{jk} + \gamma_{ki}. \tag{7.20}$$

Under BTM, $\tau_{ijk} = 0$. Triples with large values of τ_{ijk} will indicate the directions in which BTM is inadequate.

The parameters (7.20) indeed measure exactly the differences between Babington Smith and BTM. Let the θ_i's be the BTM parameters, given by

$$\theta_i = \frac{1}{m} \sum_{j \neq i} \gamma_{ij}. \tag{7.21}$$

Check that if the BTM model holds, (7.21) is correct assuming $\sum \theta_i = 0$. We claim that in general, knowing the θ_i's and the τ_{ijk}'s is equivalent to knowing the γ_{ij}'s. It is clear that the former are functions of the latter. Conversely, it is not hard to verify (Exercise 7.3) that

$$m\gamma_{ij} = \sum_{k \neq i,j} \tau_{ijk} + m(\theta_i - \theta_j). \tag{7.22}$$

There are potentially $m(m-1)(m-2)$ such τ_{ijk}'s. However, each set of three objects has essentially only one parameter. That is, for the three objects a, b, c,

$$\tau_{abc} = \tau_{bca} = \tau_{cab} = -\tau_{acb} = -\tau_{cba} = -\tau_{bac}. \tag{7.23}$$

Thus it is enough to consider unordered triples. We should note that even with the restrictions (7.23), there are more τ_{ijk}'s than degrees of freedom for the difference between the Babington Smith and Bradley–Terry/Mallows models, $\binom{m}{3}$ versus $\binom{m-1}{2}$. Thus some τ_{ijk}'s can be obtained as linear combinations of others.

Example 7.5. Gulf War: Nontransitivities
Fitting the Babington Smith model to the Gulf War data from Example 2.7, we find $LRS = 33.27$ on 10 df, while the BTM has $LRS = 20.50$ on 4 df. Thus the Bradley–Terry/Mallows model appears significant, but there is some evidence that the Babington Smith model fits better, i.e., there are some nontransitivities in the Bradley–Terry/Mallows parameters. The $\widehat{\theta}$ from (7.21) is $(0.2674, -0.3037, -0.2979, -0.0667, 0.4009)$, with standard errors from around 0.14 to 0.15. Thus the most popular options are Kuwait and Defense, then Multinational, Invade, and Out Now.

There are $\binom{5}{3} = 10$ triples of options to consider. Of those, two have parameter estimates over twice their standard errors. The triple {*Kuwait, Multinational, Out Now*} has $\widehat{\tau}_{KMO} = -5.2359$, with standard error 1.8720. The Babington Smith parameters are $\widehat{\gamma}_{KM} = -2.0156, \widehat{\gamma}_{MO} = -1.9540$, and $\widehat{\gamma}_{OK} = -1.2662$. Thus even though Kuwait is much preferred to Multinational, and Multinational is much preferred to Out Now, Out Now is somewhat preferred to Kuwait! The last fact is puzzling, since overall Kuwait is the favorite and Out Now the least favorite. The other triple, {*Invade, Kuwait, Out Now* }, has a similar profile: $\widehat{\gamma}_{IK} = -1.2719$, $\widehat{\gamma}_{KO} = -1.2662$, and $\widehat{\gamma}_{OI} = 0.5536$. I suspect that neither the Bradley–Terry/Mallows nor the Babington Smith model is capturing the essence of the data. □

7.5 Basic results for orthogonal contrast ϕ models

This section treats three types of random structures: rank vectors (Y), values of contrasts ($C(Y)$), and the Jonckheere–Terpstra statistics for contrasts ($d(C(Y))$. Given a set of nested orthogonal contrasts $C^{(1)}, \ldots, C^{(T)}$ and associated parameters $\theta' = (\theta_1, \ldots, \theta_T)$, define the ϕ model on $\mathcal{S}_m$ to have density (7.10) with respect to $Uniform(\mathcal{S}_m)$, written

$$Y \sim Phi(\theta; C^{(1)}, \ldots, C^{(T)}). \tag{7.24}$$

Recall a contrast $C = (\mathcal{O}_1, \ldots, \mathcal{O}_K)$ carries along with it the subset $\mathcal{O}^C$ of objects, a $K \times 1$ vector $\lambda (\equiv \lambda^C)$, $\lambda_k = \#\mathcal{O}_k$, and $\lambda_+ =$

$\sum_i \lambda_i$. Definition 5.4 gives the space of $C(y)$ as $\mathcal{V}(\lambda)$. See (2.69). For an element in $v \in \mathcal{V}(\lambda)$, $v = (v_1, \ldots, v_K)$ where each $v_k \subset \{1, \ldots, \lambda_+\}$, the Jonckheere–Terpstra statistic $d(v)$ from (5.29) is

$$d(v) = \sum\sum_{1 \leq k < l \leq \lambda_+} \sum_{i \in v_k} \sum_{j \in v_l} I[i > j]. \qquad (7.25)$$

Let $\mathcal{U}_\lambda$ denote the space of $d(v)$. In Corollary 7.1, we will show that

$$\mathcal{U}_\lambda = \{0, \ldots, Max_\lambda\} \text{ where } Max_\lambda = \overline{\lambda}_+ - \sum_{k=1}^{K} \overline{\lambda}_k. \qquad (7.26)$$

Recall from (3.3) that $\overline{k} = \binom{k}{2}$.

The distributions on $\mathcal{V}(\lambda)$ of interest are given by

$$V \sim Mallows(\theta; \lambda), \qquad (7.27)$$

where V has density

$$g_\theta(v; \lambda) = \exp(\theta d(v) - \psi(\theta; \lambda)) \qquad (7.28)$$

with respect to $Uniform(\mathcal{V}(\lambda))$. The ψ will be defined below. Now, let $Kendall(0; \lambda)$ be the distribution induced on $\mathcal{U}_\lambda$ through $d(V)$ where $V \sim Uniform(\mathcal{V}(\lambda))$. Note that $Kendall(0; 1_m)$ is the same as $Kendall_m$, given in Section 3.4.2 as the distribution of $d(Y, e_m)$ when $Y \sim Uniform(\mathcal{S}_m)$. The nonuniform distributions are

$$U \sim Kendall(\theta; \lambda), \qquad (7.29)$$

where U has density

$$h_\theta(u; \lambda) = \exp(\theta u - \psi(\theta; \lambda)), \qquad (7.30)$$

with respect to $Kendall(0; \lambda)$. Now we are ready for the main results.

Lemma 7.2
In the densities (7.28) for Mallows$(\theta; \lambda)$ *and (7.30) for Kendall* $(\theta; \lambda)$,

$$\psi(\theta; \lambda) = \psi(\theta; \lambda_+) - \sum_{k=1}^{K} \psi(\theta; \lambda_k), \tag{7.31}$$

where $\psi(\theta; q) = \sum_{i=1}^{q} \phi(\theta; i)$, *and* $\phi(\theta; i) = \log((1-\exp(\theta i))/(i(1-\exp(\theta)))$. *See (5.17).* □

Lemma 7.3
If $Y \sim$ *Phi*$(\theta; C^{(1)}, \ldots, C^{(T)})$, *then in (7.10),*

$$\psi(\theta; C^{(1)}, \ldots, C^{(T)}) = \sum_{t=1}^{T} \psi(\theta_t; \lambda^{(t)}), \tag{7.32}$$

where $\lambda^{(t)}$ *is the* λ *for* $C^{(t)}$. *In addition,* $C^{(1)}(Y), \ldots, C^{(T)}(Y)$ *are independent,* $C^{(t)}(Y) \sim$ *Mallows*$(\theta_t; \lambda^{(t)})$; *hence* $d(C^{(1)}(Y)), \ldots, d(C^{(T)}(Y))$ *are independent,* $d(C^{(t)}(Y)) \sim$ *Kendall*$(\theta_t; \lambda^{(t)})$. □

The proofs of Lemmas 7.2 and 7.3, as well as other results, are collected in the next section. Note that the distributions *Mallows* $(\theta; \lambda)$ and *Kendall*$(\theta; \lambda)$ depend on λ only through the values in λ, not their order.

One advantage the orthogonal contrast ϕ model has over the full Babington Smith and other contrast models is computational efficiency. If m is large, it is difficult to calculate the normalizing constant and its derivatives for Babington Smith since one must add up over all $m!$ elements of $\mathcal{S}_m$. In addition, the information matrix will be $\overline{m} \times \overline{m}$, requiring inversion of a large matrix. For orthogonal contrast models, the likelihood breaks into T components based on the independence of the $d(C^{(t)}(Y))$'s, the tth component depending only on θ_t. Furthermore, the ψ for each component is a linear combination of terms $\phi(\cdot; q)$, where there are at most $m-1$ such terms. Thus, with $T = m - 1$, calculation is in the order of m^2 rather than $m!$.

7.6 Details for the orthogonal contrast ϕ models

Our goal is to find the null ($\theta = 0$) and non-null densities of the *Phi, Mallows,* and *Kendall* distributions. Our main tool is to decompose Y into a set of independent *Mallows*'s, and $d_{\text{Ken}}(y, e_m)$ into a sum of independent *Kendall*'s. We start with some machinery.

7.6.1 *Preliminaries*

Definition 7.1
Let C and D be contrasts (5.24) and (5.27) for a given set $\mathcal{O}$ of objects such that for some l, $\mathcal{O}^C = \mathcal{O}_l^*$. Define the composed contrast $D(C)$ by

$$D(C) = (\mathcal{O}_1^*, \ldots, \mathcal{O}_{l-1}^*, \mathcal{O}_1, \ldots, \mathcal{O}_K, \mathcal{O}_{l+1}^*, \ldots, \mathcal{O}_L^*). \quad \square \quad (7.33)$$

Note that the definition requires, by Definition 5.3, that C and D are orthogonal, but is more restrictive since orthogonality does not demand equality as in (7.33). For example, we can compose $C = (\{a\}, \{b\}, \{c\})$ and $D = (\{a, b, c\}, \{d, e, f, g\})$ but not $(\{b\}, \{c\})$ and D. Equation (7.33) creates a new contrast by replacing the group $\mathcal{O}_l^*$ in$D$ with the entire contrast C, so that $D(C)$ $= (\{a\}, \{b\}, \{c\}, \{d, e, f, g\})$. Some facts about compositions are contained in the next lemma.

Lemma 7.4
Suppose C, D, and E are contrasts such that $E = D(C)$, where E has M groups.

(a) $\mathcal{O}^E = \mathcal{O}^D$, $\lambda_+^E = \lambda_+^D$, $M = K + L - 1$, and

$$\lambda^E = \lambda^D(\lambda^C) = (\lambda_1^D, \ldots, \lambda_{l-1}^D, \lambda_1^C, \ldots, \lambda_K^C, \lambda_{l+1}^D, \ldots, \lambda_L^D). \quad (7.34)$$

(b) Define the function $a : \mathcal{V}(\lambda^E) \to \mathcal{V}(\lambda^C) \times \mathcal{V}(\lambda^D)$ by $a(v) = (C(y^{(v)}), D(y^{(v)}))$, where $y^{(v)}$ is any element of $\mathcal{S}_m$ with $E(y^{(v)}) = v$. Then a is one-to-one and onto.

(c) For any $y \in \mathcal{S}_m$, $d(E(y)) = d(C(y)) + d(D(y))$. $\square$

Proof
Part (a) follows from the definition (7.33). For part (b), first check that the function a is well-defined. By definition, the range of $E(y)$ as y ranges over $\mathcal{S}_m$ is $V(\lambda^E)$, and Lemma 5.1 implies the range of $(C(y), D(y))$ is $\mathcal{V}(\lambda^C) \times \mathcal{V}(\lambda^D)$. Thus a is onto. It is also one-to-one since the domain and range have the same number of elements. (To see this, note that $\#\mathcal{V}(\lambda) = \binom{\lambda_+}{\lambda_1, \ldots, \lambda_K}$.) Part (c) is left to Exercise 7.4. $\square$

Next we consider more extensive compositions.

Definition 7.2
Given a set $\mathcal{C}$ of T orthogonal contrasts, suppose there exists a contrast C and a sequence of sets of orthogonal contrasts,

$$\mathcal{C} \equiv \mathcal{C}^{(1)} \rightarrow \mathcal{C}^{(2)} \rightarrow \cdots \rightarrow \mathcal{C}^{(T)} = \{C\}, \tag{7.35}$$

where $\mathcal{C}^{(t+1)}$ is obtained from $\mathcal{C}^{(t)}$ by a single composition, i.e., for some $D^{(t)}, E^{(t)} \in \mathcal{C}^{(t)}$,

$$C^{(t+1)} = (C^{(t)} - \{D^{(t)}, E^{(t)}\}) \cup \{E^{(t)}(D^{(t)})\}. \tag{7.36}$$

Then $\mathcal{C}$ is a decomposition of C, written $C = [\mathcal{C}]$. □

Example 7.6. Sports: Composing Contrasts
Let $\mathcal{C}$ be the set of six orthogonal contrasts in (5.25). There is a sequence $\mathcal{C}$ (7.35): Let $C_2' = C_2(C_3)$ and $C_5' = C_5(C_6)$, then $C_1' = C_1(C_2')$ and $C_4' = C_4(C_5')$, and finally $C = C_1'(C_4')$. Thus $C = [\mathcal{C}] = (\{B\}, \{F\}, \{K\}, \{T\}, \{C\}, \{S\}, \{J\})$. □

The next result follows from Lemma 7.4 by iterating over (7.35).

Lemma 7.5
Suppose $[\mathcal{C}] = C$ *as in Definition 7.2, where* $\mathcal{C} = \{\mathcal{C}^{(1)}, \ldots, C^{(T)}\}$.

(a) Define the function $a : \mathcal{V}(\lambda^C) \rightarrow \mathcal{V}(\lambda^{(1)}) \times \cdots \times \mathcal{V}(\lambda^{(T)})$, *where* $\lambda^{(t)}$ *is the ordered partition for* $C^{(t)}$, *by* $a(v) = (C^{(1)}(y^{(v)}), \ldots, C^{(T)}(y^{(v)}))$, *where* $y^{(v)}$ *is any element of* $\mathcal{S}_m$ *with* $C(y^{(v)}) = v$. *Then a is one-to-one and onto.*

(b) For any $y \in \mathcal{S}_m$, $d(C(y)) = d(C^{(1)}(y)) + \cdots + d(C^{(T)}(y))$. □

The construction above brings us to the following important result.

Theorem 7.1
Suppose $\mathcal{C}$ *is a set of contrasts such that* $C = [\mathcal{C}]$, *and* $Y \sim$ *Uniform*$(\mathcal{S}_m)$. *Set* $U_0 = d(C(Y))$ *and* $U_t = d(C^{(t)}(Y))$ *for* $1 \leq t \leq T$. *Then* $U_1, \ldots, U_T$ *are independent, and*

$$U_0 = U_1 + \cdots + U_T. \quad □ \tag{7.37}$$

Proof
The independence of the U_t's follows from Lemma 5.1, and the summation (7.37) is Lemma 7.5 (b). □

7.6.2 Null distribution for Kendall

For any set of contrasts, $Phi(0; C^{(1)}, \ldots, C^{(T)}) = Uniform(\mathcal{S}_m)$ since $f_0 = 1$ in (7.10). Similarly, $Mallows(0; \lambda) \equiv Uniform(\mathcal{V}(\lambda))$. The challenging distribution is that of $Kendall(0; \lambda)$, which has been well-studied in nonparametrics as the Jonckheere–Terpstra statistic. The next lemma will be very useful.

Lemma 7.6
Suppose $U \sim Kendall(0; \lambda)$. Then there exist $U_1, \ldots, U_K$, where $U_k \sim Kendall_{\lambda_k}$ and $U_0 \sim Kendall_{\lambda_+}$, as in (7.29), such that

$$U_0 = U_1 + \cdots + U_K + U, \tag{7.38}$$

and $U_1, \ldots, U_K, U$ are independent. □

Proof
We find a set of contrasts and apply Theorem 7.1. Start with $Y \sim Uniform(\mathcal{S}_{\lambda_+})$, and let $\mathcal{C} = \{C^{(1)}, \ldots, C^{(K)}, D\}$, where $C^{(1)} = (\{1\}, \ldots, \{\lambda_1\})$, $C^{(2)} = (\{\lambda_1 + 1\}, \ldots, \{\lambda_1 + \lambda_2\})$, ..., $C^{(K)} = (\{\lambda_1 + \cdots + \lambda_{K-1}\}, \ldots, \{\lambda_+\})$, and $D = (\mathcal{O}^{(1)}, \ldots, \mathcal{O}^{(K)})$, where $\mathcal{O}^{(k)}$ is the set of objects in $C^{(k)}$. Then $C \equiv [\mathcal{C}] = (\{1\}, \ldots, \{\lambda_+\})$. Thus Theorem 7.1 shows that U_1, ..., U_K, U are independent, where $U = d(C(Y))$. By construction, the $U_0, \ldots, U_K, U$ have the distributions stated in the Lemma; hence it is proven. □

The next result follows from Lemma 7.6, recalling that the range of $Kendall_m$ is $\{0, \ldots, \overline{m}\}$, and the mean and variance are as in Table 3.3.

Corollary 7.1
Suppose $U \sim Kendall(0; \lambda)$. Then its range is $\{0, \ldots, Max_\lambda\}$ from (7.26),

$$E(U) = \frac{1}{2}(\overline{\lambda}_+ - \sum_{k=1}^{K} \overline{\lambda}_k), \tag{7.39}$$

and

$$Var(U) = \frac{\lambda_+(\lambda_+ - 1)(2\lambda_+ + 5)}{72} - \sum_{k=1}^{K} \frac{\lambda_k(\lambda_k - 1)(2\lambda_k + 5)}{72}. \square \tag{7.40}$$

We end this subsection with an algorithm for calculating the exact null density of $U \sim Kendall(0;\lambda)$. Gibbons and Chakraborti (1992) exhibits an efficient algorithm for the case of two groups, i.e., the Mann–Whitney/Wilcoxon two-sample statistic, which can be extended to the K-group case. Burr (1960) and Valz and Thompson (1993) have an algorithm that is even more general, working also when there are ties. We suggest an approach that seems a little simpler to invoke, although it is not as efficient. Recall Section 3.4.2, in which we used convolutions to obtain the distribution of d_{Ken}. Here we devolve. That is, given random variables Z on $\{0,\ldots,a+b\}$, V on $\{0,\ldots,a\}$, and W on $\{0,\ldots,b\}$, where $Z = V + W$ and V and W are independent, we find the distribution of V from those of Z and W. To apply this method to U, we use Lemma 7.6 to first find the distribution of U_0 in (7.38), then devolve the distributions of the U_k, $k = 1,\ldots,K$, leaving us with U. Since $U_0,\ldots,U_K$ are $Kendall_q$'s, we can find their distributions as in Section 3.4.2. In fact, we can simplify a bit by noticing that finding the distribution of U_0 involves convolving $Uniform(\{0,\ldots,q\})$'s for $q = 1,\ldots,\lambda_+ - 1$, and devolving U_1 therefrom is the same as devolving the *Uniforms* for $q = 0,\ldots,\lambda_1 - 1$. [Devolving when W is $Uniform(\{0,\ldots,b\})$ is easy:

$$P[V = i] = (b+1)P[Z = i] - \sum_{j=\max(0,i-b)}^{i-1} P[V = j], i = 0,\ldots,a.] \tag{7.41}$$

Since we are left with the convolution of the uniforms for $q = \lambda_1,\ldots,\lambda_+$, we may as well begin with the latter convolution, then devolve the $Kendall_{\lambda_k}$'s for $k = 2,\ldots,K$ from that.

For large m, asymptotic results are needed. Imagine a sequence of contrasts as $m \to \infty$. It could be that the number of groups in the contrast stays fixed and the number of objects in each contrast increases, or the number of groups increases while the number of objects in each group stays fixed, or both the number of groups and the number of objects in each group increase. We use a fairly general set-up. Suppose $C^{(m)}, m = 1, 2, \ldots$ is a sequence of contrasts with $C^{(m)}$ depending on m objects. Let K_m be the number of groups in contrast $C^{(m)}$, and let $\lambda^{(m)}$ be the vector of group sizes, so that $\lambda_+^{(m)} = m$. Assume that

$$\frac{1}{m^3} \sum_{k=1}^{K_m} (\lambda_k^{(m)})^3 \longrightarrow \gamma \tag{7.42}$$

for some γ. Then setting $U^{(m)} \sim Kendall(0; \lambda^{(m)})$,

$$\frac{1}{m^{3/2}}(U^{(m)} - E(U^{(m)})) \longrightarrow N(0, \frac{1-\gamma}{36}). \tag{7.43}$$

See Exercise 7.5. It could be that $\gamma = 1$, for example, if $C^{(m)} = (\{1\}, \{2, \ldots, m\})$, in which case (7.43) is not of use.

7.6.3 Non-null distribution for Phi, Mallows, and Kendall

The proofs of Lemmas 7.2 and 7.3 are next. The first proof is essentially the same as (6.3).

Proof of Lemma 7.2
First note that the definition of $Kendall(\theta; \lambda)$ as $d(V)$ for $V \sim Mallows(\theta; \lambda)$ implies that the $\psi(\theta; \lambda)$ is the same for both densities (7.28) and (7.30). Thus it is enough to prove (7.30) for $Kendall(\theta; \lambda)$. Equation (6.3) proves that $\psi(\theta; q)$ is the constant for $Kendall_q(\theta)$, i.e.,

$$E(\exp(\theta U^*)) = \exp(\psi(\theta; q)) \text{ for } U^* \sim \ Kendall_q. \tag{7.44}$$

Now the constant (7.31) for $Kendall(\theta; \lambda)$ is defined by exp $(\psi(\theta, \lambda)) = E(\exp(\theta U))$ for $U \sim Kendall(0; \lambda)$. Define $U_0, \ldots, U_K$ as in Theorem 7.1. Independence implies that

$$E(\exp(\theta U_0)) = E(\exp(\theta U)) \prod_{k=1}^{K} E(\exp(\theta U_k)). \tag{7.45}$$

Since $U_0 \sim Kendall_{\lambda_+}$ and each $U_k \sim Kendall_{\lambda_k}$, Equation (7.44) inserted in (7.45) proves (7.31). □

Using (7.31), we have the following analog of Corollary 7.1.

Theorem 7.2
For $\theta \neq 0$,

$$E(Kendall(\theta;\lambda)) \equiv \delta(\theta;\lambda) = \delta_{\lambda_+}(\theta) - \sum_{k=1}^{K} \delta_{\lambda_k}(\theta) \qquad (7.46)$$

and

$$Var(Kendall(\theta;\lambda)) \equiv \sigma^2(\theta;\lambda) = \sigma^2_{\lambda_+}(\theta) - \sum_{k=1}^{K} \sigma^2_{\lambda_k}(\theta), \qquad (7.47)$$

where $\sigma_q^2 = Var(Kendall(\theta; 1_q))$. The moments for $\theta = 0$ are given in Corollary 7.1, and are limits as $\theta \to 0$ of (7.46) and (7.47). □

Proof of Lemma 7.3
Take $Y \sim Phi(\theta; C^{(1)}, \ldots, C^{(T)})$. Then

$$\exp(\psi(\theta; C^{(1)}, \ldots, C^{(T)})) = E(\exp(\sum_{k=1}^{K} \theta_k d(C^{(k)}(X)))), \qquad (7.48)$$

where $X \sim Uniform(\mathcal{S}_m)$. Now Lemma 5.1 shows that the $d(C^{(t)}(X))$'s are independent, and $d(C^{(k)}(X)) \sim Kendall(0; \lambda^{(k)})$ by definition. Thus (7.32) follows from (7.48), proving part (a). Next, let $V = (V_1, \ldots, V_T)$, where $V_t = C^{(t)}(y)$, and $\mathcal{V}$ be the space of V, i.e., $\mathcal{V} = \mathcal{V}(\lambda^{(1)}) \times \cdots \times \mathcal{V}(\lambda^{(T)})$. Then for $v \in \mathcal{V}$,

$$P[V = v; \theta] = \sum_{y \in \Omega(v)} P[Y = y; \theta], \qquad (7.49)$$

where

$$\Omega(v) = \{y \in \mathcal{S}_m \mid C^{(1)}(y) = v_1, \ldots, C^{(T)}(y) = v_T\}. \qquad (7.50)$$

Now

$$\sum_{y\in\Omega(v)} P[Y = y\ ;\ \theta]$$

$$= \sum_{y\in\Omega(v)} \exp(\sum_{t=1}^{T}(\theta_t C^{(t)}(y) - \psi(\theta_t; \mathcal{C}^{(t)})))P[Y = y\ ;\ 0]$$

$$= \prod_{t=1}^{T} \exp(\theta_t v_t - \psi(\theta_t; \lambda^{(t)}))P[Mallows(0; \lambda^{(t)}) = v_t], \tag{7.51}$$

the final equality following from the independence of the $C^{(t)}(Y)$'s under uniformity of Y. Thus (7.49) and (7.51) show that the $V^{(t)}$'s are independent, $V^{(t)} \sim Mallows(\theta_t; \lambda^{(t)})$. □

7.6.4 Decomposing Mallows' ϕ model; Proofs of Theorems 6.3 and 6.5

Often, such as in Example 6.7, Mallows' ϕ model will give a good first impression, but also show definite lack of fit. We would like to be able to embed the model in a larger one in an attempt to learn more about the data. One approach is to use Fligner and Verducci's (1986) ϕ-component model, which we discuss in Sections 5.6.2 and 7.9. Mallows' ϕ model is always a submodel of the Babington Smith model, but it might be more informative to find a model between Mallows' ϕ and Babington Smith, e.g., an orthogonal contrast model. For which sets $\mathcal{C}$ of orthogonal contrasts does the *Phi* model contain Mallows' ϕ model? Suppose Y is distributed as Mallows' ϕ model with modal ranking ω and corresponding ordering w. Any set $\mathcal{C}$ with $C \equiv [\mathcal{C}] = (\{w_1\}, \ldots, \{w_m\})$ provides a supermodel. By the Lemma 7.5 (b),

$$d_{\text{Ken}}(y, \omega) = d(C(y)) = \sum_{t=1}^{T} C^{(t)}(y). \tag{7.52}$$

(Exercise 7.6 checks the first equality.) Now suppose $Y \sim Phi$ $(\gamma 1_T; \mathcal{C})$, i.e., all θ_t's are equal to γ. Then (7.10) and (7.52) show that $Y \sim Phi(\theta; C)$, i.e., Y follows Mallows' ϕ model with modal ranking ω. Letting θ be unconstrained, the $Phi(\theta; \mathcal{C})$ model can be seen to contain the Mallows ϕ model. See Example 7.8 in the next section.

At last we are ready to prove Theorems 6.3 and 6.5.

Proof of Theorem 6.3
Start with $Y \sim Phi(\theta; e_m)$, and for fixed $i < j$ consider the contrasts $C = (\{i\}, \{i+1, \ldots, j\})$ and $D = (\{i+1, \ldots, j-1\}, \{j\})$. If $j = i+1$, then D can be ignored. We will henceforth assume $j > i+1$. Exercise 7.7 gives two more contrasts, A and B, such that $\mathcal{C} = \{A, B, C, D\}$ is a set of orthogonal contrasts with $[\mathcal{C}] = (\{1,\}, \ldots, \{m\})$. Thus as above, $Phi(\theta; e_m) = Phi(\theta 1_4; \mathcal{C})$, and Lemma 7.3 implies that $C(Y)$ and $D(Y)$ are independent with $U \equiv d(C(Y)) \sim Kendall(\theta; (1, \epsilon))$ and $V \equiv d(D(Y)) \sim Kendall(\theta; (\epsilon - 1, 1))$, where $\epsilon \equiv j - i$. Some reflection shows that $Y_i < Y_j$ if and only if $U + V < \epsilon$. From Lemma 7.2 and (7.30), the distributions of U and V are given by

$$\begin{aligned} P_\theta[U = u] &= \exp(\theta u - \phi(\theta; \epsilon+1))/(\epsilon+1), \ u = 0, \ldots, \epsilon, \\ P_\theta[V = v] &= \exp(\theta v - \phi(\theta; \epsilon))/\epsilon, \ v = 0, \ldots, \epsilon - 1, \end{aligned} \tag{7.53}$$

where ϕ is given in (5.17). Then (7.53) shows, with some manipulations, that

$$\begin{aligned} P[Y_i < Y_j] &= \sum_{u=0}^{\epsilon-1} P_\theta[U = u] P_\theta[V < \epsilon - u] \\ &= \frac{1 - (\epsilon+1)\exp(\theta\epsilon) + \epsilon\exp(\theta(\epsilon+1))}{(1 - \exp(\theta\epsilon))(1 - \exp(\theta(\epsilon+1)))}. \end{aligned} \tag{7.54}$$

Further manipulations show that (6.26) and (7.54) are equal, which proves the theorem. □

Proof of Theorem 6.5
Again, $Y \sim Phi(\theta; e_m)$, but now we decompose Y into the U_i's as in (3.44), i.e., the set of orthogonal contrasts is $C^{(1)}, \ldots, C^{(m-1)}$, where $C^{(i)} = (\{i\}, \{i+1, \ldots, m\})$. Then

$$U_i \equiv d(C^{(i)}(Y)) \sim Kendall(\theta; (1, m-i)), \tag{7.55}$$

where the U_i's are independent. The vector y can be reconstructed from the values $u_i = C^{(i)}(y)$. Start with e_m. Then u_1 is the number of pairwise interchanges it takes to move y_i from its slot in e_m to the first position. It takes u_2 more interchanges to move y_2 to the second slot, u_3 to place y_3 correctly, etc. The process can be described using matrices. Let $\Gamma_i(u_i)$ be the $(m-i+1)\times(m-i+1)$ permutation matrix that has a 1 in the $(u_i+1)^{st}$ slot of the first row, 1's in the $(j-1)^{st}$ slots of rows $j=2,\ldots,u_i$, and 1's in the diagonals for $j=u_i+2,\ldots,m-i+1$. For example, suppose $m=4$ and $y=(3,2,4,1)$. Then $u_1=2$, $u_2=1$ and $u_3=1$. The matrices are

$$\Gamma_1(2)=\begin{pmatrix}0&0&1&0\\1&0&0&0\\0&1&0&0\\0&0&0&1\end{pmatrix},$$
$$\Gamma_2(1)=\begin{pmatrix}0&1&0\\1&0&0\\0&0&1\end{pmatrix}\quad\text{and}\quad\Gamma_3(1)=\begin{pmatrix}0&1\\1&0\end{pmatrix}. \tag{7.56}$$

Now letting

$$\Gamma_i^*(u_i)=\begin{pmatrix}I_{i-1}&0\\0&\Gamma_i(u_i)\end{pmatrix}, \tag{7.57}$$

we have that $y=Q^{(y)}e_m$, where

$$Q^{(y)}=\Gamma_{m-1}^*(u_{m-1})\Gamma_{m-2}^*(u_{m-2})\cdots\Gamma_1^*(u_1). \tag{7.58}$$

Since the U_i's are independent,

$$M=E_\gamma(Q_Y)=\prod_{i=1}^{m-1}E_{\gamma_i}(\Gamma_i^*(U_i)). \tag{7.59}$$

Now $P_\theta[U_i=k]=g_{i,k}$ of (6.28); hence it can then be shown that $E(\Gamma_i(U_i))=A_\theta^{(i)}$; hence $E(\Gamma_i^*(U_i))=B_\theta^{(i)}$, for A and B from (6.30) and (6.31). Thus Theorem 6.5 follows from (7.59). □

7.7 Examples of orthogonal contrast models

Example 7.7. Draft Lottery: Orthogonal Contrast Model

For the 1969 Draft Lottery, Example 3.5, we found that there was indeed a downward trend. The description of the mixing process, which added each month's capsules separately, raises the question of whether the nonuniformity is due to just a between-month trend, or to the possibility there are also within-month trends. We attack the question by using a set of thirteen orthogonal contrasts. The main contrast compares the months: $C^{(Bet)} = (\{Jan.1, \ldots, Jan.\ 31\}, \ldots, \{Dec.\ 1, \ldots, Dec.\ 31\})$. The other twelve contrasts look within the months, so that we have $C^{(Jan)} = (\{Jan.\ 1\}, \ldots, \{Jan.\ 31\}), \ldots, C^{(Dec)} = (\{Dec.\ 1\}, \ldots, \{Dec.\ 31\})$. Note that the collection of sets of contrasts $\mathcal{C} = \{\ C, C^{(Jan)}, \ldots, C^{(Dec)}\ \}$ satisfies $[\mathcal{C}] = Brack(\{1, \ldots, 366\})$ as in Definition 7.2. Thus Lemma 7.5 (b) shows that

$$\begin{aligned} \textit{Year Trend} &= \textit{Between Month Trend} + \textit{Within January Trend} \\ &\quad + \cdots + \textit{Within December Trend}. \end{aligned} \tag{7.60}$$

To test for trends within months, we use $d(C^{(Month)}(y))$. Table 7.5 exhibits the monthly statistics and Z's. There appear to be no significant trends anywhere. In the last row, we have the statistic from Table 3.4, which considers the entire year. The penultimate row contains the results for $C^{(Bet)}$. The statistic is obtained as in Corollary 7.1 by subtracting the sum of the first twelve rows from the last row. We see that the between-month statistic $d(C^{(Bet)}(y))$ is very significant, with a Z almost exactly that for the total year. Thus it appears that the nonuniformity is indeed restricted to a between-months trend.

Now suppose Y follows Mallows' ϕ distribution with parameter θ and modal ranking e_{366}. The induced distributions on the $C^{(t)}(Y)$'s are independent and follow their respective orthogonal contrast ϕ models, that is, $d(C^{(Bet)}(Y)) \sim Kendall(\theta; \lambda^{(Bet)})$ for between months, where $\lambda^{(Bet)} = (31, 29, 31, \ldots, 31)$, and $Y^{(Month)} \sim Mallows(\theta,\ Brack(e_{Month}))$ for the within-month vectors. From Table 7.5, it is fairly clear that the parameter value for the between-month contrast is large, while those for the within-month contrasts are practically zero, suggesting that the overall Mallows' ϕ model is not correct. We can test $Phi(\theta; Brack(e_{366}))$ model versus the $Phi(\gamma; \mathcal{C})$ model, where in the latter the parameters γ for the thirteen contrasts are not constrained to be equal. The LRS's for the two models are 18.24 (1 df) and 32.35 (13 df), so the difference is 14.09 on 12 df, not a significant difference. Thus the two models are comparable. However, we can take the submodel of the lat-

Table 7.5 *Between- and within-month contrasts for the Draft Lottery.*

Contrast	Statistic	Z	$\widehat{\theta}^{(t)}$	se	LRS
January	260	0.93	0.0321	0.0345	0.88
February	172	−1.16	−0.0444	0.0385	1.36
March	202	−1.04	−0.0357	0.0346	1.08
April	195	−0.80	−0,0289	0.0361	0.65
May	212	−0.70	−0.0238	0.0343	0.49
June	215	−0.09	−0.0032	0.0357	0.01
July	237	0.15	0.0052	0.0340	0.02
August	195	−1.27	−0.0442	0.0350	1.64
September	186	−1.12	−0.0407	0.0365	1.27
October	278	1.55	0.0541	0.0355	2.43
November	173	−1.59	−0.5840	0.0374	2.56
December	257	0.83	0.0285	0.0344	0.70
Between Months	35787	4.37	0.0038	0.0009	19.27
Total	38369	4.25			32.35

ter model obtained by setting the within-month parameters to 0, i.e., $\gamma_{Jan} = \cdots = \gamma_{Dec} = 0$. The LRS for that model is 19.27 on 1 df, which is the LRS for the single between-month model $Mallows(\gamma_{Bet}; C^{(Bet)})$. The difference is 13.08 on 12 df, again not significant. Thus there is clearly a significant trend through the year, but not necessarily within months. It may be puzzling that for the within-month trends, either 0 or the overall $\widehat{\theta}$ are valid parameter values. I think the explanation is that $\widehat{\theta}$ itself is very small (0.0037), too small to show a significant effect within months, each of which is based on $\approx \binom{30}{2} = 435$ comparisons, but large enough to show the effect between months, which is based on over 60,000 comparisons. □

An interesting corollary of Lemma 7.3 is that if $Y \sim Phi(\theta; Brack(w))$ for some ordering $w \in \mathcal{T}_m$, then for subvector w_0 of consecutive elements, $w_0 = (w_i, w_{i+1}, \ldots, w_j)$ for some $i < j$, then $(Y_{w_i}, \ldots, Y_{w_j}) \sim Phi(\theta; Brack(w_0))$. That is, if Y follows Mallows' ϕ model, then a subvector follows Mallows' ϕ model with the same parameter if the subvector is made of consecutive elements of w. To show this fact, one finds a set $\mathcal{C}$ that contains $Brack(w_0)$ and satisfies $[\mathcal{C}] = Brack(w)$, then invokes Lemma 7.3. If the subvector does not consist of consecutive elements, then the result does not hold.

Remark 7.1

How is the set of orthogonal contrasts chosen? Contextual considerations are most important. For example, in the Political Goals example, the Materialist/Post-materialist dichotomy guided the researchers in their choice of items. In the Sports example, Team versus Non-team is a natural contrast to use. If there is no *a priori* set, the data itself can help. For example, a cluster diagram (Figure 2.10) can provide assistance. A more systematic approach is to use the maximum likelihood estimator, that is, maximize over the set $\mathcal{C}$ of contrasts as well as the parameter θ. Recall Sections 6.3 and 6.4, where the modal ranking was estimated. When searching over all sets of orthogonal contrasts, we can restrict to those sets with $m-1$ contrasts, each having two groups. See Exercise 7.8. There can be quite a few such sets, $1 \times 3 \times 5 \cdots \times (2m-3)$, in fact. Thus for large m, it would be difficult to perform a complete search. Fortunately, since the likelihood function decomposes into the individual terms for the contrasts, an algorithm is available that is practical for at least moderate m (around 10–12).

We assume each contrast has only two groups. For each contrast C, let $l(C)$ be the LRS based on the model $Phi(\theta; C)$, and for any set of orthogonal contrasts $\mathcal{C}$, let $l(\mathcal{C})$ be the LRS based on the model $Phi(\theta; \mathcal{C})$. Thus $l(\mathcal{C}) = \sum_{C \in \mathcal{C}} l(C)$. For any subset of objects $\mathcal{O}' \subset \mathcal{O}$, let $C(\mathcal{O}')$ denote the collection of sets of orthogonal contrasts that contain exactly $\#\mathcal{O}' - 1$ contrasts, each with two groups, so that all the objects in the contrasts are contained in $\mathcal{O}'$. Also, define $B(\mathcal{O}')$ to be the maximum LRS over all sets of orthogonal contrasts in $C(\mathcal{O}')$: $B(\mathcal{O}') = \max_{\mathcal{C} \in C(\mathcal{O}')} l(\mathcal{C})$. Note that for any $\mathcal{C} \in C(\mathcal{O}')$, there is one contrast that contains all the objects in $\mathcal{O}'$, i.e., there is an $A \subset \mathcal{O}'$ such that $(A, \mathcal{O}' - A) \in \mathcal{C}$. Thus we have that

$$B(\mathcal{O}') = \max_{A \subset \mathcal{O}'} [l((A, \mathcal{O}' - A)) + B(A) + B(\mathcal{O}' - A)]. \tag{7.61}$$

The search for the best overall set, i.e., the $\mathcal{C} \in C(\mathcal{O})$ that achieves $B(\mathcal{O})$, proceeds iteratively. First, find B for all the two-object subsets, which means finding $l(\{a\}, \{b\})$ for each pair of objects a, b. Next, consider the subsets $\mathcal{O}'$ with three objects. If $\mathcal{O}' =$

$\{a, b, c\}$, then the maximum in (7.61) is over $A = \{a\}, \{b\}, \{c\}$. Note that in this case, $l(A) = 0$. Then find B for all subsets of four, then five, etc., until finishing with $\mathcal{O}$.

The algorithm needs $l(C)$ for all two-group contrasts, of which there are about $3^m/2$. [Identify the contrasts (A, B) and (B, A).] For $m = 10$, this number is almost 30,000. Once we have the $l(C)$'s, the algorithm runs swiftly since there are only additions to perform. In comparison, an exhaustive search runs over 34 million sets of orthogonal contrasts.

Example 7.8: Occupations: Orthogonal Contrast Model
The algorithm found the best set of orthogonal contrasts to yield an $LRS = 1395.40$. See Marden (1992). We tinkered with the set a little, creating a set with $LRS = 1394.71$, a trivial difference. The tinkering was to take advantage of the Technical/Managerial dichotomy in the occupations. That is, the occupations *Fac, Sci, OR, IE, ME,* and *Tech* were deemed to be Technical, and *Own, Mgr, Sup,* and *For* to be Managerial. The first two contrasts split the objects into the top three, $A \equiv \{Fac, Own, Sci\}$, the middle five, $B \equiv \{OR, IE, Mgr, ME, Sup\}$, and the other two, $D \equiv \{Tech, For\}$. These groups are suggested by the large gaps in the sample means between *Sci* and *OR*, and between ME and $Tech$. We have $C^{(1)} = (A, B \cup D)$ and $C^{(2)} = (B, D)$. The next three contrasts make the Technical/Managerial splits within the above groups. The next contrast compares OR to the engineers, IE and ME. The remaining two are the leftover pairwise contrasts. Table 7.6 contains the contrasts and the statistics from the model, where $Stat = \sum d(C^{(t)}(y^{(i)}))$'s.

First, note that the preponderance of the LRS is in the first two contrasts, so that the separation of the occupations into three prestige groups is the main feature. The Technical versus Managerial contrasts show significant preference for the former, except in the Fac versus Sci case. Of the other four, Fac and Sci are equally preferred, while IE is much preferred to ME, and less so is OR preferred to the engineers, and *Mgr* to *Sup*.

One might also start with the first contrast comparing the six Technical occupations to the four Managerial. We found the best such set of contrasts has $LRS = 1287.02$, which is high but over 100 points less than the overall best. □

Table 7.6 *Occupations: orthogonal contrast ϕ model.*

Contrast	Stat	$\widehat{\theta}$	se	$100 \cdot \widehat{p}_{ij}$	$l(C^{(t)})$
$(A, B \cup D)$	350	−0.615	0.035	65	572.00
(B, D)	42	−1.697	0.130	85	660.54
$(\{Fac, Sci\}, \{Own\})$	140	−0.032	0.102	51	0.09
$(\{OR, IE, ME\}, \{Mgr, Sup\})$	310	−0.286	0.051	57	33.54
$(\{Tech\}, \{For\})$	47	−0.714	0.178	67	17.14
$(\{OR\}, \{IE, ME\})$	110	−0.353	0.106	59	11.54
$(\{Fac\}, \{Sci\})$	59	−0.353	0.170	59	4.39
$(\{Mgr\}, \{Sup\})$	43	−0.844	0.182	70	23.36
$(\{IE\}, \{ME\})$	23	−1.652	0.228	84	72.10
Total					1394.71

Remark 7.2
The ϕ models are based on Kendall's distance. It is natural to wonder if orthogonal contrast models can be based on other distances. The key is to extend the definition of the distance to values of contrasts. With Kendall, the Jonckheere–Terpstra statistic worked. Critchlow (1985) and Alvo and Cabilio (1991) have suggestions for other distances. Having defined $d(C(y))$, the $d(C^{(t)}(y))$'s are used as sufficient statistics in the exponential family. The individual components are then independent since the independence of the $C^{(t)}(Y)$'s has nothing to do with the distance used. We have been unable to find distances that retain the composition results used to obtain simple formulas for the ψ's, but the problem is an interesting one. □

7.8 Mixed orthogonal contrast/Babington Smith models

In Section 7.4.3, we find the difference between the Bradley–Terry/ Mallows and Babington Smith models. Joe and Verducci (1993) introduce an interesting model that expands the orthogonal contrast ϕ model while staying within the Babington Smith. They concentrate on one contrast with two groups, but arbitrary sets of orthogonal contrasts can be used. The idea is, given a set C

Table 7.7 *Occupations: mixed model.*

Source	χ^2	df	p-value
$(A, B \cup D)$	572.00	1	0
(B, D)	660.54	1	0
Babington Smith:			
Within A	4.91	3	0.18
Within B	151.24	10	0
Within D	17.14	1	0
Total	1495.83	16	0
Table 7.6	1394.71	9	0

of orthogonal contrasts, to add small Babington Smith models on the undivided groups. These little Babington Smith models can be fit independently of the contrast components, thereby reducing the computational complexity of the full Babington Smith model without having to restrict to the number of parameters found in the orthogonal contrast models.

Example 7.9: Occupations: Mixed Orthogonal Contrast/Babington Smith Model
Consider again Example 7.8. We would like to compare the model in Table 7.6 to the full Babington Smith model, but we are computationally unable to fit the latter. Instead, we take the two large contrasts in Table 7.6, $(A, B \cup D)$ and (B, D), then place Babington Smith models within the resulting groups, A, B, and D. This model includes that in Table 7.6. The Table 7.7 contains the likelihood ratio statistics.

We see the mixed model improves on the orthogonal contrast model significantly, increasing the LRS by almost 100 using only 7 more parameters. Since the third group, D, has only two objects, the Babington Smith and orthogonal contrast models are the same for it. The first group, which compares *Fac, Own,* and *Sci*, does not show much significance in either model. The main difference is in group B. Table 7.8 exhibits the estimates of the p_{ij}'s under the two models for these occupations.

The major differences are in IE versus Mgr and ME versus Mgr, and less so in OR versus Sup. Babington Smith has Manager

Table 7.8 *Middle occupations'* $100 \cdot \hat{p}_{ij}$*'s from Babington Smith and orthogonal contrast models.*

i	OR	OR	IE	IE	ME	ME	OR	OR	IE	Mgr
j	Mgr	Sup	Mgr	Sup	Mgr	Sup	IE	ME	ME	Sup
BabSm	56	63	41	56	68	59	57	58	89	67
OC	57	57	57	57	57	57	59	59	84	70

preferred to IE, ME strongly preferred to Manager, and IE even more strongly preferred to ME. A point to know is that the people surveyed are Industrial Engineers, so that even though there is not too much difference in the overall rankings of IE and ME, 5 and 7, respectively, the vast majority are sure to place IE somewhat higher. Manager is ranked between the engineers, but opinions are not as uniform, so that Babington Smith has to adjust its parameters in a somewhat uncomfortable manner. □

7.9 Multistage models and patterns of ties

Useful models in which contrasts are defined on the ranks rather than the objects include the ϕ models for vectors with ties as in Barton, David, and Mallows (1958), and the multistage ϕ-component models of Fligner and Verducci (1988). This section starts with the general models, then specializes to the above.

7.9.1 Patterns of ties

Partial rankings are very common, whether built into the design of an experiment or occurring by chance. Chapter 11 discusses the subject more deeply. Here, we are concerned with rank vectors that may have ties. A tie occurs when a judge fails to distinguish between two objects. The development parallels that for orthogonal contrasts, but the roles of ranks and objects are interchanged.

Definition 7.3
A *pattern of ties* Γ is a contrast defined on $\{1, \ldots, m\}$. □

We should mention that usually λ^{Γ} rather than Γ is termed the pattern of ties. See Diaconis (1988, 1989) or Critchlow (1985). The

λ^Γ is enough if we restrict to patterns of the form

$$(\{1,\ldots,\lambda_1\},\{\lambda_1+1,\ldots,\lambda_1+\lambda_2\},\ldots,\{\lambda_1+\cdots+\lambda_{K-1}+1,\ldots,m\}), \tag{7.62}$$

where $\lambda_+ = m$. However, we need the more general notion.

For example, $\Gamma = (\{1\},\ldots,\{m\})$ is a pattern with no ties, i.e., it represents a complete ranking. The pattern $\Gamma = (\{1,\ldots,k\},\{k+1,\ldots,m\})$ for some k represents splitting the objects into the top k and bottom $m-k$. A q-permutation as in Silverberg (1980) has pattern $(\{1\},\ldots,\{q\},\{q+1,\ldots,m\})$, which asks for the first, second, ..., q^{th} favorites. See Section 8.4. In practice, patterns are monotone in that if $\Gamma = (\mathcal{R}_1,\ldots,\mathcal{R}_K)$, $i \in \mathcal{R}_k, j \in \mathcal{R}_l$, and $k < l$, then $i < j$. This property need not hold. For example, $\Gamma = (\{1,m\},\{2,\ldots,m-1\})$ separates the extremes from the rest, but does not reveal which extreme is best and which worst.

Definition 7.4
If Γ is a pattern of ties, and $w \in \mathcal{T}_m$ is an ordering of $\mathcal{O}$, then the value of Γ at w is a ***partial ordering***, given by

$$\Gamma(w) = (w[R_1],\ldots,w[R_K]),\ \ w[R] = \{w_i | i \in \mathcal{R}\}. \ \ \square \tag{7.63}$$

Note that $\Gamma(w)$ is in fact a contrast of the objects. For example, if $\Gamma = \mathit{Brack}(e_m)$ of (5.30), then $\Gamma(w) = \mathit{Brack}(w)$, essentially the ordering itself. If $\Gamma = (\{1,2,3\},\{4\},\{5,6\})$ and $w = (D,B,A,F,C,E)$, then $\Gamma(w) = (\{A,B,D\},\{F\},\{C,E\})$. Note that the meaning of the partial ordering depends on Γ.

Using a pattern of ties to go from an ordering to a partial ordering is the reverse of using a contrast to go from a ranking to a value of the contrast. That is, $\Gamma(w)$ is a contrast, and for contrast C, $C(y)$ is a generalized pattern of ties. In fact, if w is an ordering, and y is the corresponding ranking, then when $\lambda_+^\Gamma = m$, $C = \Gamma(w)$ implies that $C(y) = \Gamma$. Note that $\lambda^C = \lambda^\Gamma$.

One approach to multistage modeling is to start with a crude partial ordering and refine it at each stage, the result being the complete or the partial ordering of interest. The stages can be represented by orthogonal patterns of ties, where orthogonality is the same as for contrasts in Definition 5.3. For example, in ordering four objects, one may have the orthogonal patterns $\Gamma^{(1)} = (\{1,2\},\{3,4\})$, $\Gamma^{(2)} = (\{1\},\{2\})$ and $\Gamma^{(3)} = (\{3\},\{4\})$. The first stage, $\Gamma^{(1)}$, asks to separate the objects into the top two and bottom two, e.g., $(\{A, D\}, \{B, C\})$. The second stage refines the top two into the first and second, possibly $(\{A\}, \{D\})$. Then the third distinguishes the bottom two, e.g., $(\{C\}, \{B\})$. Then the final ordering is (A, D, C, B).

A general multistage model is then a set $\mathcal{G}$ of T orthogonal patterns such that $\Gamma = [\mathcal{G}]$ (Definition 7.2), where Γ is the desired final pattern. To define the analog of the orthogonal contrast ϕ model, we need to define $d(\Gamma^{(t)}(w))$'s. The natural definition is to set

$$d(\Gamma^{(t)}(w)) \equiv d(C(y)), \text{ where } C = \Gamma^{(t)}(w), \qquad (7.64)$$

and y is the ranking corresponding to the ordering w. In order for (7.64) to make sense, the index attached to each object has to be numerically meaningful; for example, the modal ranking should be e_m. See Section 7.9.3.

7.9.2 *Mallows' ϕ model for tied rankings*

Barton, David, and Mallows (1958) developed the one-contrast ϕ model as well as the one-pattern ϕ model. The idea of the latter is to extend Mallows' ϕ model to tied rankings, as defined next.

Definition 7.5
Suppose λ is a pattern of ties, i.e., $\lambda = \lambda^{\Gamma}$ for pattern Γ of the form (7.62). Define a *tied ranking* with pattern λ to be an $m \times 1$ vector a containing λ_1 values of 1, λ_2 values of 2, ..., λ_T values of T. Let $\mathcal{S}_m(\lambda)$ denote the set of such tied rankings. □

Suppose w is an ordering. Then the partial ordering $\Gamma(w) = (\mathcal{O}_1, \ldots, \mathcal{O}_K)$ has corresponding tied ranking a, where $a_i = k$ if $i \in \mathcal{O}_k$. For example, if $\Gamma = (\{1,2,3\}, \{4\}, \{5,6\})$ and $w = (D, B, A, F, C, E)$, then $C = (\{A, B, D\}, \{F\}, \{C, E\})$ and $a = (1, 1, 3, 1, 3, 2)$.

Given a modal ranking $\omega \in \mathcal{S}_m$ (without ties), the Mallows' ϕ model on $\mathcal{S}_m(\lambda)$ has density

$$f_\theta(a; \omega) = \exp(\theta \cdot d_{\text{Ken}}(\omega, a) - \psi(\theta; \lambda)) \qquad (7.65)$$

with respect to *Uniform*$(\mathcal{S}_m(\lambda))$. Although d_{Ken} (2.29) is originally defined on complete rankings, the definition immediately extends to tied rankings as in Definition 7.5. The verification that ψ is the correct normalizing constant comes in Section 11.7.4. Note that if A has density (7.65), then $d_{\text{Ken}}(\omega, A) \sim Kendall(\theta; \lambda)$.

7.9.3 Multistage models: ϕ-component

Fligner and Verducci (1988) developed the first multistage ϕ model using the set of orthogonal patterns $\Gamma_t = (\{t\}, \{t+1, \ldots, m\})$, for $t = 1, \ldots, m-1$. Thus Γ_1 requires choosing the best from $\mathcal{O}$, Γ_2 requires choosing the best out of those not chosen in stage 1, ..., Γ_t requires choosing the best among those not chosen best in any of the first $t-1$ stages, etc. The ϕ model (5.18) is called the ϕ-component model since it reduces to Mallows' ϕ model when the parameters are equal. The model also is useful "upside-down," that is, at each stage choose the worst: $\Gamma_t = (\{1, \ldots, t-1\}, \{t\})$.

The ϕ-component model uses d in (7.64) and modal ranking ω, so that for $w \in \mathcal{T}_m$,

$$U_t = d(\Gamma^{(t)}(w)) = \sum_{i=t+1}^{T} I[\omega_{w_t} > \omega_{w_i}], \tag{7.66}$$

where ω is the modal ranking. These U_t's are illustrated in Table 5.1.

Example 7.10. Sports: ϕ-Component Model

We fit the ϕ component model using both directions. In both cases the modal ranking is that determined by the average ranks. The model that chose the best at each stage had $LRS = 40.97$, and that which chose the worst had $LRS = 49.91$. Both have 6 degrees of freedom. In the latter model, only the first two components were significant. Choosing the worst out of all had $LRS = 40.07$, and choosing the worst out of the top 6 had $LRS = 4.67$. Those two stages explain quite a lot. Recall that Bradley–Terry/Mallows had $LRS = 37.84$, again using 6 df, while Babington Smith had $LRS = 62.82$ on 10 df. The stages, of course, model mostly people ranking Jogging and Football low. □

The ϕ component model is closely related to the Plackett–Luce model. The latter model has the same stages, but the probability of choosing the best at each stage is proportional to the parameter ν_i assigned to the objects remaining. In the ϕ component model, the probability depends on the objects only through their relative rank. The advantage of the ϕ component model is, as for the orthogonal contrast models, its ease of calculation. It breaks into $m-1$ one-dimensional problems. The Plackett–Luce model has a 2^m-dimensional sufficient statistic. See Chapter 8.

7.10 Free models and contingency tables

The models in this chapter so far have been submodels of the Babington Smith model, in fact, models that place linear restrictions on the Babington Smith parameters. For some data sets, Babington Smith has more than enough flexibility. For others, e.g., Political Goals and Voting in Table 7.2, there is more structure than Babington Smith can handle. We will consider two paths for extending some of these models. In the next section, we look at the inversions of McCullagh (1993a). Babington Smith assumes that all the structure resides in the pairwise comparisons. Inversions allow higher-order comparisons, e.g., based on triples of objects, or quadruples, etc.

Fligner and Verducci (1988) relaxed the "ϕ-ness" of the ϕ component models, allowing the individual statistics U_i to have any distribution marginally, but still be independent. The same idea can be applied to the orthogonal contrast ϕ models. We also relax the independence of contrasts or components. Start with a one-contrast ϕ model, $Y \sim Phi(\theta; C)$. We can embed this model in the model that has sufficient statistic $d(C(y))$ but is otherwise unrestricted, which in turn can be embedded in the model with sufficient statistic $C(y)$, but otherwise unrestricted. The former, called the "d-free" model, is multinomial with categories defined by the possible values of $d(C(y))$, i.e., $\{0, 1, \ldots, Max_{\lambda^C}\}$ of (7.26). The "free" model is multinomial with categories defined by the values of the contrasts, $v \in \mathcal{V}(\lambda^C)$ of Definition 5.4.

Example 7.11. Political Goals: Orthogonal Contrast Models
Recall the Materialist/Post-materialist contrast in (7.9), and the corresponding ϕ model in Table 7.4. The estimates of the parameters θ and p show strong preference of the Materialist goals. The ϕ model assumes that this preference is monotone, that is, the likelihood decreases as $d(D_1(y))$ increases. To see whether the model does conform to the data, we collect the numbers of people with various values of d, and compare them to the expected numbers under the ϕ model, as in the top half of Table 7.9. The LRS for the ϕ model is 1510.72, as given in Table 7.4. For the d-free model,

Table 7.9 *Political Goals: observed counts and expected counts under the* ϕ, *d-free and free orthogonal contrast models. The value "ij" for* D_1 *means* $(\{i,j\}, \mathcal{O} - \{i,j\})$.

$d(D_1(y)) \rightarrow$	0	1	2		3	4
Observed	1188	417	358		126	173
ϕ	1031.90	517.06	518.17		129.82	65.05
Uniform	377	377	754		377	377
$D_1(y) \rightarrow$	12	13	14	23	24	34
Observed	1188	417	184	174	126	173
d-**Free**	1188	417	179	179	126	173
ϕ	1031.90	517.06	259.09	259.09	129.82	65.05
Uniform	377	377	377	377	377	377

it is 1732.21. The difference is 221.49, on $4 - 1 = 3$ degrees of freedom. Thus the ϕ model does not fit. To see where it errs, we see the biggest discrepancies in the $d = 2$ and $d = 4$ categories. The latter category consists of people very favorable to the Post-materialist goals. The ϕ model vastly underrepresents them. Thus the monotonicity of the ϕ model hides an important aspect of the data: although most people favor the Materialist Goals, there is a distinct group of Post-materialists. That group $d = 2$ is overestimated by the ϕ model shows there are fewer fence-sitters than the ϕ model allows.

The difference between the d-free model and the free model is that in the latter, there are two categories corresponding to $d = 2$. To find the expected counts for those under the ϕ or d-free models, just divide the $d = 2$ category evenly. As seen in the table, the observed counts of the two categories are very close, 184 and 174. The *LRS* for the free model is 1732.49; hence the difference between it and the d-free model is a negligible 0.28. □

Moving to sets of orthogonal contrasts, we have a d-free or free categorization for each contrast, which defines a contingency table for the data. Each factor corresponds to a contrast, and the levels within each factor are given by the categories for that contrast. Recall the contrasts for Political Goals in Example 7.3. The first contrast is discussed in Example 7.11 above. The other two compare within the Materialist and Post-materialist goals. Since they are paired comparisons, the ϕ, d-free, and free models are the same. The d-free contingency table is thus $5 \times 2 \times 2$, while the free contingency table is $6 \times 2 \times 2$. In general, there is no reason not to use d-free for some contrasts and free for others.

Once we have a contingency table, the usual hierarchical models are available. See, e.g., Bishop, Fienberg, and Holland (1975), or Agresti (1990). In particular, the independence model assumes the categorizations are independent. This model is a direct generalization of the orthogonal contrast ϕ model since it keeps the independence but relaxes the restrictions on the marginal distributions. The independence model using the d-free marginals is then contained in the independence model using the free marginals.

If either of the independence models fail to hold, more complicated contingency table models are in order, e.g., models with two-way interactions, three-way interactions, etc. If the set of contrasts composes to an ordering, i.e., $[\mathcal{C}] = (\{w_1\}, \ldots, \{w_m\})$ for some $w \in \mathcal{T}_m$, then the sequence of models based on the free categorizations is complete in that the saturated model for the contingency table is equivalent to the usual saturated model. That is, the cells in the contingency table are in one-to-one correspondence with the set of rankings $\mathcal{S}_m$; hence no information is lost by arranging the data into such a table.

Example 7.12. Political Goals: Contingency Table

Table 7.10 arranges the Political Goals data into a contingency table based on the free categorizations of the contrasts in Example 7.3. The corresponding d-free contingency table combines the third and fourth columns. The LRS for the independence model from Table 7.10 is 1775.92, on 7 degrees of freedom. Since the LRS for the saturated model is 1838.47 on 23 df, the $GOF = 62.55$ on 16 df, which shows distinct lack of fit. For the d-free table, the independence model has $LRS = 1621.94$ on 6 df, which is substantially worse than the free model's. Thus, although marginally the two models are very similar, when comparing the joint distributions of the contrasts, the free model is superior.

We fit all the usual hierarchical models to the free table. None, except the saturated model, fit. The model with all two-way interactions had $GOF = 19.46$ on 5 df, and that with just the $D_1 \times D_2$ and $D_1 \times D_3$ interactions had $GOF = 21.50$ on 6 df. We will ex-

Table 7.10 *Political Goals: free contingency table model. The value "ij" for* D_1 *means* $(\{i, j\}, \mathcal{O} - \{i, j\})$.

D_1: **Materialist vs. Post-Materialist** →	**12**	**13**	**14**	**23**	**24**	**34**
D_3 : **Say preferred to Freedom**						
D_2 : **Order preferred to Prices**	309	137	29	48	23	33
D_2 : **Prices preferred to Order**	330	117	69	61	55	59
D_3 : **Freedom preferred to Say**						
D_2 : **Order preferred to Prices**	255	93	52	30	21	29
D_2 : **Prices preferred to Order**	294	70	34	35	27	52

Table 7.11 *Political Goals: conditional independence model. Entries are percentage within* D_1*'s groups. The value "ij" for* D_1 *means* $(\{i, j\}, \mathcal{O} - \{i, j\})$.

D_1: **Materialist vs. Post-materialist** →	**12**	**13**	**14**	**23**	**24**	**34**
D_2 : **Order preferred to Prices**	47	55	44	45	35	36
D_3 : **Say preferred to Freedom**	54	61	53	63	62	53

plore the latter model first, then compare it to the observed data to see where they deviate.

The model with the $D_1 \times D_2$ and $D_1 \times D_3$ interactions has conditional independence of D_2 and D_3 given D_1. Thus it is enough to explore the dependencies between D_1 and the other two. See Table 7.11. The D_2 row shows that overall, Prices are preferred to Order, but the preference is stronger among the Post-materialists, i.e., those with $D_1 = \{3, 4\}$. For D_3, generally Say is preferred to Freedom, although the preference bounces from around 53% to around 63% with no apparent pattern.

There is some evidence of three-way interaction. The expected values under the conditional independence model are very close to the observed values except for the four cells in which $D_1(y) = (\{1, 4\}, \{2, 3\})$, which separate the two Materialist goals as much

Table 7.12 *Four offending cells in which* $C^{(1)}(y) = (\{1,4\}, \{2,3\})$. *Entries are observed (expected).*

	Say preferred to Freedom		Freedom preferred to Say	
Order preferred to Prices	29	(43.13)	69	(54.86)
Prices preferred to Order	52	(37.86)	34	(48.14)

Table 7.13 *Living Places: free component model.*

	City People		Suburbanites		Country Folk	
Favorite of top two →	**Urban**	**Rural**	**Urban**	**Rural**	**Urban**	**Rural**
Least Favorite ↓						
City	204	81	299	126	125	152
Suburb	23	8	4	4	1	0
Country	210	111	22	45	10	14

as possible. Table 7.12 presents the observed and expected counts for these cells.

The people in the first row are those who place Order first and Prices last. They prefer Freedom to Say much more than would be expected. Those in the second row, who place Prices first and Order last, are more favorable to Say than expected. Thus Order and Freedom go together, and Prices and Say go together. I do not know why. □

Example 7.13. Living Places: Free Component Model
Consider the multistage model for Example 2.14 wherein one first chooses the least favorite living place, then the favorite of the remaining two. See Duncan and Brody (1982). The resulting contingency tables appear in Table 7.13. The rows represent which place is chosen least favorite, and the columns indicate whether of the remaining two, the favorite is the more urban or the more rural.

Testing independence for the models yielded GOF's of 3.14, 34.87, and 1.71, each on 2 df. Thus only for the Suburbanites is independence untenable. The marginal percentages for people's least favorite are, in the order (*City, Suburb, Country*), (45%, 5%, 50%) for City People, (85%, 2%, 13%) for Suburbanites, and (92%, 0%, 8%) for Country Folk. Thus, as seen before, few people dislike the Suburbs, Suburbanites and Country Folk dislike the City, and City People are split on whether they dislike the City or the Country. The percentages who prefer the more urban of their top two are 69%, 65%, and 45%. The City People and Suburbanites prefer the more urban option about 2:1, while the Country Folk slightly prefer the more rural. Although there are no new surprises, these models effectively summarize what we have previously seen.

Looking more closely at the Suburbanites, it is easy to see where independence fails. Suburbanites are happy people. Whether they dislike the City or the Country, they would rather be in the Suburbs. □

7.11 Inversion models

Section 2.6.2 introduces the statistics based on comparisons of subsets of more than just two objects. The Babington Smith model has the paired comparisons as the sufficient statistics. Thus understanding the paired comparison is enough to understand the data, if the model holds. If the model does not hold, then there must be some interaction among one or more groups of three objects, or four objects, or more. The idea of the inversion models of McCullagh (1993a) is to ascertain whether extra interaction exists, and if so, among which groups of objects.

The model based on sets of h objects has sufficient statistic $X'N$ where the columns of X consist of the vectors $x^{(s,r)}$ as in (2.83). The statistic counts the number of observations at each ranking of each subset of h objects. As in Section 2.6.2, we can reduce the number of statistics by looking at just inversions. The h-object model would use the $x^{(s,r)}$'s for $r \in \mathcal{I}_s$ for $s \leq h$, i.e., only vectors for which r is an inversion, but include vectors for all subsets of objects with less than or equal to h objects.

For the model based on three-object inversions, we can reexpress the statistics in a more intuitive way. In the case $m = 3$, replace the two three-object inversion statistics with the following:

$$I[y_1 = 2] \text{ and } I[y_2 = 2]. \tag{7.67}$$

Table 7.14 *Political Goals: inversions. "bet." means "between."*

Effect	$\hat{\theta}$	Z	Effect	$\hat{\theta}$	Z
O bet. S and P	**0.0161**	**1.13**	**O bet. P and F**	**0.0267**	**1.75**
S bet. O and P	**−0.6390**	**−4.84**	**P bet. O and F**	**0.0689**	**4.89**
P bet. O and S	**0.0478**	**3.76**	**F bet. O and P**	**−0.0956**	**−6.60**
O bet. S and F	**−0.1233**	**−8.58**	**S bet. P and F**	**0.0670**	**4.86**
S bet. O and F	**0.0188**	**1.32**	**P bet. S and F**	**−0.0486**	**−3.58**
F bet. O and S	**0.1045**	**7.62**	**F bet. S and P**	**−0.0184**	**−1.30**

The two together indicate which object is ranked second. They are clearly linearly independent, and under $Uniform(\mathcal{S}_3)$ are independent of each of the two-inversions, $I[Y_i > Y_j]$, $i < j$. (They are not independent of the set of inversions, however.) See Exercise 7.9. For $m > 3$ and $k = 3$, we replace (2.83) with

$$I[rank(y_i, y_j, y_k)_l = 2] \text{ for } 1 \leq i < j < k \leq m, \text{ and } l = 1, 2, \tag{7.68}$$

that is, indicate for each triple of objects which object is ranked between the other two.

Example 7.14. Political Goals: Inversions
In Table 7.2 we see that although Babington Smith explains 85% of the nonuniformity in the Political Goals data, there is still a significant amount to be explained. Example 7.12 used the orthogonal contrast free models to find some extra structure. Here we will add the three-object inversions to Babington Smith. Fitting the extended model has LRS of 1810.81 on 14 df, the 14 including the 6 for Babington Smith and $\binom{4}{3}3!(1/2! - 1/3!) = 8$ for the three-object inversions. The difference between this model and Babington Smith is 250.48 on 8 df, quite a substantial improvement. The $GOF = 27.65$ on 9 df, significant but much reduced.

To determine in which directions the new models differs from Babington Smith, we investigate the estimates of the parameters for the statistics in (7.68). For symmetry, we add to each set of statistics that for $l = 3$, then require the three parameters sum to zero to retain identifiability. Table 7.14 exhibits the estimated effects for each object being ranked between two others. A positive value for θ indicates that the first object is more likely to be between the other two than expected under the Babington Smith model. A negative value means it is less likely. Note that these effects are to be interpreted after conditioning on Babington Smith. That is, since Order and Prices are much more popular than Say, under Babington Smith one would rarely expect Say to come between Order and Prices. The fact that the effect is negative in Table 7.14 says that Say is even less likely to split the others than expected.

Perusing the table, we note that all the significant negative effects involve either a Materialist trying to split two Post-materialists, or vice versa. All the other effects, except the very last, are positive, meaning it is more likely any goal will come between one Materialist and one Post-materialist. A similar effect is revealed in Example 7.11, where people are more likely to prefer both Post-materialist goals than expected from the ϕ model. □

For higher-way inversions, interpretation seems to be more difficult. The next example decomposes the Voting data into χ^2 statistics. McCullagh (1993a) contains a more thorough discussion of these data.

Example 7.15. Voting: Inversions
We decompose the Voting Data into components due to the levels of inversions. The models and the LRS's and degrees of freedom for the consecutive differences in the models are Babington Smith: 189.60 on 10 df; Triples: 1281.41 on 20 df; Quadruples: 187.66 on 45 df; and $GOF = 58.85$ on 44 df. What is most notable is that the three-object inversions explain almost 75% of the nonuniformity. The effects are generally negative when some candidate tries to come between either A and C, or D and E, the Researchers and Clinicians, respectively. Note also that the four-inversions are significant. Interestingly, the only group of four that did not show significant effects is $ACDE$. Fortunately, there is no significant five-inversion effect. □

7.12 Exercises

7.1. Show that (7.4) gives the relationship between the p_{ij}'s in the Babington Smith models (5.7) and (7.3).

7.2. Show that $\widehat{\theta}_{MB} = \log(7)$ and $\widehat{\theta}_{FB} = \log(6/7)$ maximizes the likelihood l^* in Example 7.1.

7.3. Use (7.20) and (7.21) to show (7.22).

7.4. Use the definition in (7.25) to prove Lemma 7.4 (c).

7.5. Prove (7.43). [Consider Lemma 7.6, where the U's are indexed by m. Then $U_0^{(m)} = U^{(m)} + V^{(m)}$, where $V^{(m)} = U_1^{(m)} + \cdots + U_K^{(m)}$. Subtract the mean from each random variable, and divide the result by $m^{3/2}$. The left-hand side is asymptotically $N(0, 1/36)$ by Theorem 3.4 applied to Kendall's distance. See Table 3.3. The $V^{(m)}$ term needs two cases. If $\sum(\lambda_i^{(m)}/m)^3 \to 0$, then $Var(V^{(m)})/m^3 \to 0$, and $U^{(m)}$ and $U_0^{(m)}$ are asymptotically equivalent. Otherwise, apply Theorem 3.4 to $V^{(m)}$ by breaking each $U_i^{(m)}$ into $\lambda_i^{(m)}$ components.]

7.6. Prove the first equality in (7.52).

7.7. Show that $[\mathcal{C}] = (\{1\}, \ldots, \{m\})$, where $\mathcal{C} = \{A, B, C, D\}$, $A = \{1\}, \ldots, \{i-1\}, \{i, \ldots, j\}, \{j+1\}, \ldots, \{m\})$, $B = (\{i+1\}, \ldots, \{j-1\})$, $C = (\{i\}, \{i+1, \ldots, j\})$ and $D = (\{i+1, \ldots, j-1\}, \{j\})$.

7.8. Show that any orthogonal contrast ϕ model is a submodel of an orthogonal contrast ϕ model with $m-1$ contrasts, each with two groups. [The smaller model is obtained from the larger one by equating some parameters, and setting others to 0.]

7.9. Given $Y \sim \mathcal{P}_3$, show that $(I[Y_1 = 2], I[Y_2 = 2])$ is independent of $I[Y_i < Y_j]$ for any pair i, j. (It is not independent of all three $I[Y_i < Y_j]$'s simultaneously.)

CHAPTER 8

Plackett–Luce, Logistic, and Vase Models

8.1 The first-order model

In Section 5.6.1, the Plackett–Luce model is given as a multistage model, which we extend in this chapter. Section 5.13.1 notes that it is also a Thurstonian model. The basic model is based on the positive constants ν_i, $i = 1, \ldots, m$, where ν_i is associated with Object i. The larger the parameter, the more preferred is the object. The ν_i's are proportional to the probability that Object i is ranked first:

$$P[W_1 = i] = \frac{\nu_i}{\nu_1 + \cdots + \nu_m}, \tag{8.1}$$

where $W \in \mathcal{T}_m$ is a random ordering. The key assumption for the *first-order model* is that the conditional probability that Object j is ranked k given the Objects ranked 1 through $k-1$ is also proportional to the same parameters:

$$P[W_k = j \mid W_1 = i_1, \ldots, W_{k-1} = i_{k-1}] = \frac{\nu_j}{\sum_{l \mid l \neq i_1, \ldots, i_{k-1}} \nu_l}. \tag{8.2}$$

The probability of any ordering is then easily obtained in stages, the same stages as in the ϕ component model of Section 5.6.2, but with different probabilities. Thus, as in (5.13),

$$\begin{aligned} P[W = w] =& \frac{\nu_{w_1}}{\nu_{w_1} + \cdots + \nu_{w_m}} \times \frac{\nu_{w_2}}{\nu_{w_2} + \cdots + \nu_{w_m}} \times \\ & \cdots \times \frac{\nu_{w_{m-1}}}{\nu_{w_{m-1}} + \nu_{w_m}} \times \frac{\nu_{w_m}}{\nu_{w_m}}. \end{aligned} \tag{8.3}$$

Note that the parameters are identifiable only up to a multiplicative constant. This model assumes no interaction between the objects, as formalized by the Luce choice axiom in Section 5.13.1. In the Voting data, for example, this property is likely not to hold, since people who rank Candidate A first are more prone to rank Candidate C high than are those who rank Candidates D or E first. See Example 2.15.

It is useful to use Silverberg's (1980) *vase* model metaphor. Imagine a vase containing an infinite number of balls, each ball associated with one of the objects. (Since infinite numbers are needed, vases are used. Urns can hold only a finite number of balls.) The proportion of balls in the vase for Object i is proportional to ν_i. The stages are draws from the vase. First, randomly draw one ball. Its object is then W_1, the best object. At the second stage, draw another ball. If it is different than W_1, then it is W_2. If you happen to draw W_1 again, then replace it and draw another. Continue making draws until you have one distinct from W_1. Stage three is the same: keep drawing until obtaining a ball distinct from W_1 and W_2, and assign it to W_3. The other stages follow, until the mth stage where there is only one object you have not chosen. It is ranked last. Note that the conditional probabilities of obtaining an object, given the objects already chosen, are given in (8.2). Pictorially, the vase model is

$$\overbrace{Vase}^{Stage\ 1} \rightarrow \overbrace{Vase}^{Stage\ 2} \rightarrow \cdots \rightarrow \overbrace{Vase}^{Stage\ m} . \tag{8.4}$$

This model is not a regular exponential family model. Though there are $m-1$ free parameters, the sufficient statistic is of dimension $2^m - 2$. The likelihood function is a fairly simple function of the sufficient statistics and parameters, so that at least for moderate m, the calculations are reasonably easy. For the likelihood calculations, we will set $\nu_m = 1$ for identifiability, and $\theta_i = \log(\nu_i)$ to prevent the estimates from straying outside the range. For one observation, the loglikelihood from (8.3) is thus

$$l(\theta; w) = \sum_{i=1}^{m} \theta_i - \sum_{i=1}^{m} \log(\sum_{j=i}^{m} \exp(\theta_{w_j})). \tag{8.5}$$

The first term on the right-hand side is independent of the ordering w. Each summand in the second term has components depending on the objects ranked as the k worst, $k = 1, \ldots, m$. In a sample, there is one statistic for each possible such subset, of which there are $2^m - 1$. Consequently, for any nonempty subset $\mathcal{O}' \subset \mathcal{O}$ with $k = \#\mathcal{O}'$, let

$$\gamma(\mathcal{O}') = \sum_{i \in \mathcal{O}'} \exp(\theta_i) \text{ and}$$
$$S(\mathcal{O}') = \#\{w^{(j)} | w_l^{(j)} \in \mathcal{O}', m - k + 1 \le l \le m\}, \tag{8.6}$$

where $w^{(1)}, \ldots, w^{(n)}$ is a sample of orderings. Thus $S(\mathcal{O}')$ is the number of orderings that rank the objects in $\mathcal{O}'$ at the bottom. Then the likelihood for the sample is

$$l(\theta) = n \sum_{i=1}^{m} \theta_i - \sum_{\mathcal{O}' \subset \mathcal{O}} S(\mathcal{O}') \log(\gamma(\mathcal{O}')). \tag{8.7}$$

(Ignore the empty set in the summation.) A slight simplification arises by noting that $S(\mathcal{O}) = n$. The first and second derivatives of l are

$$\frac{\partial l}{\partial \theta_i} = n - \sum_{\mathcal{O}' \subset \mathcal{O} | i \in \mathcal{O}'} S(\mathcal{O}') \frac{\exp(\theta_i)}{\gamma(\mathcal{O}')}, \tag{8.8}$$

$$\frac{\partial^2 l}{\partial \theta_i^2} = - \sum_{\mathcal{O}' \subset \mathcal{O} | i \in \mathcal{O}'} S(\mathcal{O}') [\frac{\exp(\theta_i)}{\gamma(\mathcal{O}')} - \frac{\exp(2\theta_i)}{\gamma(\mathcal{O}')^2}], \tag{8.9}$$

and

$$\frac{\partial^2 l}{\partial \theta_i \partial \theta_j} = \sum_{\mathcal{O}' \subset \mathcal{O} | i,j \in \mathcal{O}'} S(\mathcal{O}') \frac{\exp(\theta_i) \exp(\theta_j)}{\gamma(\mathcal{O}')^2}, i \neq j, \tag{8.10}$$

for $1 \le i, j \le m - 1$. The Newton–Raphson algorithm then proceeds in a straightforward manner. In order to prevent the estimates from approaching infinity, remove an object if it is always ranked first or always ranked last, and apply the algorithm to the reduced set. (Be sure that no object is always ranked first or last in that set, too.)

Example 8.1. Feed Flavors: Plackett–Luce
Recall Example 5.1, in which cows were forced to use a multistage process to rank five flavorings for their feed. One experiment had $n = 6$ cows. The flavors are Control, Sucrose, HCl, Urea, and NaCl. Three cows had the ranking 21534, and one each had 12435, 21435, and 23541. Even though the sample size is small, the model detects some effects. The $LRS = 28.64$ on 4 df. (The $GOF = 13.90$ on 115 df is meaningless.) The estimated ν_i's are, respectively, 35.67%, 59.24%, 0.26%, 4.26%, and 0.54%. (We normalize so that they add to 100%.) Clearly, the Sucrose is the favorite, followed by the plain Control. The others are not liked at all. □

The multistage set-up in (8.4) is a clever approach to modeling noninteraction of objects since, although the object chosen second clearly cannot be independent of that chosen first, at least the relative chances of the remaining objects being second given the first choice are the same as the relative (unconditional) chances of being first. Silverberg extends this model, preserving the noninteraction, by allowing the vases to depend on the stages:

$$\overbrace{Vase_1}^{Stage\ 1} \rightarrow \overbrace{Vase_2}^{Stage\ 2} \rightarrow \cdots \rightarrow \overbrace{Vase_m}^{Stage\ m}. \tag{8.11}$$

The vases are allowed to have different proportions of balls, but the vase drawn from at each stage depends only on the stage, not on which objects have already been drawn. This model is appropriate if judges' opinions change over time. For example, it may be that judges work very hard to choose their top three, but after that the judgments are less careful. Then the vases after the third one may be much more uniform than the first three. We present this model in Section 8.2.

In many data sets, we can expect interactions, e.g., if a judge chooses a Team sport as #1, then it is likely a Team sport will also be chosen #2, and similarly for Non-team sports. Plackett's solution is to allow the vase at each stage to depend on the previously chosen objects. The most general such model is

$$\overbrace{Vase}^{Stage\ 1} \rightarrow \overbrace{Vase(w_1)}^{Stage\ 2} \rightarrow \overbrace{Vase(w_1, w_2)}^{Stage\ 3} \rightarrow \cdots \rightarrow \overbrace{Vase(w_1, \ldots, w_{m-1})}^{Stage\ m}. \tag{8.12}$$

The Plackett–Luce model sets all vases equal. Plackett's logistic models require $Vase(w_1, \ldots, w_k)$ to depend only on the last fixed number of elements of $(w_1, \ldots, w_k)$. See Section 8.3. One advantage of the vase models is that they are immediately applicable to patterns of ties in which one gives the first, second, ..., and q^{th} favorite objects for some q, so that the pattern of ties as in Definition 7.3 is $(\{1\}, \{2\}, \ldots, \{q\}, \{q+1, \ldots, m\})$. Silverberg (1980) calls such partial rankings "q-permutations." The models proceed in stages, but stop after q stages. See Section 8.4.

8.2 Extended vase models: No interaction

Consider the vase model represented by (8.11). Each vase has its own set of proportions of balls; hence has $m - 1$ free parameters. Since the last vase's parameters are irrelevant, the dimension is $(m-1)^2$. The parameters are $\nu_j^{(i)}$ for $1 \leq i \leq m-1$ and $1 \leq j \leq m$, where i denotes the stage and j denotes the object. For each i, the parameters $\nu_1^{(i)}, \ldots, \nu_m^{(i)}$ are identifiable up to a multiplicative constant. The density is then

$$P[W = w] = \frac{\nu_{w_1}^{(1)}}{\nu_{w_1}^{(1)} + \cdots + \nu_{w_m}^{(1)}} \times \frac{\nu_{w_2}^{(2)}}{\nu_{w_2}^{(2)} + \cdots + \nu_{w_m}^{(2)}} \times \cdots \times \frac{\nu_{w_{m-1}}^{(m-1)}}{\nu_{w_{m-1}}^{(m-1)} + \nu_{w_m}^{(m-1)}}. \tag{8.13}$$

Although there are $(m - 1)^2$ parameters, the density breaks into $m - 1$ pieces, each depending on one of the vases; hence the likelihood calculations are reasonably simple.

Table 8.1. *Political Goals: vases. Entries are $\widehat{\nu}_j^{(i)}$'s normalized to 100% within each vase.*

Object i	$Vase_1$	$Vase_2$	$Vase_3$
Order	**38.68**	**36.77**	**24.56**
Say	**12.33**	**14.02**	**18.17**
Prices	**40.41**	**38.53**	**41.87**
Freedom	**8.58**	**10.67**	**15.40**

As in Section 8.1, we set $\nu_m^{(i)} = 1$ for each i, and $\theta_j^{(i)} = \log(\nu_j^{(i)})$. Recall $S(\mathcal{O}')$ from (8.6), and let

$$\gamma^{(i)}(\mathcal{O}') = \sum_{j \in \mathcal{O}'} \exp(\theta_j^{(i)}) \tag{8.14}$$

be the γ for the ith vase. Also, recall the Marginals matrix $\widehat{M}$ from (2.10), so that $n\widehat{M}_{ij} = \#\{w^{(k)} | w_j^{(k)} = i\}$. The loglikelihood from (8.13) is

$$l(\theta) = \sum_{i=1}^{m-1} \left(\sum_{j=1}^{m} n\widehat{M}_{ij}\theta_j^{(i)} - \sum_{\{\mathcal{O}' | \#\mathcal{O}' = m-i+1\}} S(\mathcal{O}') \log(\gamma^{(i)}(\mathcal{O}')) \right). \tag{8.15}$$

The summands for the different vases can be maximized separately, using equations very similar to (8.8), (8.9), and (8.10).

Example 8.2. Political Goals: Vases
The difference between the LRS's for the Plackett–Luce and vase model (8.13) for the data in Example 2.3 is 87.55 on 6 df, so that even though the latter model does not fit, it improves significantly on Plackett–Luce. The individual LRS's for the first three vases are 832.23, 654.98, and 123.76 for stages 1, 2, and 3, respectively. The differences among the objects do seem to diminish as the stages progress. Table 8.1 exhibits the estimates.

The main feature is that as stages progress, Order becomes less popular and Say and Freedom become more popular, while Prices remains stable. At the first two stages, Order and Prices are roughly equally likely to be chosen. If both are still left at the third stage, Prices is about 1.7 times more likely to be chosen. That is, if people rank the Post-materialist goals first and second, then they are more likely to prefer Prices to Order. The relative preference between the two Post-materialist goals stays fairly constant. The Post-materialist goals become relatively more popular, although still much below the Materialist goals in absolute terms. For example, Order is about three times as likely to be ranked #1 as Say, but if they are the only two choices left at the third stage, then Order is only 1.35 times as likely to be chosen third. □

It is easy to imagine models between Plackett–Luce (8.4) and (8.11) obtained by equating some of the vases.

8.3 Extended Plackett models

Plackett (1975) presents a hierarchy of models, each level of the hierarchy defined by the number of previous choices the vases depend upon. The first-order model (8.4) uses the same vase at each stage irrespective of the results at previous stages. The "Lag 1" model allows the kth vase to depend upon the object ranked $k-1$:

$$\overbrace{Vase}^{Stage\ 1} \rightarrow \overbrace{Vase(w_1)}^{Stage\ 2} \rightarrow \overbrace{Vase(w_2)}^{Stage\ 3} \rightarrow \cdots \rightarrow \overbrace{Vase(w_{m-1})}^{Stage\ m}. \quad (8.16)$$

Thus from (8.12), $Vase(w_1, \ldots, w_k) = Vase(w_k)$. There are $m+1$ vases. The first one is always used at Stage 1. The others can appear at any subsequent stage, depending on the ordering so far. The first vase has $m-1$ free parameters. Each of the others has $m-2$ free parameters, since the number of balls for Object i in $Vase(i)$ is irrelevant. The total number of free parameters is thus $(m-1)+m(m-2) = m^2 - m - 1$.

The likelihood function again decomposes into one term for each vase. Letting $\nu_j^{(i)}$ be the parameter for object j in $Vase(i)$, we have

$$P[W=w] = \frac{\nu_{w_1}}{\nu_{w_1}+\cdots+\nu_{w_m}} \times \frac{\nu_{w_2}^{(w_1)}}{\nu_{w_2}^{(w_1)}+\cdots+\nu_{w_m}^{(w_1)}} \times \cdots \times \frac{\nu_{w_{m-1}}^{(w_{m-2})}}{\nu_{w_{m-1}}^{(w_{m-2})}+\nu_{w_m}^{(w_{m-2})}}. \quad (8.17)$$

Again, set $\nu_m = 1 = \nu_l^{(i)}$ for all i, where l is the largest index not i, and let $\theta_j^{(i)} = log(\nu_j^{(i)})$. The parameters and statistics in (8.6) also receive some superscripts:

$$\gamma^{(i)}(\mathcal{O}') = \sum_{j \in \mathcal{O}'} \exp(\theta_j^{(i)}) \text{ and} \tag{8.18}$$

$$S^{(i)}(\mathcal{O}') = \#\{w^{(j)} | w_{m-k}^{(j)} = i \text{ and } w_l^{(j)} \in \mathcal{O}', m-k+1 \leq l \leq m\}, \tag{8.19}$$

where $\#\mathcal{O}' \leq m-1$ since these vases are used only after at least one object has been chosen. In addition, we need the statistics corresponding to the numerators $\nu_j^{(i)}$ in the likelihood, i.e.,

$$T_j^{(i)} = \#\{w^{(k)} | w_l^{(k)} = i \text{ and } w_{l+1}^{(k)} = j \text{ for some } l\}, \tag{8.20}$$

the number of orderings in which Object j immediately follows Object i. The loglikelihood for the sample is

$$\begin{aligned} l(\theta, \theta^{(1)}, \ldots, \theta^{(m)}) = \sum_{i=1}^{m} \theta_i n \widehat{M}_{i1} - n \log(\sum_{i=1}^{m} \exp(\theta_i)) \\ + \sum_{i=1}^{m} l^{(i)}(\theta^{(i)}), \end{aligned} \tag{8.21}$$

where

$$l^{(i)}(\theta^{(i)}) = \sum_{j \neq i} \theta_j^{(i)} T_j^{(i)} - \sum_{\mathcal{O}' \subset \mathcal{O} | i \notin \mathcal{O}'} S^{(i)}(\mathcal{O}') \log(\gamma^{(i)}(\mathcal{O}')). \tag{8.22}$$

The $l^{(i)}$ in (8.22) is similar to the likelihood for the first-order model (8.5) except that for each ordering, it takes into account only the set of objects ranked below Object i. Once the statistics are found, the individual $l^{(i)}$'s are maximized just as is l in (8.5). The first line on the right-hand side of (8.21) is even easier. It is just the multinomial model based on the objects ranked first; hence $\widehat{\nu}_i \propto \widehat{M}_{i1}$.

It is clear that the Plackett–Luce model is a submodel of the Lag 1 model, since the former equates the $m+1$ vases. Plackett suggests parameters to assess the difference between the models. One may wish to look at $\nu_j^{(i)}/\nu_j$'s to see how much the vases differ, but since the ν's are determined only up to a multiplicative constant, these ratios are not defined. Even constraining each vase to have ν's sum to 1 will not work since there are no estimates of the $\nu_i^{(i)}$'s. The relative values of the ν's are defined, so that we can estimate ratios

$$\frac{\nu_j^{(i)}/\nu_k^{(i)}}{\nu_j/\nu_k}, \quad i, j, k \text{ distinct.} \tag{8.23}$$

The denominator is the likelihood Object j is ranked first relative to Object k, and the numerator is the likelihood Object j is ranked l^{th} relative to Object k given that Object i is ranked $(l-1)^{st}$. Under the Plackett–Luce model, these relative likelihoods are equal for given j and k; hence all the terms (8.23) equal 1. Plackett takes logarithms of (8.23), and since these are ratios of odds, he calls the resulting parameters "logistic." Example 8.3 below illustrates these parameters.

For larger m, the Lag 1 model may not fit, as is the case for the Political Goals and Voting data. The general Lag L model allows the vases in (8.12) to depend on $(w_1, \ldots, w_k)$ only through the last L objects, $(w_{k-L+1}, \ldots, w_k)$. If $L > k$, then the vase depends on the entire vector. The saturated model takes $L = m - 2$, since a vase depending on the $m-1$ previous choices is unestimable. The models are nested since the Lag $L-1$ model is obtained from the Lag L model by equating some of the vases.

The Lag L model has $m(m-1)\cdots(m-L)-1$ free parameters: The first vase has $m-1$ parameters; there are m vases that condition on one object, each with $m-2$ parameters, etc., down to $m(m-1)\cdots(m-L+1)$ vases each with $m-L-1$ parameters. The number of parameters increases rapidly with L, so that in practice models with only fairly small L are practical. For example, in the Sports data $m = 7$ and $n = 130$, so that the Plackett–Luce ($L = 0$) model with 6 parameters and the Lag 1 model with 41 parameters are the only ones conceivable. The Lag 2 model has 209 parameters. The Political Goals and Voting data sets are good candidates for these models since they have a reasonable number of observations at each possible ordering.

The likelihood is again a sum of separate pieces. The principle is straightforward but the notation is a bit complicated. Let $\mathcal{W}_l$ denote the set of vectors of l of the objects. Then the vases, ν's, and θ's in the Lag L model are indexed by $u \in \mathcal{W}_l$ for $l = 0, \ldots, L$. For each u, $\nu_k^{(u)} = 1$ for $k = \max\{j | 1 \leq j \leq m, j \notin u\}$. Extend γ and S of (8.6), and T of (8.20) to $u \in \mathcal{W}_l$: for $\mathcal{O}'$ with $u_i \notin \mathcal{O}'$, $i = 1, \ldots, l$, $\#\mathcal{O}' = k$,

$$\gamma^{(u)}(\mathcal{O}') = \sum_{j \in \mathcal{O}'} \exp(\theta_j^{(u)}), \tag{8.24}$$

$$\begin{aligned} S^{(u)}(\mathcal{O}') = \#\{w^{(j)} | w^{(j)}_{m-k-l+i} &= u_i \text{ for } 1 \leq i \leq l, \\ &\text{and } w_i^{(j)} \in \mathcal{O}' \text{ for } m-k+1 \leq i \leq m\}, \end{aligned} \tag{8.25}$$

and

$$\begin{aligned} T_a^{(u)} = \#\{w^{(j)} | w^{(j)}_{i+b-1} &= u_i \text{ for } 1 \leq i \leq l \\ &\text{and } w^{(j)}_{i+b+1} = a \text{ for some } b\}. \end{aligned} \tag{8.26}$$

In addition, we need the following extension of $\widehat{M_{ij}}$:

$$M_a^{(u)} = \#\{w^{(j)} | w_i^{(j)} = u_i \text{ for } 1 \leq i \leq l \text{ and } w^{(j)}_{l+1} = a\}. \tag{8.27}$$

Thus $S^{(u)}(\mathcal{O}')$ is the number of sample orderings that end with the objects in $\mathcal{O}'$, and have $(u_1, \ldots, u_l)$ as the immediately preceding objects, and $T_a^{(u)}$ is the number that have the subsequence u followed immediately by a somewhere in the ordering. $M_a^{(u)}$ is like $T_a^{(u)}$ but demands that the first l components match u, and the next be a. The likelihood consists of a sum of terms, one for

Table 8.2. *Voting: Plackett models.*

Source	χ^2	df	*p*-value	$100 \cdot R^2$
Plackett–Luce (Lag 0)	136.67	4	0	7.96%
Lag 1 – Lag 0	997.45	15	0	58.08%
Lag 2 – Lag 1	427.24	40	0	24.88%
Lag 3 – Lag 2	151.16	60	0	8.80%
Total (Lag 3)	1717.51	119	0	

each u of length $\leq L$. Thus

$$l(\theta) = \sum\nolimits_{u \in \mathcal{W}_l, 0 \leq l \leq L-1} \left(\sum_{i \notin u} \theta_i^{(u)} M_i^{(u)} - \log(\sum_{i \notin u} \exp(\theta_i^{(u)})) \right)$$
$$+ \sum_{u \in \mathcal{W}_L} \left(\sum_{i \notin u} \theta_i^{(u)} T_i^{(u)} - \sum_{\mathcal{O}' | u_j \notin \mathcal{O}', 1 \leq j \leq L} S^{(u)}(\mathcal{O}') \log(\gamma^{(u)}(\mathcal{O}')) \right). \tag{8.28}$$

Maximizing each component proceeds as before. Care must be taken to adjust for maximum likelihood estimates approaching $\pm\infty$.

As in (8.23), Plackett defines parameters that measure the difference between the Lag $L-1$ and Lag L models. The idea is to compare the $\nu^{(u)}$'s for $u = (u_1, \ldots, u_L)$ to those based on a Lag of $L-1$, i.e., $\nu^{(u[-1])}$, $u[-1] = (u_2, \ldots, u_L)$. Since the parameters are determined only up to a multiplicative constant within each vase, we look at odds ratios

$$\frac{\nu_a^{(u)} / \nu_b^{(u)}}{\mu_a^{(u[-1])} / \nu_b^{(u[-1])}}, \quad a, b \notin u. \tag{8.29}$$

If the parameter is 1, then Object u_1 has no effect on the relative preference of a and b given the intervening $u[-1]$. As with any hierarchical system, interpretation of parameters becomes more difficult as the order of the model increases.

Table 8.3. *Voting: Lag 0 and Lag 1 parameters.*

	First→	A	B	Lag 1 C	D	E
Pair *a*,*b* ↓	**Lag 0**					
A vs. B	1.36	—	—	1.77	0.55*	0.87
A vs. C	0.65	—	2.13*	—	2.29*	1.89*
A vs. D	0.90	—	1.05	4.69*	—	0.81
A vs. E	0.93	—	0.94	3.01*	0.57*	—
B vs. C	0.48	0.88	—	—	4.20*	2.18*
B vs. D	0.66	1.48	—	2.65*	—	0.93
B vs. E	0.69	1.06	—	1.70	1.05	—
C vs. D	1.37	1.68	0.49*	—	—	0.43*
C vs. E	1.43	1.21	0.44*	—	0.25*	—
D vs. E	1.04	0.72	0.89	0.64*	—	—

The "*" 's indicate values with $|Z|$'s over 2.

Example 8.3. Voting: Plackett Models

Table 8.2 contains the likelihood ratio statistics for the sequence of Plackett models of Lag 0, 1, 2, and 3. Note first that even the Lag 2 model does not fit; hence it is necessary to look at the saturated model. Also, the Lag 1 parameters explain by far the most of the nonuniformity, 58%. Compare this result to that for the inversion model in Example 7.13. The estimates in Table 8.3 are based on the saturated model.

We will start with the Lag 0 and Lag 1 parameters. See Table 8.3. The rows are indexed by the objects whose relative preference we wish to assess, i.e., they are the a and b in (8.29). The "Lag 0" column contains the ratios from vase 1, that is, they are the ratios of numbers of people ranking the candidates #1. C is most likely to be first, followed by D, E, A, and B. The other five columns assess the relative likelihood of being second given the column candidate is ranked first *divided by* the relative odds of being first. Recall that previous analyses have revealed two groups in the electorate, the "AC" group, Researchers, and the "DE" group, Clinicians. Of the two researchers, C is more extreme; hence less palatable to the non-researchers. Of the clinicians, D is the more extreme. Look at the ratios by column. If A is ranked first, the odds ratios are not far from 1. The corresponding $|Z|$'s are all under 2. We see that

C gains the most from having A first, followed by D and E, then B. If C is ranked first, A gains the most by quite a bit, especially versus D and E. Candidate B gains second most, then E, and D always loses support.

Suppose D is ranked first. The likelihoods for B and E increase substantially relative to A and C, and A's likelihood doubles with respect to C's. If E is ranked first, A, B, and D's likelihoods stay fairly constant, while C's is cut in half. Thus, generally, if a researcher (clinician) is first, the chance that one is second increases. The effect is much more pronounced if the more extreme candidate, C or D, is first, and if a clinician is first, C becomes quite unpopular.

Finally, turn to the B, the community psychologist, column. A, D, and E are left alone, but C's support is again halved. Thus while people who rank C first are more favorable to B than D or E, those who rank B first especially dislike C. If one of the clinicians is first, then the likelihood of the other relative to B remains the same. It appears that B is closer to the DE camp than to AC, and closer to the milder people A and E than to the extremes.

There are 20 vases for the Lag 2 parameters indexed by the candidates ranked first and second, and each vase has three odds ratios. About half of these estimates are statistically significant, meaning that generally who is ranked first does affect who is ranked third, even conditioning on second place. The effects are similar in direction to the ones above. For example, take $a = A$ and $b = D$ and $u = (C, E)$ in (8.29). The coefficient is 2.25. The interpretation is that the relative likelihood of A being ranked #3 to D being ranked #3 when C and E are ranked #'s 1 and 2 is 2.25 times greater than the relative likelihood of A versus D being ranked #2 when E is ranked first. Thus the presence of C out in front causes support to shift from D to A. Taking $u = (E, C)$, the coefficient is 0.34, meaning having E out in front shifts support in the other direction, even with C intervening. Other coefficients behave similarly, that is, a researcher in first place increases the chance one will be in third, and similarly for a clinician, no matter who is ranked second. Having B in first causes interesting effects. If a researcher is second, then the B causes the other researcher's support to decrease relative to the clinicians, while if a clinician is second, then the B causes the other clinician's support to decrease relative to the researchers.

For Lag 3, there are 60 vases depending on which objects are ranked #'s 1, 2, and 3, and each vase has only one odds ratio since there are only two objects left. Ten of these have Z over 2. The story is similar, the effect of who is first being felt even at rank 4, with a researcher in first helping the other in fourth, and the same for a clinician. □

The application of the model in this example is somewhat disappointing. A good hierarchical model should capture main effects at the lower orders, exposing interesting new effects at higher orders. In this case, the major effect, i.e., that there are two strong camps, seems to dominate at every lag.

8.4 q-permutations

Silverberg (1980) defines a q-permutation to be a partial ordering that gives the judge's #1 through #q choices, distinguishing between them. Thus a tied ranking has one each of $1, \ldots, q$, and $m - q$ of $q + 1$'s. The advantage of the general vase models (8.11) and (8.12) is that they extend immediately to such patterns of ties. One stops after drawing from q vases. Models can be fit whether or not all the observations in the data set are q-permutations for the same q, but care must be taken if the judges themselves decide on the q. It may be that different types of people choose different q's, as in the Voting example.

Example 8.4. Voting: Vase Model for Tied Rankings
Diaconis (1989) presents ballots from the APA election that rank only the top one, two, or three candidates, in addition to the fully ranked data we have seen so often. (There are 5141, 2462, and 2108 observations, respectively, for $q = 1, 2$, and 3.) We fit the above unordered vase model as in Example 8.3 to the additional three sets of data. For those who give only their favorite q candidates, the model has q vases. We first compare the first vases for the four groups. Table 8.4 contains the percentages for the vases. The test of homogeneity yields $\chi^2_{12} = 182.38$, showing significant differences between the groups.

Note that the people who give their top three are quite even-handed. The people who give two seem to be more likely to be in the AC camp than are the other groups. The group of full rankings and those with $q = 2$ are quite similar, although the latter shifts some support from C to B.

Table 8.4. *Voting: percentage of candidates ranked #1, categorized by q.*

Candidate	$q = 5$	$q = 3$	$q = 2$	$q = 1$
A	18.35	20.11	21.57	17.41
B	13.51	16.27	11.78	17.14
C	28.04	20.11	31.88	23.30
D	20.43	21.82	16.61	22.27
E	19.68	21.68	18.16	19.88

The second vases can be compared only for $q = 2, 3$, and 5. Each model has five possible vases, one for each possible choice for rank #1. For each candidate, we test homogeneity of the vases across q-groups, obtaining $\chi^2_{30} = 205.26$, quite significant. Generally, the vases for the full rankings and $q = 3$ are similar, and the vases for B, D, and E are similar across groups. The main differences appear in the A and C vases for those who give their top two. We saw above that these people are more prone to vote for A and C. They also show more dependence between A and C. Among the $q = 5$ and 3 groups, 41% and 37% who rank A first rank C second, while for the $q = 2$ group the percentage is 57%. If C is ranked first, the percentages that rank A second are 50%, 46%, and 70% for $q = 5, 3$, and 2, respectively.

Only the full and $q = 3$ rankings have third vases. Testing for homogeneity yields $\chi^2_{20} = 47.32$. (These are vases depending on who are the top two choices, unordered.) The biggest difference is in $Vase(\{B, C\})$. The percentages who rank (A, D, E) third are $(46\%, 28\%, 25\%)$ and $(64\%, 19\%, 17\%)$ for $q = 5$ and 3. Thus the $q = 3$ people favor A more highly, perhaps indicating more AC interaction than for the full rankings.

The vase model here does give a good handle for comparing the different groups. We see mainly that, compared to the people who give full rankings, the $q = 2$ group is more fiercely researcher-oriented, while the $q = 3$ group is more mild in its differences but to a slight extent shows a stronger connection between A and C. □

CHAPTER 9

Marginals and ANOVA-Type Loglinear Models

9.1 Introduction

Chapter 7 presented models based on paired and higher-order comparisons of objects, and Chapter 8 looked at multistage models in which the choice of object at a given stage could depend on the results of previous stages. In both chapters, hierarchies of models resulted from more complicated comparisons or dependencies. This chapter starts with the Marginals model, that is, the exponential family model with the Marginal counts, the numbers of judges who give rank j to Object i for each i, j, as sufficient statistic. The model can also be approached as a quasi-independence model analogous to the independence model in contingency table analysis. The objects are the factors, and the ranks are the levels within each factor. The "quasi" is due to the presence of many structural zeroes in the resulting contingency table.

We will consider two extensions of the Marginals model, depending on the approach. A natural extension of the Marginals model uses sufficient statistics that take into account more object/rank combinations, such as the number of judges giving ranks j_1, j_2, and j_3 to Objects i_1, i_2, and i_3. These models use the statistics in the spectral decompositions of Diaconis (1988, 1989) as in Section 2.6.1. In Sections 9.3, 9.4, and 9.5, we show how the quasi-independent model extends to other models analogous to the usual hierarchical set of loglinear models for contingency tables. See Silverberg (1980) and Croon (1989a).

9.2 Submodels of the Marginals model

Verducci (1982) explores several submodels of the Marginals model, including the distance models using Hoeffding distances and the Bradley–Terry/Mallows model. The basic Marginals model has

Table 9.1 *Idea: Spearman, Bradley–Terry/Mallows, and Marginals.*

Source	χ^2	df	p-value	R^2
Spearman	428.65	1	0	80.96%
BTM – Spearman	22.25	3	0	4.20%
Marginals – BTM	33.71	12	0.0007	6.37%
Residual	44.81	103	—	8.46%
Total	529.43	119	—	

sufficient statistic $n\widehat{M}$, and parameters $\theta_j^{(i)}$, $1 \le i, j \le m$, where parameter $\theta_j^{(i)}$ is associated with statistic $n\widehat{M}_{ij}$. The Bradley–Terry/Mallows model places the linear restrictions on the parameters $\theta_j^{(i)} = j\theta_i^{(\mathrm{BTM})}$, and Spearman's distance model places the further restriction $\theta_j^{(i)} = -2j\omega_i\theta^{(\mathrm{Spear})}$, where ω is the modal ranking.

Example 9.1. Idea: Marginals Model
Fligner and Verducci (1986) analyze data from the Graduate Record Examination Board in which 98 students rank five words according to their association with the word "Idea." The five words are Thought, Play, Theory, Dream, and Attention. The average ranks for the words are, respectively, 1.21, 4.67, 2.37, 2.69, and 4.05. The modal ordering is fairly clear: Thought is first, Theory and Dream are next, then Attention, and finally Play. Table 9.1 gives the results of fitting Spearman with modal ranking $(1,5,2,3,4)$, Bradley–Terry/Mallows, and the Marginals model.

The Spearman model explains over 80%, but the Bradley–Terry/Mallows model is significantly better, showing that the modal ordering is strong, but the words are not equally spaced. (Recall that the Bradley–Terry/Mallows model fits the average ranks exactly.) Note that the Marginals model explains more than the Bradley–Terry/Mallows. We will try another model between Spearman and Marginals, suggested by looking at the difference between the observed Marginals matrix and expected under Spearman.

In Table 9.2, the words are listed in their modal ordering. The Spearman model underrepresents Theory at #1, since it tries to

Table 9.2 *Idea: observed – expected Marginals, under Spearman.*

Rank → **Word↓**	**1**	**2**	**3**	**4**	**5**
1. Thought	**15.02**	**−14.73**	**−2.00**	**−0.28**	**1.99**
2. Theory	**−15.73**	**1.84**	**10.49**	**3.69**	**−0.28**
3. Dream	**0.00**	**13.49**	**0.02**	**−10.51**	**−3.00**
4. Attention	**0.72**	**−1.31**	**−6.51**	**5.84**	**1.27**
5. Play	**−0.01**	**0.72**	**−2.00**	**1.27**	**0.02**

space the words somewhat equally, but the interesting feature is the symmetry of the entries, so that Spearman overrepresents Thought at #2 to about the same degree as it overrepresents Theory at #1, etc. The symmetry model is defined by symmetrically equating the parameters, i.e., $\theta_j^{(i)} = \theta_i^{(j)}$, where the words are numbered as in Table 9.2. The dimension of the resulting model is 10 in this case, since there are $(m-1)^2$ free parameters in the Marginals model, and we are setting $\binom{m-2}{2}$ independent linear restriction among them. The $GOF = 55.83$, so that the difference between the Marginals and symmetric models is 11.02 on 6 df, yielding a p-value of 0.088. Table 9.3 gives the expected Marginals under the symmetric model, and the difference between the observed and expected. Note that these quantities are easy to calculate from the observed Marginals by averaging or differencing the appropriate pair of cells.

The observed – expected table does not show anything interesting. The expected table smooths out the observed Marginals. We see that Thought is clearly #1, and Theory and Dream are #2 and #3, but with a reasonable chance they are switched, and similarly Attention and Play are #4 and #5, with some chance they are switched. □

Other situations might call for other submodels of the Marginals model. The general d_p distances can be written

$$d_p(\omega, y) = \sum_{\delta=0}^{m-1} \delta^p \sum_{i=1}^{m} I[|y_i - \omega_i| = \delta]; \qquad (9.1)$$

Table 9.3 *Idea: expected and observed − expected, under symmetry.*

	Expected				
Rank → **Word↓**	**1**	**2**	**3**	**4**	**5**
Thought	85.0	8.5	3.0	0.5	1.0
Theory	8.5	53.0	31.5	4.5	0.5
Dream	3.0	31.5	51.0	11.0	1.5
Attention	0.5	4.5	11.0	57.0	25.0
Play	1.0	0.5	1.5	25.0	70.0

	Observed − Expected				
Rank → **Word↓**	**1**	**2**	**3**	**4**	**5**
Thought	0	−0.5	1.0	0.5	−1.0
Theory	0.5	0	1.5	−2.5	0.5
Dream	−1.0	−1.5	0	2.0	0.5
Attention	−0.5	2.5	−2.0	0	0
Play	1.0	1.0	−0.5	0	0

hence the distance model can be extended to a submodel of the Marginals model by inserting a parameter γ_δ for each δ:

$$\sum_{\delta=0}^{m-1} \delta^p \gamma_\delta \sum_{i=1}^{m} I[|y_i - \omega_i| = \delta]. \tag{9.2}$$

Thus the distance model equates the γ's, while model (9.2) is obtained from the Marginals model by setting $\theta_{\omega_j}^{(i)} = \gamma_\delta$ if $|i - \omega_j| = \delta$. Model (9.2) will result in a symmetric expected Marginals table with constant counts down each parallel to the diagonal.

We already know that Hamming's distance produces an expected Marginals table wherein the diagonal counts are all equal, and the off-diagonal counts are all equal. This model can be extended by allowing the chance that the object receives the correct rank to depend on the object, so that the exponent in the density is

Table 9.4 *Political Goals: tests for extended Marginals models.*

Null →	Uniform	$M^{(3,1)}$	$M^{(2,2)}$	$M^{(2,1,1)}$
Alternative →	$M^{(3,1)}$	$M^{(2,2)}$	$M^{(2,1,1)}$	Saturated
LRS	1596.77	219.61	21.48	0.61
Projection	1536.82	762.66	25.04	0.21
df	9	4	9	1

$\sum_{i=1}^{m} \alpha_i \; I[y_i = \omega_i]$. The Hamming model equates the α's, while the Marginals model equates the off-diagonal $\theta^{(i)}_{\omega_j}$'s.

Remark 9.1
Verducci (1989) presents an interesting method for calculating the fits of the Marginals model that uses neither Newton–Raphson nor iterative proportional fitting. It is based on finding subsets ("Latin sets") of $\mathcal{S}_m$ over which the product of the Marginals model's density is constant no matter what the values of the parameters. Fitting involves iteratively spreading mass equally over the various Latin sets. □

9.3 Extended Marginals models

In Section 2.6.2 we have extended Marginal statistics (2.72). Any of these can be used as sufficient statistics in an exponential family model to generalize the Marginals model. Thus if X is a matrix containing the basis vectors of U^λ in (2.75) for some partition λ, then we have the exponential family model with statistic $X'N$, which is equivalent to $\widehat{M}^\lambda$ in Definition 2.5. Two such models will be nested if their corresponding subspaces (2.75) are. Thus we can test hypotheses about such subspaces. In fact, the dimension of the reduced subspace V^λ in (2.77) is the degrees of freedom for testing the λ model versus the model combining all submodels of λ. Note that the uniform model has $\lambda = (m)$ and the saturated model has $\lambda = 1_m$.

Example 9.2. Political Goals: Extended Marginals Model
In Example 2.21 we find the projection on the subspaces, which are completely nested (2.79). Table 9.4 contains the χ^2 statistics

for testing each model versus the next larger. The LRS row shows that the Marginals model is overwhelmingly significant, as is the $\lambda = (2,2)$ model. Thus the unordered pairs effects we see in Table 2.11 are real. Testing $(2,2)$ versus $(2,1,1)$ is testing whether there is an effect of ordered pairs beyond the unordered pairs. It does appear statistically significant, but much less in magnitude than the previous effects. The final column shows that the ordered pairs model fits fine. The "Projection" row contains the T^2's as in Theorem 3.1 based on the projections of $\widehat{P}$. (They are $m! \times SS/n$'s for SS/n in Table 2.10.) They have the same asymptotic distributions as the LRS's under the *Uniform*. Note that, except for the second column, they are reasonably close to the LRS's. The advantage of the LRS's is that the statistics are valid under the relevant null hypothesis rather than just under the *Uniform*. That is, under the unordered pairs model, the LRS for testing the ordered pairs versus the unordered pairs is asymptotically χ^2_9, while that of the projection is more complicated. □

Example 9.3. Sports: Extended Marginals Models
The Marginals model for the Sports data has $LRS = 81.73$ on 36 df, out of a total of 963.12. The unordered pairs model increased the LRS by 319.48 using 225 degrees of freedom, but I doubt that that means much. Since we have previously observed some distinctions between Team and Non-team sports, as in Table 2.12, we simplify the unordered pairs model by allowing the parameters to depend only on whether the sports are Team or not. That is,

$$\theta^{\{i,j\}}_{\{k,l\}} = \beta^{\{t(i),t(j)\}}_{\{k,l\}}, \tag{9.3}$$

where $t(i) = T$ if Sport i is Baseball, Football, or Basketball, and $t(i) = N$ otherwise. Fitting this model yielded $LRS = 175.74$, which is 94.01 more than the Marginals model. The degrees of freedom are not obvious. We first construct the vectors that span the subspace, then let the computer figure out the dimension by Gaussian elimination. In this case, we obtain 50, so that the increase in the LRS is due to 14 degrees of freedom. □

It is possible to go beyond the pairs to groups of three, four, etc. So far, I have not come across a situation where it is necessary. When $m = 4$, there is only one degree of freedom between the ordered-pairs model and the saturated model, and for $m > 4$ the sample sizes needed to fit even the ordered-pairs model become quite large. The Voting data would be a good candidate, but for the ordered-pairs model, its $GOF = 37.33$ on 42 df, which fits remarkably well.

9.4 ANOVA-type loglinear models

Contingency table analysis has been extremely well-served by loglinear models, especially those analogous to the hierarchical models in analysis of variance. This section assumes knowledge of such models at the level of Agresti (1990) or Bishop, Fienberg, and Holland (1975). A contingency table consists of a number of factors, each factor having a finite number of levels. The ANOVA-like loglinear models provide a framework for testing independence of factors, conditional independence of factors, two- and higher-level interactions, etc. Quasi-independence models arose to treat cases such as social mobility tables, in which the data consist of father/son pairs, and the table has two factors being the social status of the father and the social status of the son. Independence is obviously not going to fit, but quasi-independence may, which basically says that, ignoring the cells that correspond to the status being the same, the father's and son's status are independent. See Goodman (1969).

Imagine applying these ANOVA-type models to ranking data, where the factors are the objects, and the levels are the cells. See Section 5.9. It is clear that most cells are impossible since one cannot give the same level to two different factors. In fact, the proportion of non-zero cells is $m!/m^m$. However, the model is still easily defined and not-quite-so-easily fit. The models are defined on the logs of the probabilities for the cells that are not structurally 0 , i.e., the $\log(P_y)$'s for $y \in \mathcal{S}_m$. The quasi-independence model is

$$\log(P_y) = \alpha^{(0)} + \alpha^{(1)}_{y_1} + \cdots + \alpha^{(m)}_{y_m}, \tag{9.4}$$

where $\alpha^{(i)}_k$ is the effect for the Object i being ranked k. In fact, (9.4) is the Marginals model. The ordered-pairs model is the model

$$\log(P_y) = \alpha^{(0)} + \sum_{i=1}^{m} \alpha^{(i)}_{y_i} + \sum_{i \neq j} \alpha^{(ij)}_{y_i y_j}, \tag{9.5}$$

Table 9.5 *Voting: adding pairs to the Marginals model.*

Model	*GOF*	df	*p*-value
Marginals	**973.81**	**16**	**0**
+*AC*	**352.23**	**27**	**0**
+*BE*	**145.75**	**38**	**0**
+*DE*	**104.06**	**49**	**0.0051**
+*AB*	**66.82**	**60**	**0.2262**
+*AD*	**41.09**	**71**	**0.7495**
Ordered Pairs	**37.32**	**77**	**0.6760**

where $\alpha_{kl}^{(ij)}$ is the interaction effect for the two Objects i and j being ranked k and l, respectively. In general, the model with all interactions of degree $l \leq k$, where the l-way interaction effects are denoted $\alpha_{k_1 \cdots k_l}^{(i_1 \cdots i_l)}$, is the extended Marginals model corresponding to $\lambda = (m-k, 1, 1, \ldots, 1)$, the ordered k-tuples. Other models from the previous section are obtained by equating appropriate α's. For example, the unordered-pairs model sets $\alpha_{ij}^{(ab)} = \alpha_{ij}^{(ba)}$.

Example 9.4: Voting: Two-Way Interaction ANOVA-Type Models Start with the ordered-pairs model applied to the Voting data in Example 2.15. The model with all the ordered pairs had $GOF =$ 37.32 on 42 df. The unordered-pairs model has $GOF = 106.39$ on 78 df, for a difference of 69.07 on 36 df. Thus there is still some information in the ordered pairs not captured by the unordered pairs. The ordered-pairs model contains all the Marginals parameters, $\theta_k^{(i)}$, plus all the two-way parameters, $\alpha_{kl}^{(ij)}$. We tried to find a loglinear model between the Marginals and ordered-pairs model by starting with the Marginals model and sequentially adding the parameters for pairs of candidates. At each stage, we added the pair that provided the largest reduction in the GOF. Table 9.5 gives the results.

The first model is the regular Marginals model, the second has the Marginals parameters plus the $\alpha_{kl}^{(AC)}$'s, the third has the Marginals, AC and $\alpha_{kl}^{(BE)}$'s, etc. The model with the Marginals plus the AC, BE, DE, and AB parameters is the smallest one that fits very well. We will look at this model, although adding AD does fit marginally better.

There is no simple name for this model. The model does imply that B, C, and D are quasi-conditionally independent given A and E, and further that C and D are quasi-conditionally independent given A, B, and E. Recall that A is the milder of the researchers A and C, and E is the milder of the clinicians D and E. The model suggests that the relationships among the candidates are governed by these two milder representatives. We can see best what these relationships are by inspecting the sufficient statistics. To see the effect of the pairs, we look at the numbers of people who give ranks (k, l) to Candidates (i, j) minus the numbers expected under the Marginals model. Call these differences $D_{kl}^{(ij)}$. We have four pairs (i, j), each with 20 possible (k, l)'s. To organize the numbers, for each unordered pair of rankings we find the sum and difference of the two possible unordered pairs. In Table 9.6, the first panel contains the sums, i.e., $D_{kl}^{(ij)} + D_{lk}^{(ij)}$, and the second the differences $D_{kl}^{(ij)} - D_{lk}^{(ij)}$.

The first panel reiterates what Diaconis (1988, 1989) found: A and C tend to be ranked either both high or both low, as do D and E. Candidate B seems to be ranked low if A is high and vice versa. The relation between B and E is less clear. They are less likely to be ranked 2 and 3, or 4 and 5, than expected.

The largest number in the second panel is the -111 for AC being 1 and 2. The negative indicates that when A and C are ranked at the top, C is much more likely to be in the #1 spot. This trend emphasizes that C is the more extreme researcher. Otherwise, it is hard to see any other major pattern. □

9.5 Paired and multisample models

The contingency table paradigm is easily extended to paired and multisample situations. In the latter case, we have G samples, each one consisting of a set of rankings. The contingency table adds the "sample" factor, so that it is now $m \times m \times \cdots \times m \times G$. With paired rankings, each observation consists of two rankings of the same objects, e.g., before and after some treatment. There are $2m$ factors corresponding to the Before and After rankings for each of m objects. Useful models can be defined using the loglinear parameters from previous sections. These contingency tables can

Table 9.6 *Voting: effects of the selected pairs, rounded to integers.*

Sums $D^{(ij)}_{kl} + D^{(ij)}_{lk}$

$(k,l) \rightarrow$ $(i,j) \downarrow$	12	13	14	15	23	24	25	34	35	45
AB	−109	−2	18	92	80	13	15	−1	−77	−31
AC	444	−96	−170	−177	−113	−163	−168	98	111	235
BE	18	−43	12	13	−108	66	23	55	97	−133
DE	295	−31	−145	−118	2	−163	−133	44	−14	265

Differences $D^{(ij)}_{kl}$-$D^{(ij)}_{lk}$

$(k,l) \rightarrow$ $(i,j) \downarrow$	12	13	14	15	23	24	25	34	35	45
AB	−14	36	−17	−5	55	−35	−35	56	35	5
AC	−111	53	55	3	−4	−45	−62	11	38	21
BE	38	13	−30	−21	69	−58	27	30	52	−57
DE	6	−52	−2	47	−3	13	−3	11	−67	22

be huge; hence most of the models have to be very low-order. The next two examples show some possibilities.

Example 9.5. Literary Criticism: Loglinear Model

Ye and McCullagh (1993) consider models for the paired comparison data described in Example 4.3. Our treatment here is similar to theirs. The first-order model is the quasi-independence of all eight rankings, implying in particular that the Before and After rankings for the students are independent. The dimension of the model is 18 since the Before and After each have $(m-1)^2 = 9$ free parameters. With only 38 observations, it will be difficult to assess any higher-order models; hence we first reduce the model by using Bradley–Terry/Mallows parameters for each rank. We denote an observation by (z, y), where $z, y \in \mathcal{S}_m$, z is the Before ranking, and

y is the After ranking. Our first model is

$$\log(P_{z,y}) = \alpha^{(0)} + \sum_{i=1}^{m} \beta_i z_i + \sum_{i=1}^{m} \alpha_i y_i, \tag{9.6}$$

where $P_{z,y}$ is the probability of observing the Before/After pair (z, y). The model has dimension 6 since each of the Before and After models has $m - 1 = 3$. The model of no difference between Before and After sets $\beta_i = \alpha_i$ for $i = 1, 2, 3, 4$. The LRS for model (9.6) is 34.01, and for the model with no difference it is 22.95, so that there does appear to be a change in preferences. However, the assumption that Before and After rankings are independent is too strong to accept blindly. There are several ways to model nonindependence. The simplest is to include two-way interaction parameters $\theta_{kl}^{(ii)}$, which indicate when Object i receives rank k Before and rank l After. To trim the dimension of the model, we will use the following submodel, which includes an overall parameter for each object, plus parameters for possible values of the differences $z_i - y_i$:

$$\log(P_{z,y}) = \alpha^{(0)} + \sum_{i=1}^{m} \gamma_i (z_i + y_i) + \sum_{d=-(m-1)}^{(m-1)} \delta_d^{(i)} I[z_i - y_i = d]. \tag{9.7}$$

If $\delta_d^{(i)} = 0$ for all i and d, then we have the no-difference model from above. If $\delta_d^{(i)} = d\delta^{(i)}$ for some $\delta^{(i)}$, then we have model (9.6). Both of these models assume independence between Before and After, but with general δ_d, there is dependence. Model (9.7) has $LRS = 61.45$ on 26 df, to compare with the 34.01 on 6 df of model (9.6). The difference appears substantial, but remember that there are only 38 observations.

We next try to find a model between the no-difference model and (9.7) by removing the $\delta_d^{(i)}$'s for some of the Objects i. The best model of these appears to be that with only the Comparative differences included. It has $LRS = 45.77$ on 9 df, not worse than the full model, and significantly better than no-difference. Adding two or three objects increased the LRS by only around five or six points, using six additional df. Table 9.7 has the final model, and Table 9.8 contains the expected values of the averages under

Table 9.7 *Literary Criticism: loglinear model in which only the Comparative rankings change.*

Source	χ^2	df	p-value	R^2
Model	45.77	9	0	21.30%
Residuals	169.16	566		
Total	214.93	575		

Table 9.8 *Literary Criticism: expected rankings.*

Style↓	Before	After	Before − After
Personal	2.01	2.30	-0.29
Comparative	2.68	1.87	0.82
Authorial	2.95	3.18	-0.22
Textual	2.36	2.64	-0.29

this model. These values are very close to the observed means in Example 4.3.

Basically, the main effect seems to be that Comparative's rank improved, becoming closer to the teacher's. However, the teacher's favorite, Personal, declined. Thus it unclear from this analysis whether there is movement towards the teacher's preferences. □

Example 9.6. Ethnic Preferences: Comparisons
Consider the data in Example 2.6. The model we use here will be much like that in the previous example. Within each group we assume a Bradley–Terry/Mallows model, so that with l representing group and i representing choice, we have

$$\log(P_y^{(l)}) = \alpha^{(l)} + \sum_{i=1}^{m} \alpha_i^{(l)} y_i. \tag{9.8}$$

Table 9.9 *Ethnic Preferences: Bradley–Terry/Mallows models.*

Source	χ^2	df	p-value	R^2
Model (9.9)	94.90	5	0	13.53%
Model (9.8) – Model (9.9)	294.38	20	0	41.99%
Residuals	311.87	3570		
Total	701.15	3595		

The $\alpha^{(l)}$ is the normalizing constant for each group, and $\alpha_i^{(l)}$ is the Bradley–Terry/Mallows parameter for choice i in group l. Note that the model has 25 free parameters since each of the five represented groups has an $(m-1) = 5$ parameter Bradley–Terry/Mallows model. We explore some submodels of (9.8). One model, a straw man, equates the parameters between the groups:

$$\alpha_i^{(l)} = \alpha_i^{(k)} \text{ for all } i, k, l. \tag{9.9}$$

Table 9.9 presents the fits of the two models. Clearly the general model fits much better than the model that equates the parameters. It may be that preferences are the same between groups except that each group feels most comfortable with itself. That is, set

$$\alpha_i^{(l)} = \alpha_i^{(k)} \text{ if } i \neq k, i \neq l, \tag{9.10}$$

and allow $\alpha_k^{(k)}$'s to range freely. We obtain an LRS of 362.18 on 10 df, which means this model is substantially better than the model with equal parameters, but still does not fit as well as the model with arbitrary parameters. We also try the model that equates the parameters in (9.10) over i as well, which declares all preferences equal across group and choice except for the choice corresponding to the group:

$$\alpha_i^{(l)} = \alpha_j^{(k)} \text{ if } i \neq l, j \neq k. \tag{9.11}$$

The LRS = 259.05 on 5 df. It does not fit, but does abstract a feature of the data.

To inspect the models, we look at the expected mean ranks of the choices within groups, and their differences as the models be-

Table 9.10 *Ethnic Preferences: expected Means.*

	Model (9.11)				
Choice↓ Group →	Asian	Black	Latino	Jewish	White
Asian	1.16	4	3.83	3.76	4
Black	3.97	1	3.83	3.76	4
Latino	3.97	4	1.83	3.76	4
Jewish	3.97	4	3.83	2.20	4
Native American	3.97	4	3.83	3.76	4
White	3.97	4	3.83	3.76	1
	Model (9.10) – Model (9.11)				
Choice↓ Group →	Asian	Black	Latino	Jewish	White
Asian	0.00	0.44	0.54	0.63	0.13
Black	0.38	0.00	0.40	0.49	-0.01
Latino	0.32	0.26	0.00	0.41	-0.08
Jewish	0.59	0.54	0.65	0.00	0.24
NativeAmerican	0.12	0.07	0.11	0.18	-0.28
White	-1.41	-1.31	-1.69	-1.71	0.00
	Observed – Model(9.10)				
Choice↓ Group →	Asian	Black	Latino	Jewish	White
Asian	0.00	0.44	0.21	0.01	-0.34
Black	0.06	0.00	-0.40	-0.25	0.15
Latino	-0.03	-1.70	0.00	1.43	0.75
Jewish	-0.06	0.84	0.10	0.00	-0.45
NativeAmerican	0.22	-0.07	0.06	-0.74	-0.11
White	-0.19	0.49	0.03	-0.45	0.00
	Observed				
Choice↓ Group →	Asian	Black	Latino	Jewish	White
Asian	1.16	4.88	4.58	4.40	3.79
Black	4.41	1.00	3.83	4.00	4.14
Latino	4.25	2.56	1.83	5.60	4.68
Jewish	4.50	5.38	4.58	2.20	3.79
NativeAmerican	4.31	4.00	4.00	3.20	3.61
White	2.38	3.19	2.17	1.60	1.00

come more complex. The first panel in Table 9.10 gives the expected mean ranks in (9.11), where only the rank of the choice corresponding to the group matters. In all cases, people tend to rank their own group high, but the Latino and Jewish groups less so. Next, we use model (9.10), which relaxes the equality between choices. The second panel contains the difference between the expected means under (9.10) and the previous means. The main feature here is the presence of large negative numbers for the choice White, which suggests that in general people rank White fairly high, after adjusting for their own group. The third panel contains the observed means minus those from the model (9.10). The largest effects show Blacks more favorable toward Latinos, and less towards Jewish, than expected, and Jewish and White groups are less favorable towards Latinos than expected. The observed means are given in the last panel. □

Remark 9.2
The models in Chapters 7 and 8 can also be extended to paired and multisample situations by writing the models in terms of the exponential family parameters. In particular, the orthogonal contrast contingency table models extend immediately to the multisample case by adding the group as a new factor. □

CHAPTER 10

Latent Class and Unfolding Models

10.1 Introduction

Almost all data sets we have investigated show clustering of judges to some extent. The Political Goals data has the Materialist/Post-materialist dichotomy; the Living Places data splits into groups based on the degree of urbanness the judges enjoy; Occupations are Managerial or Technical; there are Team and Non-team Sports. In many cases, the experiment is set up to reveal groupings, and other times they arise naturally. Some models are more adept at accommodating such grouping than others. The distance models, Plackett–Luce, and Bradley–Terry/Mallows models presume a certain degree of unimodality in the judges' rankings. More complicated models, such as the (extended) Marginals and Babington Smith models, and the orthogonal contrast ϕ and Free models, are flexible enough to capture additional dimensionality. By the same token, they are somewhat less simple in their interpretation.

One effective approach to extending the simpler models (or any model) is to explicitly assume a small number of groups of judges, within each of which a simple model obtains. Typically, the same type of model is assumed for each group, but the parameters are allowed to differ. Croon (1989b), Croon and Luijkx (1993), Stern (1993), Böckenholt (1993), and McCullagh (1993a) have successfully used such models for ranking data. A great advantage of these models is their easy interpretability.

From a different perspective, psychologists have developed scaling and unfolding techniques for assessing observers' (judges') responses to stimuli (objects). Interest may center on the objects, on the judges, or on their relation. Coombs (1964) develops an overall theory for such data, which includes as a major component rank data. The theory is based on situating the judges and the objects in a common space, usually Euclidean space. A judge's ranking is defined through the distances of the objects to the observer; the closer the object the more it is preferred. A placement of objects in a space will typically limit the possible rankings, called *admissible* rankings, that can be observed. For example, if three objects are placed on a line, there are only four possible rankings. In particular, the object in the middle can never be ranked last. See (2.46). Just as typical, given a sample of rankings and a fixed dimension for the space, there will be no placement of objects such that all observed rankings are admissible. Thus there must be some allowance for deviation from the admissible rankings. One solution is, given a sample, place the objects so that the observed rankings are as close as possible to admissible rankings. We will follow the approach of van Blokland-Vogelesang (1989, 1993), who uses Kendall's distance to find the placement and Mallows' ϕ model to model the observations. The connection with latent class models is the following: given a set of admissible rankings, assume there is one group associated with each admissible ranking, and assume the observations follow Mallows' ϕ model with that ranking as the mode. The parameter θ is assumed equal for the groups, although that restriction can be relaxed.

Section 10.2 discusses the general latent class models. Section 10.3 goes into detail about the EM algorithm for fitting these models. The one-dimensional unfolding model is presented in Section 10.4.

10.2 Latent class models

Assume there are L latent groups, $G_1, \ldots, G_L$, in the population. We observe rankings, but not from which groups the rankings arise. Denote the density for rankings of judges sampled from group G_l by $f_l(y)$, and the proportion of judges in the population in group l by π_l. The overall density of an observation y is then

$$f(y; \pi) = \sum_{l=1}^{L} \pi_l f_l(y). \tag{10.1}$$

The individual densities f_l can be anything, but usually they are from the same parametric family.

There are numerous problems, statistical and computational, that arise. One is identifiability. If $f_1 = f_2$, say, then π_1 and π_2 are not identifiable, and if any $\pi_l = 0$, then clearly any parameters unique to f_l are not estimable. Thus to remain a regular model, in addition to the assumptions on the individual f_l's, we must restrict the π_l's to be positive, and remove from consideration parameter values that allow two f_l's to be equal. One often wishes to test two models against each other, where the smaller model has fewer groups. Unfortunately, we need to set some π_l's to zero in the larger model, which violates the regularity conditions in Section 5.14. Consequently, the likelihood ratio test between such models is not asymptotically χ^2. See Everitt and Hand (1981) for a general introduction to mixture models, Lindsay (1983a,b) for geometric insight into their likelihoods, and McLachlan and Basford (1988) for ideas on assessing the likelihood ratio statistics.

In principle, if the f_l's are smooth enough, Newton–Raphson can be used to maximize the likelihood. However, unlike the regular exponential family models, the likelihood function may be multi-modal; hence there is no guarantee the algorithm will proceed to the correct maximum. In addition, especially if some π_l's are close to zero, the algorithm can be quite unstable. The EM algorithm is a very stable and conceptually simple approach, at least when individual f_l's are easy to maximize. It may approach only a local, not global, maximum, or possibly a saddle point. It also converges very slowly. However, it has worked very well in the examples we have tried. The next section presents more details, but first some examples are given.

Example 10.1. Sports: Latent Class Models
Recall from Example 2.17 that the Sports data clustered around two orderings. The respective rankings are $\omega^{(1)} = (6,7,4,3,2,1,5)$ and $\omega^{(2)} = (1,2,3,4,5,6,7)$, indicating Non-team and Team sports groups. Consider the latent class model (10.1) with the f_l's being Mallows' ϕ model:

$$f(y;\pi,\theta_1,\theta_2,\omega^{(1)},\omega^{(2)}) = \pi_1 \exp(\theta_1 d_{\text{Ken}}(\omega^{(1)},y) - \psi(\theta_1;m)) + \pi_2 \exp(\theta_2 d_{\text{Ken}}(\omega^{(2)},y) - \psi(\theta_2;m)), \tag{10.2}$$

where $\pi_1 + \pi_2 = 1$. We find $LRS = 143.79$ on 3 (two θ's plus one π) degrees of freedom. Compare this result to that for the regular one-group Mallows' ϕ model, which has $LRS = 27.08$ using modal ranking $(4,6,3,5,1,2,7)$. The $\widehat{\pi}_l$'s for the two groups are 43.77% and 56.23%, respectively, indicating a majority in the Team sports camp. The Kendall's τ's for the two groups are estimated to be

Table 10.1 *Occupations: latent class orthogonal contrasts'* $100 \cdot \tau$*'s.*

	One Group	Two Groups	
Group →		Technical	Managerial
$100 \cdot \hat{\pi}$ →	100%	66.56%	33.44%
Contrast ↓			
$(A, B \cup D)$	77	79	71
(B, D)	94	91	100
$(\{Fac, Sci\}, \{Own\})$	2	26	−46
$(\{OR, IE, ME\}, \{Mgr, Sup\})$	28	57	−30
$(\{Tech\}, \{For\})$	34	49	4
$(\{OR\}, \{IE, ME\})$	23	25	19
$(\{Fac\}, \{Sci\})$	17	−3	59
$(\{Mgr\}, \{Sup\})$	40	44	32
$(\{IE\}, \{ME\})$	68	59	86

0.50 and 0.40, respectively. Thus the Non-team group is clustered a bit more tightly. The three-group model, using the three rankings $(1,2,3,4,5,6,7)$, $(6,7,5,3,1,2,4)$ and $(7,3,1,5,4,2,6)$, obtained as in Example 2.17, has an insignificantly lower LRS of 143.71. There must exist a better set of three modal rankings, but it is unlikely to be substantially better.

We also looked at the Bradley–Terry/Mallows model, which allows one parameter for each object (minus 1). The one-group model has $LRS = 37.84$ on 6 df, and the two-group model has $LRS = 148.85$ on 13 df. Thus two groups are better than one, and one BTM group is better than one Mallows' ϕ group, but two BTM groups is not substantially better than two Mallows' ϕ groups, considering the former has 10 more parameters. □

Example 10.2. Occupations: Latent Class Models

In Example 7.8 we obtain a good orthogonal contrast ϕ model, exhibited in Table 7.6. Using the same set of orthogonal contrasts, we try a two-group model. Now $LRS = 1447.98$ on 19 df, which is an improvement of 53.27 on 10 df. Table 10.1 contains the estimated Kendall's τ's for each group and each contrast. We define Kendall's τ for contrast C to be $1 - d(C(y))/E_0(d(C(Y)))$, where the expected value is over $Y \sim \mathit{Uniform}(\mathcal{S}_m)$. Positive τ's indicate preference of the first-named subset of occupations in the contrast. The major differences between the two groups in the two-group model occur in the second set of contrasts, i.e., those that compare the Technical to the Managerial occupations. The first group is relatively more favorable to the Technical occupations; hence we call it the Technical group. The only other notable difference is in the Faculty versus Scientist contrast, where the Technical group is more favorable towards the Scientist.

Marden (1992) proceeds further, finding four interesting latent groups. □

10.3 The EM algorithm for latent class models

Dempster, Laird, and Rubin (1977) present a complete accounting of the EM algorithm in general, and Croon and Luijkx (1993) give the details for latent class models for rank data. We summarize their explanations for latent class models in the exponential family case.

Consider the complete but only partially observed data $x^{(1)}, \ldots, x^{(n)}$, a sample from a K-parameter exponential family density with

$$h(x;\phi) = \exp(\phi' S(x) - \psi(\phi)), \tag{10.3}$$

where ϕ and S are $K \times 1$. We assume that the sample space $\mathcal{X}$ is finite, and the density is with respect to $\mathit{Uniform}(\mathcal{X})$, but the development works for other cases as well. The observed data are $y^{(1)}, \ldots, y^{(n)}$, where $y^{(i)} = u(x^{(i)})$ for some known function u. Let $\mathcal{Y}$ be the space of the y's. Then the density of the $y^{(i)}$'s with respect to $\mathit{Uniform}(\mathcal{Y})$ is

$$f(y;\phi) = c \sum_{x|u(x)=y} h(x;\phi) = c \cdot \exp(-\psi(\phi)) \sum_{x|u(x)=y} \exp(\phi' S(x)), \tag{10.4}$$

where $c = \#\mathcal{Y}/\#\mathcal{X}$. The loglikelihood based on the observed $y^{(i)}$'s is then

$$l(\phi) = \sum_{i=1}^{N} \log\left(\sum_{x|u(x)=y^{(i)}} \exp(\phi' S(x)) \right) - n \cdot \psi(\phi) + n \cdot \log(c). \tag{10.5}$$

Differentiating with respect to the elements of ϕ yields

$$\nabla l(\phi) = \sum_{i=1}^{n} \frac{\sum_{x|u(x)=y^{(i)}} S(x)\exp(\phi' S(x))}{\sum_{x|u(x)=y^{(i)}} \exp(\phi' S(x))} - n \cdot \nabla\psi(\phi). \quad (10.6)$$

We know the derivatives of ψ are the expectations of the S_k's, but notice also that the quantities in the summation are the *conditional* expectations given $y^{(i)}$, so that

$$\begin{aligned} \nabla l(\phi) &= \sum_{i=1}^{n} E_\phi(S(X_i)|u(X_i) = y^{(i)}) - nE_\phi(X) \\ &\equiv C(\phi; y^{(1)}, \ldots, y^{(n)}) - U(\phi). \end{aligned} \quad (10.7)$$

Thus solving the likelihood equation is equivalent to setting the conditional and unconditional expectations, C and U, equal.

If we know C, then solving $C = U(\phi)$ is the same as finding the MLE of the complete data model (10.3) using C as the sufficient statistic. On the other hand, if we knew ϕ, it is straightforward to find the conditional expectations C. Unfortunately, we do not know either ϕ or C to begin with. The EM algorithm solves the problem by iterating between the two calculations. That is, suppose we start with a guess for ϕ, $\phi^{(1)}$. The "E" (expectation) steps find C based on the current ϕ, and the "M" (maximization) steps find ϕ based on the current C:

E-Step. $C^{(j)} = C(\phi^{(j)}; y^{(1)}, \ldots, y^{(n)})$;

M-Step. Find $\phi^{(j+1)}$ to solve $C^{(j)} = U(\phi^{(j+1)})$.

Dempster, Laird, and Rubin (1977) and Wu (1983) give conditions under which this procedure indeed converges to a solution to the likelihood equation. The procedure is especially useful when the two steps are computationally quick. If they are not, the EM procedure can be quite slow since often many iterations are necessary, although there are techniques for accelerating the algorithm. See Tanner (1993). Note that the information matrix for f in (10.4) can be found by differentiating (10.6) again, yielding the conditional information matrix minus the unconditional one.

Turn to the special case of latent classes. The observed data $y^{(1)}, \ldots, y^{(n)}$ are the observed rankings. Underlying these are the complete data $x^{(1)} = (g_1, y^{(1)}), \ldots, x^{(n)} = (g_n, y^{(n)})$, where g_i denotes the unobserved group membership of the ith observation. Thus $u(x) = u(g, y) = y$. We assume that the density for group l is a regular exponential family density with respect to $Uniform(\mathcal{Y})$,

$$f_l(y; \theta^{(l)}) = \exp(\theta^{(l)\prime} S^{(l)}(y) - \psi_l(\theta^{(l)})), \tag{10.8}$$

where each density may or may not have a different dimension and set of sufficient statistics $S^{(l)}$. The density with respect to $Uniform(\mathcal{X})$ for a complete observation $x = (g, y)$ is

$$f^*(x; \theta^{(1)}, \ldots, \theta^{(L)}, \pi) = L\pi_g f_g(y; \theta^{(g)}). \tag{10.9}$$

(Note that $\#\mathcal{X}/\#\mathcal{Y} = L$ in this case.) We need to write g as a large exponential family model. The parameter ϕ contains the $\theta^{(l)}$'s as well as parameters corresponding to the π_l's. Call these $\rho_1, \ldots, \rho_{L-1}$. (We drop the last one for identifiability.) Then

$$\phi = (\theta^{(1)}, \ldots, \theta^{(L)}, \rho_1, \ldots, \rho_{L-1}). \tag{10.10}$$

The sufficient statistic S in (10.3) contains the individual $S^{(l)}$'s as well as indicators for the groups g. Thus

$$\begin{aligned} S(g, y) = &(I[g = 1]S^{(1)}(y), \ldots, I[g = L]S^{(L)}(y), \\ &I[g = 1], \ldots, I[g = L - 1]). \end{aligned} \tag{10.11}$$

Exercise 10.1 gives $\psi(\phi)$ and π in terms of the ψ_l's and ρ_l's. Next we handle the two steps.

E Step. Given y, the $S^{(l)}(y)$'s are fixed; hence the only random parts of S are the group indicators $I[g = l]$'s. Now, using Bayes'

Theorem,

$$
\begin{aligned}
E_\phi(I[G=l]|Y=y) = & P_\phi(G=l|Y=y) \\
= & \frac{P_\phi(Y=y|G=l)P_\phi(G=l)}{\sum_{k=1}^{L} P_\phi(Y=y|G=k)P_\phi(G=k)} \\
= & \frac{\pi_l f_l(y;\theta^{(l)})}{\sum_{k=1}^{L} \pi_k f_k(y;\theta^{(k)})}.
\end{aligned}
\tag{10.12}
$$

Thus

$$
\begin{aligned}
C(\phi) &= \sum_{i=1}^{n} E(S(G,Y)|Y=y^{(i)}) \\
&= (C^{(1)},\ldots,C^{(L)},n^{(1)},\ldots,n^{(L-1)}),
\end{aligned}
\tag{10.13}
$$

where

$$
\begin{aligned}
C^{(l)} &= \sum_{i=1}^{n} P_\phi(G=l|Y=y^{(i)})S^{(l)}(y^{(i)}), \\
n^{(l)} &= \sum_{i=1}^{n} P_\phi(G=l|Y=y^{(i)}).
\end{aligned}
\qquad \square \tag{10.14}
$$

M Step. At this point it is easier to revert to the parametrization (10.9). We then need to solve

$$
\frac{C^{(l)}}{n^{(l)}} = \nabla\psi_l(\widehat{\theta}^{(l)}),\ \widehat{\pi}_l = \frac{n^{(l)}}{n},\ l=1,\ldots,L. \qquad \square \tag{10.15}
$$

The first equation in (10.15) is the likelihood equation for the single group l, at least if there are no relationships between $\theta^{(l)}$'s from different groups. In the next section, the $\theta^{(l)}$'s are equal.

One problem is where to start. It is usually not wise to set all $\theta^{(l)}$'s to 0, since then all groups look the same. In particular cases, one may be able to make judicious guesses. For example, in the distance models as in Example 10.1, one might start with all $\theta^{(l)} = -0.5$ to ensure that clusters grow around the given modal rankings. For the orthogonal contrast Example 10.2, we randomly generated values $P[G = g|Y = y]$, then plugged these into the M Step. It is important to try several different starting values, since there may be many modes to the likelihood.

10.4 Unidimensional unfolding models

The use of Mallows' ϕ model for the errors in unfolding models was pioneered by van Blokland-Vogelesang (1989, 1993). We will take this approach to modeling because it is sensible and fits in well with the rest of this book.

We start with a *joint* scale, that is, a unidimensional space (line) on which the m objects are placed. Each judge is assumed to have an *ideal point*, which is a point on this line that is the judge's favorite. See Section 5.11. A canonical example is sweetness of cookies. Several cookies are arranged on the line according to their sweetness. A judge has an ideal sweetness, that being the judge's ideal point. The judge's preferences for the cookies then depends on how far the cookies' sweetnesses are from the judge's ideal. An ordering arises by taking the cookies by distance, the closest being ranked first. Note that the model makes a number of assumptions. The judges should agree on the placement of the objects. For example, some judges may judge cookies based on the number of raisins rather than sweetness. Even if everyone agrees on which characteristic to arrange the objects, they might not agree on the arrangement. Imagine trying to order several chameleons based on color. The model also assumes that preferences are monotone in the distance. It may be that some people dislike moderate positions. In a war, it is reasonable to have first and second choices be "Bomb them back to the stone age" and "Get out." If an unfolding model does make sense, and fits the data, then it can be an extremely effective means for interpreting the data.

A joint scale for the objects can be given by an $m \times 1$ vector z, where z_i is the location of Object i. We will assume that the z_i's are distinct. We also take all ideal points to be different from any midpoint of two z_i's in order to prevent ties in the ideal rankings. The ranking associated with ideal point z^* is $rank(|z_1 - z^*|, \ldots, |z_m - z^*|)$. See Definition 2.3. For example, if $z = (1, 2, 4, 5)$ and $z^* = 2.7$, then the ranking is $rank(1.7, 0.7, 1.3, 2.3) = (3, 1, 2, 4)$. The name "unfolding" is suggested by folding the scale at the ideal point, in which case the objects are ordered on the resulting ray according to the ranking. For each z, let $\mathcal{A}(z)$ be the set of admissible rankings, i.e., those rankings that correspond to some ideal point.

The number of distinct rankings in $\mathcal{A}(z)$ depends on the spacing of the z_i's. The relative preference a judge has for a pair of objects depends on which side of the midpoint of the two objects the ideal point is located. In the example above, Objects 1 and 2 have midpoint 1.5. Since the ideal point is 1.7, Object 2 is preferred. The midpoints of the objects divide the line into a number of disjoint segments, one segment for each admissible ranking. The number of segments is one more than the number of distinct midpoints. Thus if all midpoints are distinct, we obtain the maximum $\binom{m}{2}+1$ admissible rankings. If the objects are equally spaced, e.g., $z \in \mathcal{S}_m$, then there are only $2m-3$ distinct midpoints: $1.5, 2, 2.5, \ldots, m-0.5$; hence the minimum $2m-2$ admissible rankings. A wider class of rankings is available if we only regard the joint scale as *qualitative*, that is, specifying the order of the objects but not their spacing. If ω is the qualitative ranking, then the set of admissible rankings is the union of all $\mathcal{A}(z)$ for z with $rank(z) = \omega$.

The van Blokland-Vogelesang model allows actual observations to deviate from the admissible set of rankings. Each judge has a modal ranking in the admissible set, but the observation may be slightly off, the error given by Mallows' ϕ model. (Other distances can also be used.) Given the joint scale z and the Mallows' parameter θ, the density is

$$f_\theta(y; z) = \sum_{\omega \in \mathcal{A}(z)} \pi_\omega \exp(\theta d_{\text{Ken}}(\omega, y) - \psi(\theta; m)), \tag{10.16}$$

where π_ω is the proportion in the population whose ideal point leads to the ranking ω. Compare (10.16) to the distance model in Example 10.1. The main difference is that in the example, the two or three modal rankings were not chosen for any internal consistency. Also, here we assume that the same θ obtains for each mode.

Given z, the EM algorithm can be invoked to estimate θ and π. Because the $\theta^{(l)}$'s are all equal to θ, the M step (10.15) is even easier: solve

$$\frac{\sum_{l=1}^{L} C^{(l)}}{n} = \psi'(\widehat{\theta}), \ \widehat{\pi}_l = \frac{n^{(l)}}{n}, \ l = 1, \ldots, L. \tag{10.17}$$

The difficult part is finding z or $\mathcal{A}(z)$. Note that z is very unidentifiable, since there is a finite number of possible $\mathcal{A}(z)$'s. Van Blokland-Vogelesang and van Blokland (1990) have an algorithm to find a good joint scale. The approach is to find z to minimize

$$\sum_{i=1}^{n} Min_{\omega \in \mathcal{A}(z)}\, d_{\text{Ken}}(\omega, y^{(i)}). \tag{10.18}$$

A simplification is to restrict to equally spaced z, so that one can exhaustively search if m is not too large, say $m \leq 7$. In the examples below, we iteratively search starting with random z and sequentially add and subtract fixed amounts to the individual components of z. We vary the size of the fixed amounts interactively.

Example 10.3. Living Places: Unfolding
The results from Example 2.14 strongly suggest an unfolding model along the Urban/Rural continuum. There are only three sets of admissible rankings when $m = 3$, defined by which object is never ranked last. For these three data sets, Suburbs are rarely ranked last, so the objects have positions $z = (1, 2, 3)$ on the joint scale. [In fact, any values with Suburbs in the middle yields the same $\mathcal{A}(z)$.] Table 10.2 contains the fits of the unfolding models to the three data sets.

The χ^2's show that the unfolding model captures essentially everything. The estimated percentages of people associated with each admissible ordering repeat the story we know, but very clearly. See Table 2.7. Kendall's τ's are the estimated average τ over the group, where each judge's τ is that relative to that judge's modal ranking. We see they are all very high, but increase as we go from the City to the Country, suggesting that City People are the most variable, and Country Folk the least. □

Example 10.4. Occupations: Unfolding Models
Van Blokland-Vogelesang (1989) fit the unidimensional unfolding model to the Occupations data, finding the overall prestige ranking as well as the Managerial/Technical split. Our results are somewhat different since we use a less sophisticated search technique, and reversed the three outlying observations (see Example 2.4), while van Blokland-Vogelesang removed them. We should also mention that the occupations are named differently here and in van

Table 10.2 *Living Places: unfolding models.*

	Estimated Percentages $100 \cdot \hat{\pi}_l$		
Modal Ordering	**City People**	**Suburbanites**	**Country Folk**
(*Cty, Sub, Cntry*)	38.60	4.80	3.61
(*Sub, Cty, Cntry*)	13.84	6.25	4.39
(*Sub, Cntry, Cty*)	36.00	64.85	41.58
(*Cntry, Sub, Cty*)	11.56	24.10	50.42

	Fits (χ^2's)		
Source (df)	**City People**	**Suburbanites**	**Country Folk**
Model (4)	406.78	701.52	482.08
Residuals (1)	0.49	4.03	5.32
Total (5)	407.28	705.55	487.41
Kendall's $100 \cdot \tau$	87	93	99

Blokland-Vogelesang, owing to different translations of the original names in Hebrew.

First we restricted to equally spaced values for the occupations. The best such z is $(2, 1, 3, 4, 5, 6, 7, 8, 9, 10)$, the average ranking modified by interchanging the first two occupations. There are 18 admissible rankings. The $LRS = 1202.25$, and the average Kendall's $\tau = 0.68$. Allowing the z to be arbitrary, we obtain $\binom{10}{2} + 1 = 46$ admissible rankings. Now, the $LRS = 1342.00$, with $\tau = 0.75$. Thus the latter is quite a bit better, even taking into account its 28 additional parameters. Figure 10.1 contains the "unfolding histogram" for the latter joint scale, where the bars in the histogram are situated above the ranges of ideal points corresponding to the admissible rankings, and their areas are proportional to the estimated percentage of judges with that ideal point. On the scale, the Technical occupations start with Faculty and go to the right, while the Managerial occupations start with Owner and go to the left. Thus there really are two orderings, one for each group of occupations. Within each group, the ordering is very strong, to the extent that all the judges' ideal rankings agree. The differences are based on how the two orderings fold into one. People on the right tend to rank the Technical occupations higher, and those on

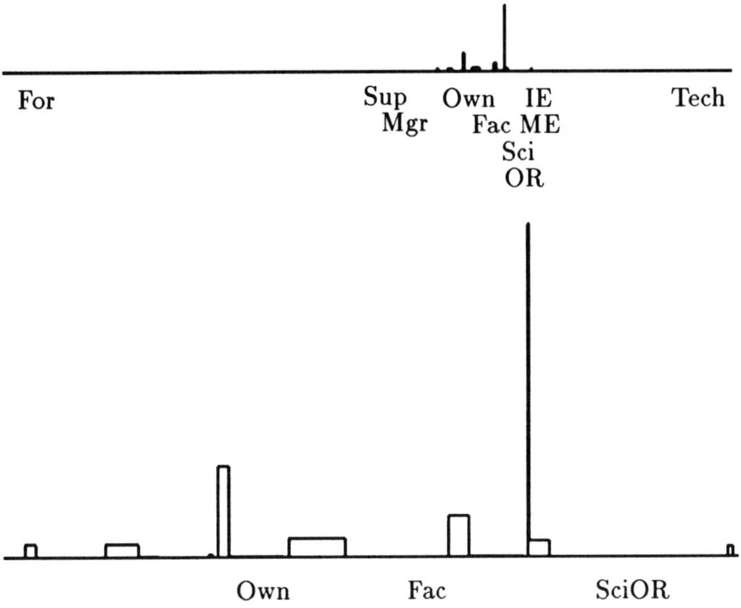

Figure 10.1 ***Occupations: unfolding histograms. In the top panel, the horizontal axis ranges from 0 to 1. The bottom panel zooms in on the dense area, its horizontal axis ranging from 0.65 to 0.75.***

the left Managerial occupations, but there is a good deal of interleaving. The unfolding model here appears to capture the essence of the data very succinctly and clearly. □

Example 10.5. Literary Criticism: Unfolding Models
The main question in the Literary Criticism data (Example 4.3) is whether the students' rankings moved toward the teacher's ranking as the course progressed. Implicit in the question is the idea that the students and the teacher perceive the choices in the same way, that is, generally agree on the placement of the choices in a space. Then the question is to assess where their ideal points are in the space. We will make the strong assumption that there is a common unidimensional joint scale, and that the scale is the same before and after the course. Since the sample size is small, we will limit ourselves to equally spaced scales. The optimal z is then $(1, 3, 2, 4)$, which translates to the ordering (A, P, C, T). Note that the teacher's ordering, (P, C, A, T), has ideal point just to the right of the P. The LRS's are 16.46 and 19.95 for Before and After rankings, respectively, each on 6 df. We fit the models separately,

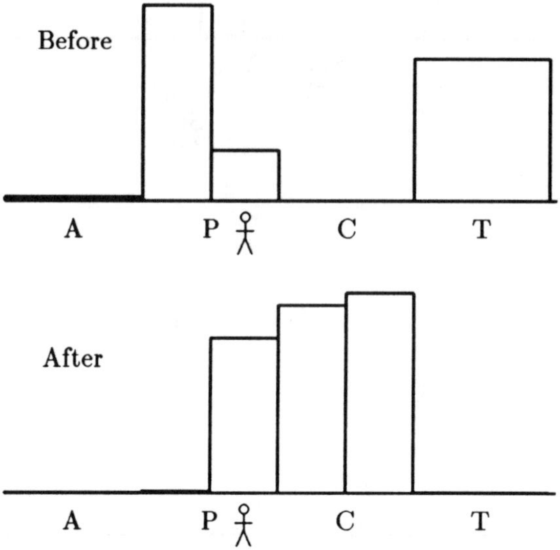

Figure 10.2 ***Literary Criticism: the unfolding histograms, before and after the course.***

so the paired-ness of the observations is not taken into account. Figure 10.2 contains the unfolding histograms for the two samples. It looks as though the mass is indeed moving towards the teacher's ideal point, indicated by the stick figure. Alternatively, as noted in Example 9.5, we can interpret the mass as moving towards choice C. In either case, the movement is clear, although we have not attached a p-value to the observed difference. □

Remark 10.1. Multidimensional Unfolding Models
Multidimensional scaling is an effective descriptive technique for finding structure in the data. It is straightforward to define the analog of the unidimensional Mallows' ϕ models when the objects and the judges' ideal points are in multidimensional space: given a placement of the objects, there is a corresponding set of admissible rankings. These are then the modes for the latent class model as in (10.16). Unfortunately, at this point the computational aspects are a bit daunting. First, a multidimensional search for the best placement is much harder than a unidimensional search. Also, given the placement, there may be too many possible rankings. Coombs (1964, Table 7.1) shows that in two dimensions, the num-

ber of admissible rankings for 5, 6, 7, 8, 9, and 10 objects can be 46, 101, 197, 351, 583, and 916, respectively. For higher dimensions, the numbers are even higher. With $m = 10$ in three dimensions, there are over ten thousand admissible rankings. I am confident that these problems can be overcome, and expect that modeling can be a useful supplement to the descriptive approach. □

Remark 10.2. Thurstonian Models
As the unfolding models, Thurstonian models assume an underlying continuum that contains actual responses, and the observed rankings are discretizations of the responses. For the unfolding models, the ranking for a particular judge arises from the distances between the judge's ideal point and the objects. In Thurstonian models, a judge's response consists of an m-dimensional continuous random variable, whose i^{th} element is attached to Object i, and the observed ordering on the objects is that of the random variables. The basis of the Thurstonian models is that the judges all have the same fundamental reaction to the objects, but that some error is induced by the judges' sensory perceptions. That is, when confronted with sweetness of cookies, the judges attempt to rank the cookies from most to least sweet, differences being due to imperfections in the taste buds. For the unfolding models, the judges rank the cookies on their preferences of sweetness, and it is expected that their preferences will be different. Brady (1989) presents a method for incorporating unfolding models into the Thurstonian framework. Objects have points in the space, judges have ideal points, and the mean of the random vector for any judge depends on the distances between the ideal point and the objects' points. This approach provides an alternative error structure to Mallows' ϕ model for unfolding. See also the "wandering ideal point" models in Böckenholt and Gaul (1986) and DeSoete, Carroll, and DeSarbo (1986).

In Section 5.3, we limited our discussion of Thurstonian models to the location-family models, that is, the random vector $Z = (Z_1, \ldots, Z_m)$ has independent components, and $Z_i \sim g(z_i - \mu_i)$ for some density g on the real line, and vector μ of means. These assumptions are clearly quite strong. It may well be that the variance of the variables depends on the objects, being lower (possibly) for objects with more moderate means, or that the observations are correlated. Correlations could arise because some judges are having a bad day, or different judges have basically different judgments. The latter case can be handled using latent class models, as in Böckenholt (1993). A strength of the Thurstonian model is that it is easy, in principle, to define models of any complexity on the underlying parameters. Thurstone (1927) suggested several possible extensions, based on means, variances, and correlations. Linear models on the means and structural equations for the covariance matrix (e.g., factor analytic) can be immediately defined. In addition, multisample or paired data can be handled by defining usual normal-theory-type models on the underlying means.

We are not going into these interesting and rich possibilities for two reasons. First, the actual modeling of the parameters takes us into fairly advanced multivariate analysis. Second, the computations for the models can be difficult. Going from the random vectors to rankings often requires $(m-1)$-fold multiple integrations. Current computational techniques appear to limit to about six the practical number of objects that can be dealt with. Yao (1994), a student of Böckenholt, is working on an intriguing use of the Gibbs sampler (Tanner, 1993) to handle larger numbers of objects. We note that, for the location-family models, there is a "quick" estimation device based on the paired comparisons suggested by Brady (1989) and Critchlow and Fligner (1993). The idea is that the Thurstonian parameters can be estimated from the paired comparisons, for which the calculations involve only the distribution function of the difference of Z_i's. □

10.5 Exercise

Consider the exponential family (10.3) with parameter and statistic given in (10.10) and (10.11), respectively. Show that

$$\exp(\psi(\phi)) = \frac{1}{L}\sum_{l=1}^{L}\exp(\rho_l)\exp(\psi_l(\theta^{(l)})),$$

and

$$\pi_l = \frac{\exp(\rho_k)\exp(\psi_k(\theta^{(k)}))}{\sum_{l=1}^{L}\exp(\rho_l)\exp(\psi_l(\theta^{(l)}))},$$

where $\rho_l = \log(\pi_l/\pi_L)$, and ψ_l is given in (10.8).

CHAPTER 11

Tied, Partial, and Incomplete Rankings

11.1 Introduction

There are many ways in which a judge could partially rank a set of objects. Some examples:

Example 11.1. Voting
In the Voting data, Examples 2.15 and 8.3, people were supposed to rank the five candidates from 1 to 5. A total of 5738 people ranked all five, but 5141 gave only their top candidate, 2462 gave their first and second choices, and 2108 gave their top three. □

Example 11.2. Desirable Qualities
In the National Opinion Research Center General Social Survey, one question presented 13 desirable qualities for a child to possess. The respondent was to give the top three choices, unordered, the bottom three choices, unordered, the top choice of the top three, and the bottom choice of the bottom three. Thus the objects are separated into five groups: #1, #'s 2 and 3, #'s 4 through 10, #'s 11 and 12, and #13. See Alwin and Jackson (1982). □

Example 11.3. Soft Drinks
Böckenholt (1992) presents a study in which judges rank eight soft drinks. However, instead of ranking all eight together, each judge was presented with two groups of four soft drinks. The judges ranked only the drinks within each group of four drinks so that preferences between drinks that were in two different groups were not obtained. The groups of four were chosen according to an incomplete block design. For example, some people were asked to rank Coke, Pepsi, Diet Coke, and Diet Pepsi, and then to rank 7-Up, Sprite, Diet 7-Up, and Diet Sprite. The purpose of using

blocks of four was to make ranking easier, as well as to reduce the number of possible rankings from 8! to 4!. □

Example 11.4. Awards

In sports, there are often many judges who must give their top choices out of a large number. For example, in the college football and basketball polls, coaches give their top 20 (?) out of the hundreds of teams. In choosing the Most Valuable Player or Cy Young Award winner in baseball, or the Heisman Trophy winner in college football, the judges give their top k choices out of the hundreds or thousands of players eligible. □

Example 11.3 is of a different type than the other examples since there are no ties, but many comparisons are not made. In the others, all objects are compared, but in some cases two or more objects are equally preferred. One can also imagine mixtures of situations, such as if someone preferred A to B and B to D and C to D, but did not compare C to either A or B. Thus C cannot be tied with A or B.

In specific cases, there are often natural representations of incomplete rankings. For any case, we can use the idea of "compatibility" as in Alvo and Cabilio (1991), which is seen as early as Kendall (Kendall and Gibbons, 1990, the latest revision of Kendall's 1948 classic), and later in Critchlow (1985) for distances. The idea is that for any incomplete ranking, there is a well-defined subset of the complete rankings $\mathcal{S}_m$ that are compatible. We go one step further in the formal definition, declaring an incomplete ranking to be a subset of $\mathcal{S}_m$.

Definition 11.1

An *incomplete ranking* v is an element of $\mathcal{W}(\mathcal{S}_m)$, the set of all subsets of $\mathcal{S}_m$. □

An *incomplete ordering* is similarly defined as a subset of the orderings. Given an incomplete ranking, the corresponding incomplete ordering is the set of orderings corresponding to the rankings in the incomplete ranking.

We probably should remove the empty set from consideration, but it does not appear to be a problem in what follows. It is up to the researcher to map actual judges' responses to such incomplete rankings, but in most cases the mapping should be fairly obvious. Two common types of incomplete rankings are tied rankings and rankings of subsets of objects, which we define in the next two subsections.

11.1.1 *Tied rankings*

In Section 7.9.1 we introduced tied rankings, or equivalently, partial orderings. A ranking with ties is represented as an $m \times 1$ vector a of integers, the integers being between 1 and K for some K such that each integer appears at least once. See Definition 7.5. Thus several objects may attain the same rank. The corresponding pattern of ties is given in (7.62), where $\lambda = (\lambda_1, \ldots, \lambda_K)$ with $\lambda_k = \#\{i|a_i = k\}$. For example, $a \equiv (1,1,3,2,3,3,4)$ means objects A and B are tied at the top, object D is alone at #3, objects C, E, and F are tied at the next spot, and object G is uniquely in last place, and $\lambda = (2,1,3,1)$. The corresponding partial ordering groups the objects in that same way: $(\{A,B\},\{D\},\{C,E,F\},\{G\})$. The incomplete ranking v consists of all rankings compatible with a, that is, all rankings that have A and B at the top, in some order, D at #3, C, E, and F next, in some order, and G at #7. There are $2! \times 3! = 12$ such rankings, i.e.,

$$v(a) = \{1243567, 2143567, 1243657, 2143657, 1253467, 2153467, \\ 1253647, 2153647, 1263457, 2163457, 1263547, 2163547\}. \tag{11.1}$$

One way to define the mapping formally is $v(a) = \{y \in \mathcal{S}_m | \, d_{\text{Ken}}(a,y) = 0\}$, where d_{Ken} is defined as in (2.29) for any $m \times 1$ vectors a and b. □

11.1.2 *Rankings of subsets: Balanced incomplete block designs*

When assessing an object is time-consuming or takes some effort, it may be unreasonable to ask every judge to rank every object. For example, in taste testing, after tasting a number of objects, a judge's taste buds may become numb, or stomach full. The quality of judgments will deteriorate. Quicker and more reliable rankings can be obtained by presenting each judge with a subset of objects. If different judges are assigned different objects systematically, the results can be as good as for full rankings. Durbin (1951), Alvo and Cabilio (1991), and Böckenholt (1993) use *balanced incomplete block designs* as in Cochran and Cox (1957). That is, each judge is presented $h(< m)$ of the objects in such a way that each object is presented to the same number of judges, and each pair of objects is presented to the same number of judges. (In design parlance, the judges are the blocks.) For example, if $m = 6$, we can obtain a balanced incomplete block design using $h = 3$ with only $n = 10$ judges by presenting each judge with one of the following subsets: $\{A, B, E\}$, $\{A, B, F\}$, $\{A, C, D\}$, $\{A, C, F\}$, $\{A, D, E\}$, $\{B, C, D\}$, $\{B, C, E\}$, $\{B, D, F\}$, $\{C, E, F\}$, $\{D, E, F\}$. See Cochran and Cox (1957), page 471, Plan 11.4. Note that each object is assigned to four judges, and each pair of objects is assigned to two judges. For given m and h, the numbers n for which there is such a design is rather limited, although if a design exists for n, then it also exists for any integral multiple of n by assigning the same number of judges to each subset of objects.

A judge's ranking when presented with h objects is an element r of $\mathcal{S}_h$, so that the corresponding compatibility set consists of all elements of $\mathcal{S}_m$ that rank the objects assigned to the judge in the same order as r. That is, suppose the judge is assigned objects $l_1 < \cdots < l_h$. Then $v = \{y \in \mathcal{S}_m \mid rank(y_{l_1}, \ldots, y_{l_h}) = r\}$ from Definition 2.3. The set v contains $m!/h!$ rankings. □

One can imagine a number of variants of the above, e.g., having tied rankings of subsets of objects, or asking a judge for rankings within two or more different subsets of objects. See Example 11.3. Paired comparisons (David, 1988) and tripled comparisons (Pendergrass and Bradley, 1960) are special cases, where each judge ranks objects within pairs or triples. Another type of incomplete ranking occurs when creating rotations of objects, as with pitchers in baseball or, in post-Tito pre-collapse-of-communism Yugoslavia, the rotation of the presidency among the republics. A judge's rotation has no beginning, so that with five objects, a possible rotation is the incomplete ordering $\{CABED$, $ABEDC$, $BEDCA$, $EDCAB$, $DCABE\}$.

Having incomplete rankings complicates the computational and mathematical analysis of the data, but more importantly makes interpretation more difficult. As long as all incomplete rankings are of the same type, interpretation is fairly straightforward. The problem comes when trying to compare or combine different types.

In the next section, we suggest a simple procedure for extending the definition of the basic descriptive statistics to incomplete rankings. In Section 11.3 we use the statistics to develop tests of uniformity. Sections 11.4 and 11.5 give details of the statistics and tests for the tied ranking and balanced incomplete block design situations, respectively. Section 11.6 briefly discusses comparing populations. In Section 11.7, we present extensions of some models to incomplete rankings.

11.2 Descriptive statistics

Describing arbitrary samples of incomplete rankings can be problematic. We will start with the general situation, creating a full-ranking statistic as in Chapter 2 out of incomplete rankings by averaging. This approach is taken by Kendall for average rankings, and Alvo and Cabilio for distances. Given any statistic z on $\mathcal{S}_m$, for incomplete ranking v define

$$z(v) = \frac{\sum_{y \in v} z(y)}{\#v}. \tag{11.2}$$

That is, it is the average of the statistic over the rankings in the incomplete ranking. If $z(y) = y$ and v is a tied ranking, then $z(v) \equiv y(v)$ is called the *midrank* vector. See Kendall and Gibbons (1990). The elements of the tied ranking are replaced by the average of the rankings they would have received had there been no ties. In the example (11.1), $y(v) = (1.5, 1.5, 5, 3, 5, 5, 7)$.

The thinking behind (11.2) is that the judge has an actual full ranking in mind, but due to the strictures of the experiment, time constraints, laziness, or such, only an incomplete ranking is reported. The true $z(y)$ is then estimated by (11.2). This procedure is conservative in that it implicitly assumes uniformity of the rankings given the incomplete ranking. If a parametric model is assumed, then nonuniform averages arise naturally. See Section 11.7.

Using the notation in Section 2.6, where $z(y) = X'\zeta^y$ (2.48), we have that $z(v) = X'\zeta(v)$, where

$$\zeta_y(v) = 1/\#v \text{ if } y \in v, 0 \text{ otherwise}. \tag{11.3}$$

Given a sample $v^{(1)}, \ldots, v^{(n)}$ of incomplete rankings, the average of the $z(v^{(i)})$'s can be used as the analog of the $\overline{z}$'s in Chapter 2. Thus the Probabilities $\widehat{P}$, Marginals $\widehat{M}$, Means $\overline{y}$, Pairs $\widehat{K}$, and distances d can be handled no matter what the form of the incomplete rankings. Not that they will always be helpful. For example, in the rotation illustration at the end of the previous section, the Means and Marginals are constants. See Exercise 11.1.

The distance d can be extended to incomplete rankings by setting

$$d(u,v) = \frac{1}{\#u\#v} \sum_{x\in u} \sum_{y\in v} d(x,y) \tag{11.4}$$

for $u, v \in \mathcal{W}(\mathcal{S}_m)$. If either v or u is a complete ranking, then the corresponding sum is over only one ranking. As Alvo and Cabilio (1991) point out, $d(u,v)$ is not a distance on $\mathcal{W}(\mathcal{S}_m)$ since $d(v,v)$ will not be 0 if v has more than two points. Critchlow (1985) uses Hausdorf distances to define a true distance, where

$$d_{\text{Haus}}(u,v) = \max\{\max_{x\in u} d^*(x,v), \max_{y\in v} d^*(u,y)\}, \tag{11.5}$$

and $d^*(x,v) = d^*(v,x) = \inf_{y\in v} d(x,y)$. We will use the Alvo/Cabilio approach since the formulas are simpler. The Hausdorf approach may reap more rewards in the future. An annoying problem with distances arises when trying to construct a coefficient as in Section 2.5: Should one use $\delta_0 = E_0(d)$ in (2.21) as if the rankings were complete, or for the incomplete rankings? We consider this question in Remark 11.1.

Some general points can be noted. For any incomplete rankings u and v, we can write

$$d(u,v) = \zeta(v)' D \zeta(u), \tag{11.6}$$

where D is the matrix of distances in (3.75). Hoeffding distances can be obtained from Q by

$$d_{\text{Hoeff}}(u,v) = \text{tr}(Q(u)\Delta Q'(v)), \; Q(v) = \frac{1}{\#v}\sum_{y\in v} Q^{(y)}; \tag{11.7}$$

hence for Spearman,

$$d_{\text{Spear}}(u,v) = (constant) - 2y'(u)y(v). \tag{11.8}$$

For Kendall's distance, by (3.73), we have that

$$d_{\text{Ken}}(u,v) = \frac{1}{2} - 2(\kappa(u) - \frac{1}{2}1_{\overline{m}})'(\kappa(v) - \frac{1}{2}1_{\overline{m}}). \tag{11.9}$$

If the number of rankings in v is large and the structure of v is ungainly, it may be challenging to find the average z in (11.2). For special types of incomplete rankings and statistics, relatively simple formulas apply.

In the next section, we consider using the statistics $z(v)$ to develop tests of uniformity as in Chapter 3. Sections 11.4 and 11.5 then apply the methods to tied rankings and balanced incomplete block designs, situations for which some of the formulas are nice.

11.3 Tests of uniformity

The previous section showed how to extend the basic descriptive statistics to incomplete rankings. In this section, we use the statistics to test the null hypothesis that $Y \sim Uniform(\mathcal{S}_m)$. With arbitrary incomplete rankings, it is not obvious how to proceed, since Y being uniform does not imply that V is uniform. In well-designed situations, the answer is usually clear. For example, if the experiment produces tied rankings with the same pattern, then it is plausible that the null hypothesis means that A is uniformly distributed over all rankings with that pattern. If each judge is presented a subset of h objects, and each judge produces a complete ranking of the subset, then again uniformity of Y implies uniformity of the observed rankings. However, when different judges show different patterns of ties, or rank different numbers of objects, then uniformity vanishes. For example, in the Voting Example 11.1, some people gave complete rankings, and others ranked only the first one, two, or three. Within each group, uniformity may be plausible, but it makes no sense to test that the probability of any full ranking is the same as any ranking of just one candidate.

To deal with arbitrary situations, we break the sample into groups depending on their type of incomplete rankings. We first formally define what we mean by type.

Definition 11.2
A *type* of incomplete ranking is a partition $\mathcal{P}(\subset \mathcal{W}(\mathcal{S}_m))$ of $\mathcal{S}_m$ such that each element contains the same number of rankings:

$$\cup_{v\in\mathcal{P}} v = \mathcal{S}_m, v \cap u = \emptyset \text{ if } u \neq v, \text{ and } \#u = \#v, \text{ for } u, v \in \mathcal{P}. \quad \square \tag{11.10}$$

Now assume we have B types of incomplete rankings, $\mathcal{P}^{(1)}, \ldots, \mathcal{P}^{(B)}$, such that each incomplete ranking v that may arise from the experiment appears in one and only one of the $\mathcal{P}^{(b)}$'s.

Example 11.5. Desirable Qualities
Consider Example 11.2. Everyone was asked for a tied ranking with the same pattern of ties: $1, 2, 7, 2, 1$. The sample space consists of all possible permutations of the tied ranking $(1,2,2,3,3,3,3,3,3,3,4,4,5)$, of which there are $\binom{13}{1,2,7,2,1}$ different ones. Then $B = 1$, and $\mathcal{P}^{(1)}$ consists of the v's corresponding to the tied rankings. Each v has $1!2!7!2!1!$ elements, and the union of the v's is $\mathcal{S}_{13}$. □

Example 11.6. Voting
In Example 11.1 there are $B = 4$ sets corresponding to the number of candidates ranked, i.e., 5, 3, 2, or 1. $\mathcal{P}^{(4)} = \mathcal{S}_m$ contains the complete rankings, $\mathcal{P}^{(2)}$ and $\mathcal{P}^{(3)}$ contain the tied rankings for patterns $(1,1,3)$ and $(1,1,1,2)$, respectively, and $\mathcal{P}^{(1)}$ contains the results of ranking just the favorite. Note that $\mathcal{P}^{(1)}$ has five elements, the ith containing the rankings with candidate i ranked first. □

Example 11.7. Balanced Incomplete Block Designs
In Section 11.1.2 we describe the balanced incomplete block design. Each judge is presented with the same number h of objects to rank, where the experimenter has only B possible subsets. For each subset, there is a $\mathcal{P}^{(b)}$, equivalent to $\mathcal{S}_h$, containing the possible incomplete rankings for that subset. □

The two sampling schemes we will deal with are exemplified by Example 11.6, where the judges themselves decide on the pattern of ties, and Examples 11.5 and 11.7, where the experiment sets the type of incomplete ranking to be obtained from each judge. In the former case, Case I, we assume $V^{(1)}, \ldots, V^{(n)}$ are independent and identically distributed, where

$$P[V = v] = \sum_{b=1}^{B} p^{(b)} P[V = v | V \in \mathcal{P}^{(b)}], \tag{11.11}$$

$p^{(b)}$ being the probability a random judge chooses a partial ranking from $\mathcal{P}^{(b)}$. In the latter case, Case II, we assume that $n^{(b)}$ observations are chosen from each $\mathcal{P}^{(b)}$, where the $n^{(b)}$'s are fixed by the experimenter and sum to n.

The null hypothesis of uniformity will now be conditional uniformity given b, that is,

$$H_0: \; P[V = v \mid V \in \mathcal{P}^{(b)}] = \frac{1}{\#\mathcal{P}^{(b)}}, \; b = 1, \ldots, B. \tag{11.12}$$

The first objective is to find the mean and covariance matrix under the uniform for given statistic z. Let $\mu^{(b)} = E^{(b)}(z(V))$ and $\Sigma^{(b)} = Cov^{(b)}(z(V))$ under $V \sim Uniform(\mathcal{P}^{(b)})$. The next theorem gives the unconditional values.

Theorem 11.1
Under the null hypothesis of uniformity (11.12), $\mu^{(b)} = E(z(Y))$ for $Y \sim Uniform(\mathcal{S}_m)$ for every b. Under Case I,

$$\Sigma(z) \equiv Cov(z(V)) = \sum_{b=1}^{B} p^{(b)} \Sigma^{(b)}. \quad \square \tag{11.13}$$

Proof
Start with $\mu^{(b)} = \sum_{v \in \mathcal{P}^{(b)}} z(v)/\#\mathcal{P}^{(b)}$, and insert $z(v)$ from (11.2) to obtain $\sum_{v \in \mathcal{P}^{(b)}} \sum_{y \in v} z(y)/(\#v \times \#\mathcal{P}^{(b)})$. The numerator is the sum of $z(y)$ over all $y \in \mathcal{S}_m$ by Definition 11.2, and the denominator is $m!$; hence the result is $E(z(Y))$. For (11.13), use the identity

$$Cov(z(V)) = E(\Sigma^{(J)}) + Cov(\mu^{(J)}), \tag{11.14}$$

where the outer E and Cov indicate J is random with $P[J = b] = p^{(b)}$. Then (11.13) follows since $Cov(\mu^{(J)}) = 0$. $\square$

To find test statistics, we use the quadratic forms as in Chapter 3. One difference is that in Case I, the $n^{(b)}$'s are random, so that $Cov(z(V))$ is unknown but has an unbiased and consistent estimator

$$\widehat{\Sigma}(z) = \sum_{b=1}^{B} \frac{n^{(b)}}{n} \Sigma^{(b)}. \tag{11.15}$$

Then the test statistic is

$$T^2 = n(\overline{z} - \mu(z))' A(\overline{z} - \mu(z)) \tag{11.16}$$

for some symmetric A, possibly estimated, where $\overline{z}$ is the average of the $z(v^{(i)})$'s in the sample and $\mu(z)$ is the common value of the $\mu^{(b)}$'s. As in Chapter 3, under the null hypothesis

$$T^2 \longrightarrow \chi^2[A^*\Sigma], \tag{11.17}$$

where $A^* = A$ in Case II, and A^* is the consistent limit of A in Case I. In Case II, the asymptotics assume $n^{(b)}/n \longrightarrow p^{(b)} > 0$ for each b.

The concordance for a sample $V^{(1)}, \ldots, V^{(n)}$ of incomplete rankings is

$$\widehat{\alpha} = \frac{1}{n^2} \sum \sum d(v^{(i)}, v^{(j)}). \tag{11.18}$$

Letting $\widehat{P} = \frac{1}{n} \sum_i \zeta(v^{(i)})$ be the estimate of the probability vector, Exercise 11.2 shows that (3.75) again holds. Also, the formula (3.69) for Hoeffding distances and (3.74) for Kendall's distance hold, where $\widehat{M}$ and $\widehat{K}$ are now the averages over the $Q(v^{(i)})$'s and $\kappa(v^{(i)})$'s, respectively. The proofs of Theorems 3.5 and 3.6 concerning one-group concordance can be extended to incomplete rankings. Thus we have that under uniformity, for $\widehat{\alpha}$ in (11.18),

$$n(\widehat{\alpha} - \delta_0) \rightarrow \chi^2[\Sigma(\zeta) D], \tag{11.19}$$

where $\delta_0 = E_0(d(Y, e_m))$. For Hoeffding distances,

$$n(\widehat{\alpha}_{\text{Hoeff}} - \delta_0) \rightarrow \chi^2[\Sigma(Q)(I_m \otimes \Delta)], \tag{11.20}$$

and for Kendall's distance,

$$n(\frac{\overline{m}}{4} - \frac{1}{2}\widehat{\alpha}_{\text{Ken}}) \longrightarrow \chi^2[\Sigma(\kappa)]. \qquad (11.21)$$

The prescriptions in this section and the previous can be applied to any type of incomplete rankings. In general, the formulas will be much more complicated than for complete rankings. The next two sections apply the results to the special cases in Section 11.1 in which nice formulas do exist: all rankings are tied rankings with the same pattern, or there are rankings of subsets arranged in a balanced incomplete block design. Alvo and Cabilio (1994, 1995b) look at more general arrangements in the latter case.

11.4 Tied rankings

Assume we have tied rankings as in Section 11.1.1, that is, we have the set $\mathcal{S}_m(\lambda)$ consisting of all tied rankings with a given (fixed) pattern λ. See Definition 7.5. Thus the tied rankings a under consideration are all permutations of each other. For $a \in \mathcal{S}_m(\lambda)$, let $v \in \mathcal{P}(\lambda)$ be its associated incomplete ranking. Let $R_k = \{i \mid a_i = k\}$ be the set of objects that receive the tied ranking k, $k = 1, \ldots, K$. Given any $y \in v$, the rankings in v consist of those that can be obtained from y by permuting the rankings within each set R_k. Thus if $a = 1132334$, $R_1 = \{1, 2\}$, $R_2 = \{4\}$, $R_3 = \{3, 5, 6\}$, and $R_4 = \{7\}$. Take $y = 1243567 \in v$, given in (11.1). Then the other elements of v are found by interchanging the ranks assigned to Objects 1 and 2 (R_1), or permuting the ranks assigned to Objects 3, 5, and 6 (R_3), or performing both a permutation and an interchange. We can express this idea mathematically by defining $\mathcal{Q}_\lambda$ to be the subset of permutation matrices of the form

$$Q_\lambda = diag\{Q_1, \ldots, Q_K\} \qquad (11.22)$$

where $Q_k \in \mathcal{Q}_{\lambda_k}$. [Here, $diag(A_1, \ldots, A_K)$ denotes the block diagonal matrix with diagonal blocks being the matrices A_k.] Taking any fixed $y \in v$, let $Q_v = Q^{(y)}$. Then

$$\{Q^{(y)} | y \in v\} = \{Q_v Q_\lambda | Q_\lambda \in \mathcal{Q}_\lambda\}. \qquad (11.23)$$

In fact, the two sets are in one-to-one correspondence. [We are actually using the notions of subgroups and quotient groups for tied rankings as presented in Diaconis (1989) and Critchlow (1985).] We next present details for some of the basic statistics.

11.4.1 Marginals and Means

Consider $Q(v)$, the average Q in v. By (11.23), taking the average over v is the same as taking the average over $\mathcal{Q}_\lambda$; hence

$$Q(v) = \frac{1}{\#v}\sum_{y\in v} Q^{(y)} = Q_v E(Q_\lambda) = Q_v \overline{J}^\lambda, \qquad (11.24)$$

where the expectation is over $Uniform(\mathcal{Q}_\lambda)$, and

$$\overline{J}^\lambda = diag\{\overline{J}_{\lambda_1}, \ldots, \overline{J}_{\lambda_K}\}. \qquad (11.25)$$

See (3.3). From (11.24) we can calculate the midrank vector by

$$y(v) = Q_v \overline{J}^\lambda e_m. \qquad (11.26)$$

Note that the ith element of $\overline{J}^\lambda e_m$ for $i \in R_k$ is the average of the elements in R_k.

To obtain test statistics, we assume that we have a sample $V^{(1)}, \ldots, V^{(n)}$ from $\mathcal{P}(\lambda)$. We will need $Cov(Q(V))$ for $V \sim Uniform(\mathcal{P}(\lambda))$. For any $y \in V$, as in (11.24), $Q(V) = Q^{(y)}\overline{J}^\lambda$. Since there is the same number of rankings y in each $v \in \mathcal{P}(\lambda)$, the distribution of $Q(V)$ under uniformity on V is the same as the distribution of $Q^{(Y)}\overline{J}^\lambda$ under uniformity on Y. Thus by (3.16) and Lemma 12.3,

$$\Sigma(Q) = \frac{1}{m-1} H_m \otimes \overline{J}^\lambda H_m \overline{J}^\lambda = \frac{1}{m-1} H_m \otimes H^\lambda, \qquad (11.27)$$

where $H^\lambda = \overline{J}^\lambda - \overline{J}_m$. It is not hard to show that H^λ is idempotent, and that $\text{tr}(H^\lambda) = K - 1$. See Exercises 11.3 and 11.4. Thus we take the same T^2 for the Marginals as in Table 3.1, where now

$\widehat{M}$ is the sample average of the $Q(v^{(i)})$'s. Under uniformity, T^2 is asymptotically $\chi^2_{(m-1)(K-1)}$.

For the Means test, note that $Cov(y(V)) = Cov(Q(V)e_m)$, which by (11.27) is $(e'_m H^\lambda e_m)H_m/(m-1)$. Exercise 11.5 shows that

$$e'_m H^\lambda e_m = \|a - c_m\|^2 = \frac{m(m^2-1) - \sum_{k=1}^K \lambda_k(\lambda_k^2 - 1)}{12}, \quad (11.28)$$

where a is any vector of midranks arising from the pattern of ties λ. See Kendall and Gibbons (1990). Thus, similar to the result in Table 3.1,

$$\frac{n(m-1)}{e'_m H^\lambda e_m}\|\overline{y} - c_m\|^2 \longrightarrow \chi^2_{m-1} \quad (11.29)$$

under uniformity, where $\overline{y}$ is the average of the $y(v^{(i)})$'s.

11.4.2 Pairs

The pairs vector κ of (2.59) is easy to handle. For Objects i and j, $\kappa_{ij}(v) = I[a_i < a_j]$ if $a_i \neq a_j$, since if one object is preferred to another in a, it is in all elements of v. On the other hand, if $a_i = a_j$, then the elements are tied; hence v contains all possible ways to break the tie. Half will prefer i, the other half j. Summarizing,

$$\kappa_{ij}(v) = \begin{cases} 0 & \text{if } a_i < a_j \\ \frac{1}{2} & \text{if } a_i = a_j \\ 1 & \text{if } a_i > a_j. \end{cases} \quad (11.30)$$

It is somewhat of a challenge to find $Cov(\kappa(V))$ for $V \sim Uniform$ $(\mathcal{P}(\lambda))$. We use the conditional covariance formula, expressing the unconditional covariance of κ under the full data by conditioning on the incomplete ranking:

$$\begin{aligned} Cov(\kappa(Y)) &= Cov(E(\kappa(Y)|Y \in V)) + E(Cov(\kappa(Y)|Y \in V)) \\ &= Cov(\kappa(V)) + E(Cov(\kappa(Y)|Y \in V)), \end{aligned} \quad (11.31)$$

where the outside operators on the right are over $V \sim Uniform$ $(\mathcal{P}(\lambda))$. We know that $Cov(\kappa(Y)) = (1/12)(I_{\overline{m}} + \Gamma_m\Gamma'_m)$ from (3.23). For $Cov(\kappa(Y)|Y \in v)$, note that the only randomness is within the sets of tied objects, i.e., within sets R_k. That is,

$Cov(\kappa_{ij}(Y), \kappa_{kl}(Y)|Y \in v) = 0$ unless $a_i = a_j = a_k = a_l$. Within such sets, the covariances are the same as under the full rankings, so that for $i < j$ and $k < l$, $a_i = a_j = a_k = a_l$,

$$Cov(\kappa_{ij}(Y), \kappa_{kl}(Y)|Y \in v) = \frac{1}{12}\begin{cases} 3 & \text{if } i = j \text{ and } k = l \\ 1 & \text{if } i = j \text{ and } k \neq l \\ 1 & \text{if } i \neq j \text{ and } k = l \\ -1 & \text{if } j = k \\ 0 & \text{if } i, j, k, l \text{ are distinct.} \end{cases} \tag{11.32}$$

Take the expected value of (11.32) over V. Entries are 0 if i, j, k, l are distinct. The other cases give

$$E(Var(\kappa_{ij}(Y)|Y \in V)) = \frac{1}{4}P[A_i = A_j] = \frac{1}{4}\gamma_2, \tag{11.33}$$

and

$$\begin{aligned} E(Cov(\kappa_{ij}(Y), \kappa_{ik}(Y)|Y \in V)) \\ &= E(Cov(\kappa_{ij}(Y), \kappa_{kj}(Y)|Y \in V)) \\ &= -E(Cov(\kappa_{ij}(Y), \kappa_{jk}(Y)|Y \in V)) \\ &= \frac{1}{12}P[A_i = A_j = A_k] = \frac{1}{12}\gamma_3, \end{aligned} \tag{11.34}$$

where

$$\gamma_2 = \sum_{k=1}^{K} \frac{\lambda_k(\lambda_k - 1)}{m(m-1)} \text{ and } \gamma_3 = \sum_{k=1}^{K} \frac{\lambda_k(\lambda_k - 1)(\lambda_k - 2)}{m(m-1)(m-2)} \tag{11.35}$$

are the probabilities that two and three objects are tied, respectively. Putting (11.31) through (11.35) together, we have that

$$\Sigma(\kappa) = Cov(\kappa(V)) = \alpha I_{\overline{m}} + \beta \Gamma_m \Gamma_m', \tag{11.36}$$

where

$$\alpha = \frac{1 - 3\gamma_2 + 2\gamma_3}{12} \text{ and } \beta = \frac{1 - \gamma_3}{12}. \tag{11.37}$$

See Exercise 11.6. Except in the trivial case when all objects are tied, $\alpha > 0$, so that $\Sigma(\kappa)$ has the inverse given in Lemma 3.4.

The test statistic for the Pairs is

$$T^2 = n(a\|\widehat{K}^* - \frac{1}{2}1_{\overline{m}}\|^2 - b\|\overline{y} - c_m\|^2), \tag{11.38}$$

where $\widehat{K}$ represents the average of the $\kappa(y^{(i)})$'s, $\overline{y}$ is as above, and the a and b are obtained from Lemma 3.4 using α and β from (11.37).

11.4.3 Distances

Turn to the distances. We start with a fixed incomplete ranking u, and consider the average distance from u based on the sample from $\mathcal{P}(\lambda)$:

$$\overline{d} = \frac{1}{n}\sum_{i=1}^{n} d(v^{(i)}, u). \tag{11.39}$$

The ranking u may or may not have ties, and need not have the same pattern of ties as the $v^{(i)}$'s. The Hoeffding distances yield $\overline{d}$ = $\mathrm{tr}(\widehat{M}\Delta Q'(u))$. From (11.27) and Lemma 12.5,

$$Var(d(V,u)) = \frac{1}{m-1}\mathrm{tr}(H_m Q(u)\Delta H^\lambda \Delta Q'(u)). \tag{11.40}$$

Exercise 11.7 shows that for Hamming and Spearman, respectively, the variance is

$$\begin{gathered} \frac{1}{m-1}\mathrm{tr}(Q'(u)H_m Q(u)H^\lambda) \text{ and} \\ \frac{4}{m-1}(e'_m Q'(u)H_m Q(u)e_m)(e'_m H^\lambda e_m). \end{gathered} \tag{11.41}$$

When u is a full ranking, then $Q(u)$ is a single permutation matrix; hence $Q'(u)\ H_m\ Q(u) = H_m$. We then have that the variances in (11.41) further simplify to $\mathrm{tr}(H^\lambda)/(m-1) = (K-1)/(m-1)$ and $4(e'_m H_m e_m)\ (e'_m H^\lambda e_m)/(m-1)$, respectively. The final term can be made more explicit using Exercise 3.7 and (11.28).

Kendall's distance has a nice formula. From (11.9) and (11.30),

$$d_{\text{Ken}}(u,v) = \frac{\overline{m}}{2} - 2 \sum\sum_{\substack{i<j \\ a_i \neq a_j, b_i \neq b_j}} (I[a_i < a_j] - \frac{1}{2})(I[b_i < b_j] - \frac{1}{2}). \tag{11.42}$$

Thus we need to count only the pairs for which both incomplete rankings provide a distinction. See Exercise 11.8. There are other extensions of Kendall's distance for tied rankings. One is the straightforward one from (2.29) that counts 1 if the pair (i,j) is discordant, and 0 if it is either concordant or there is a tie in one or both of the vectors. Recall the permutation polytope-o-grams of Section 2.1. Thompson (1993a) extends them to tied rankings for samples in which all rankings have the same pattern of ties. Each tied ranking a is a point in m-dimensional space. The polytope-o-gram places a ball at each tied ranking, where the radius is a function of the frequency of observations at that tied ranking. Two points a and b are connected if $a_i = b_i$ for all but two i, say j and k, and $a_j = b_k = a_k - 1 = b_j - 1$. Then the distance between two tied rankings is the minimum number of steps it takes to go from one to the other, traveling between connected points.

To use Kendall's distance for a test statistic, note that (11.9) and (11.36) show that

$$\begin{aligned} Var(d_{\text{Ken}}(V,u)) &= 4(\kappa(u) - (1/2)1_{\overline{m}})'\Sigma(\kappa)(\kappa(u) - (1/2)1_{\overline{m}}) \\ &= 4(\alpha\|\kappa(u) - (1/2)1_{\overline{m}}\|^2 + \beta\|y(u) - c_m\|^2) \end{aligned} \tag{11.43}$$

for α and β in (11.37).

Finally, we treat one-group concordance. For Hoeffding distances, since $\Sigma(Q) = \frac{1}{m-1} H \otimes H^{\lambda}$, the limiting distribution in (11.20) is

$$\frac{1}{m-1}\chi^2[H \otimes H^{\lambda}\Delta] = \frac{1}{m-1}\sum_{i=1}^{m-1} C_i, \tag{11.44}$$

where the C_i's are independent $\chi^2[H^{\lambda}\Delta]$. Results for Hamming and Spearman follow quickly, being equivalent to the Marginals and Means tests, respectively.

In Kendall's case, Lemma 3.8 shows that the eigenvalues of $\alpha I_{\overline{m}} + \beta\Gamma_m\Gamma'_m$ consist of $m-1$ $(\alpha + \beta m)$'s, and $\binom{m-1}{2}$ α's (see

11.37); hence

$$\chi^2[\Sigma(\kappa)] = (\alpha + \beta m)\chi^2_{m-1} + \alpha\chi^2_{\binom{m-1}{2}}, \tag{11.45}$$

where the two χ^2's on the right-hand side are independent.

11.5 Rankings of subsets: Balanced incomplete block designs

Recall Section 11.1.2 for the rankings of subsets. We have B subsets of $\mathcal{O}$, say $\mathcal{O}^{(1)}, \ldots, \mathcal{O}^{(B)}$, each containing the same number of objects, say h, such that each object appears in the same number of subsets (i.e., Bh/m), and each pair of objects appears in the same number of subsets [i.e., $B\binom{h}{2}/\binom{m}{2}$.] The data consist of n/B complete rankings of objects in subset $\mathcal{O}^{(b)}$, $b = 1, \ldots, B$. (Thus n must be an integral multiple of B.) For the Pairs and Kendall's distance test statistics, we additionally assume that each triple of objects occurs the same number of times, as well as each quadruple. The development here follows Alvo and Cabilio (1991). The result for the Means test, i.e., concordance for Spearman's distance, is their Theorem 4.1. That for concordance using Kendall's distance is their Theorem 4.2. [In Alvo and Cabilio (1995a), they note that the proof in Alvo and Cabilio (1991) omitted the necessary requirement on the triples. Our approach is slightly different, and appears to need the additional requirement on the quadruples.]

11.5.1 Marginals and Means; Hamming's and Spearman's distances

Lemma 4.1 of Alvo and Cabilio (1991) finds $Q(v)$ in general. We will start with the subset $\mathcal{O}^{(o)} = \{1, \ldots, h\}$, that is, the judge is to rank Objects $1, \ldots, h$. Take the ranking of these objects to be $r^{(0)} = (1, 2, \ldots, h)$. Then the corresponding incomplete ranking, denoted $v^{(0)}$, consists of all $y \in \mathcal{S}_m$ such that $y_1 < \cdots < y_h$. For each i, j, we need to find

$$Q_{ij}(v^{(0)}) = P[Y_i = j \mid Y_1 < \cdots < Y_h], \tag{11.46}$$

where the probability is under $Y \sim Uniform(\mathcal{S}_m)$. First note that if $i > h$, i.e., Object i is not ranked, then Y_i is equally likely to be anything, i.e., $P[Y_i = j] = 1/m$. Take $i \leq h$. If $j < i$ or $j > m - h + i$, the answer is 0 by the constraints on the first h values of Y. Consider the case $i \leq j \leq m - h + i$, $i \leq h$. The number of rank vectors for which $Y_1 < \cdots < Y_h$ and $Y_i = j$ is $\binom{j-1}{i-1}\binom{m-j}{h-i}$, the number of way to choose which values less than j to assign to $Y_1, \ldots, Y_{i-1}$, times the number of ways to choose which values greater than j to assign to $Y_{i+1}, \ldots, Y_h$. Divide by the total number of possible ways to have the first h in order, $\binom{m}{h}$. Then the complete specification of (11.46) is

$$Q_{ij}(v^{(0)}) = \begin{cases} \frac{\binom{j-1}{i-1}\binom{m-j}{h-i}}{\binom{m}{h}} & \text{if } i \leq h \text{ and } i \leq j \leq m-h+i \\ 0 & \text{if } i \leq h \text{ and } (j < i \text{ or } j > m-h+i) \\ \frac{1}{m} & \text{if } i > h. \end{cases} \tag{11.47}$$

Now consider an arbitrary ranking r of the first h objects, so that $r \in \mathcal{S}_h$. With v being the corresponding incomplete ranking, the last $m - h$ rows of $Q(v)$ are as for $Q(v^{(0)})$, and the first h rows are permuted according to $Q^{(r)} \in \mathcal{Q}_h$. Finally, if the set of observed objects is an arbitrary set of h objects, $\mathcal{O}' \subset \mathcal{O}$, then we must further permute the rows so that they correspond to the objects in $\mathcal{O}'$. That is, if we observe $r = rank(Y_{l_1}, \ldots, Y_{l_h}) \in \mathcal{S}_h$, where $\mathcal{O}'$ consists of $l_1 < \cdots < l_h$, then

$$Q(v) = Q(\mathcal{O}') \begin{pmatrix} Q^{(r)} & 0 \\ 0 & I_{m-h} \end{pmatrix} Q(v^{(0)}), \tag{11.48}$$

where $Q(\mathcal{O}') \in \mathcal{Q}_m$ such that $Q(\mathcal{O}')_{l_k k} = 1$ for $k = 1, \ldots, h$.

To find $y(v)$, use Exercise 11.9, which shows that $Q(v^{(0)})e_m = ((m+1)e_h'/(h+1), (m+1)1_{m-h}'/2)'$. Then

$$y_i(v) = \begin{cases} \frac{m+1}{h+1} r_k & \text{if } i = l_k \in \mathcal{O}' \\ \frac{m+1}{2} & \text{if } i \notin \mathcal{O}' \end{cases}. \tag{11.49}$$

That is, to find the average of the complete rankings, just scale the rankings of the observed objects by $(m + 1)/(h + 1)$, and set the unobserved rankings to the average $(m + 1)/2$.

Return to the sampling scheme at the beginning of Section 11.3, and assume H_0 of (11.12) holds. For a ranking from a given subset

$\mathcal{O}^{(b)}$, note that in (11.12) the randomness is only in the $Q^{(R)}$, where here $R \sim Uniform(\mathcal{S}_h)$. Since $Cov(Q^{(R)}) = (H_h \otimes H_h)/(h-1)$,

$$Cov(Q(v)) = \frac{1}{h-1} Q(\mathcal{O}^{(b)}) \begin{pmatrix} H_h & 0 \\ 0 & 0 \end{pmatrix} Q'(O^{(b)}) \otimes S, \qquad (11.50)$$

where

$$S = Q_1'(v^{(0)}) H_h Q_1(v^{(0)}), \qquad (11.51)$$

and $Q_1(v^{(0)})$ is the upper $h \times m$ submatrix of $Q(v^{(0)})$ in (11.47). Letting $\widehat{M}$ be the sample average of all $Q(v^{(i)})$'s, we obtain

$$Cov(\widehat{M}) = \frac{1}{n}\frac{1}{h-1} G \otimes S, \qquad (11.52)$$

where

$$G = \sum_{b=1}^{B} \frac{1}{B} Q(\mathcal{O}^{(b)}) \begin{pmatrix} H_h & 0 \\ 0 & 0 \end{pmatrix} Q'(\mathcal{O}^{(b)}). \qquad (11.53)$$

Because of the balanced design, we have that $G_{ii} = (h/m)(1-1/h)$, and $G_{ij} = -(h-1)/m(m-1)$ if $i \neq j$. A little algebra yields $G/(h-1) = H_m/(m-1)$, so that

$$Cov(\widehat{M}) = \frac{1}{n}\frac{1}{m-1} H_m \otimes S. \qquad (11.54)$$

Unfortunately, S is neither invertible nor idempotent in general. Thus to perform the Marginals test, one must first find a generalized inverse of S. Also, the concordance using Hamming's distance requires finding the eigenvalues of S. The Means test does work out nicely:

$$Cov(\overline{y}) = Cov(\widehat{M} e_m) = \frac{e_m' S e_m}{n(m-1)} H_m. \qquad (11.55)$$

From (11.49), $Q_1(v^{(0)})e_m = (m+1)e_h/(h+1)$; hence

$$\frac{e'_m S e_m}{n(m-1)} = \frac{(m+1)^2 e'_h H_h e_h}{(h+1)^2 n(m-1)} = \frac{(m+1)^2 h(h-1)}{12n(m-1)(h+1)} \equiv \frac{w_{h,m}}{n}. \tag{11.56}$$

Thus, under uniformity,

$$\frac{n}{w_{h,m}} \sum_{i=1}^{m} \|\overline{y} - c_m\|^2 \longrightarrow \chi^2_{m-1}. \tag{11.57}$$

Results for the Hoeffding distances can be derived from (11.52) and Lemma 12.5, similar to the approach used to obtain (11.41) for tied rankings. Concordance for Spearman's distance is again equivalent to the Means test.

11.5.2 Pairs and Kendall's distance

Suppose we observe a complete ranking of the objects in $\mathcal{O}'$ as above. We need to find

$$\kappa_{ij}(v) = P[Y_i < Y_j \mid rank(Y_{l_1}, \ldots, Y_{l_h}) = r] \tag{11.58}$$

for $r \in \mathcal{S}_h$, where $l_1 < l_2 < \cdots < l_h$. Alvo and Cabilio (1991) show that

$$\kappa_{ij}(v) = \begin{cases} I[r_k < r_{k'}] & \text{if } i = l_k, j = l_{k'} \in \mathcal{O}' \\ 1 - \frac{r_k}{h+1} & \text{if } i = l_k \in \mathcal{O}' \text{ and } j \notin \mathcal{O}' \\ \frac{r_{k'}}{h+1} & \text{if } i \notin \mathcal{O}' \text{ and } j = l_{k'} \in \mathcal{O}' \\ \frac{1}{2} & \text{if } i, j \notin \mathcal{O}' \end{cases}. \tag{11.59}$$

See Exercise 11.10. Exercises 11.11, 11.12, and 11.13 exhibit the $Cov(\kappa(V))$ for V uniformly distributed.

To find $\Sigma(\kappa)$ as in (11.13) for the balanced incomplete block design, we assume additionally that every triple of objects appears in the same number of blocks, and every quadruple of objects appears in the same number of blocks. Let $\widehat{K}$ denote the sample average of the observed $\kappa(v^{(i)})$'s. Exercise 11.14 shows that

$$n\, Var(\widehat{K}_{ij}) = \frac{h(h-1)(2m+h+3)}{12m(m-1)(h+1)} \equiv v_{h,m} \tag{11.60}$$

and

$$n\ Cov(\widehat{K}_{ij}, \widehat{K}_{il}) = \frac{h(h-1)(m^2-h-2)}{12m(m-1)(m-2)(h+1)} \equiv c_{h,m}. \quad (11.61)$$

Thus

$$n\ Cov(\widehat{K}) = \alpha^* I_{\overline{m}} + \beta^* \Gamma_m \Gamma_m', \quad (11.62)$$

where

$$\alpha^* = v_{h,m} - 2c_{h,m} \text{ and } \beta^* = c_{h,m}. \quad (11.63)$$

The Pairs test now follows as in (11.38), but with the starred quantities (11.63). Similarly, the analogous results for Kendall's distance (11.43) and concordance using Kendall's distance (11.45) follow.

Remark 11.1. Coefficients
Normalizing distances or concordances when the observations are complete rankings has difficulties as mentioned in Chapter 3. Incomplete rankings add to the difficulties. See Alvo and Cabilio (1994). The main obstacle is deciding what a coefficient of "1" should mean. For example, suppose we have a tied ranking v with associated λ, and we wish to measure the closeness to e_m. Kendall's τ coefficient is then

$$t(v) = 1 - \frac{4}{m(m-1)} d_{\text{Ken}}(v, e_m). \quad (11.64)$$

If V is uniformly distributed over the tied rankings, then $E(\tau(V)) = 0$ as desired. However, the minimum $d_{\text{Ken}}(v, e_m)$ for such v is $(1/2)\sum \overline{\lambda}_k$, and the maximum is $\overline{m} - (1/2)\sum \overline{\lambda}_k$. (Exercise 11.15.) Thus

$$-1 + \gamma_2 \leq t(v) \leq 1 - \gamma_2, \quad (11.65)$$

where γ_2 is given in (11.33). Are the bounds desirable? To some extent it depends on what other v's are under consideration. If only v's with the same pattern are allowed, then it would be reasonable to scale $t(v)$ by dividing by $1 - \gamma_2$ so that "+1" means as close to e_m as possible. On the other hand, if different patterns of ties are expected, especially if some rankings might be complete, then scaling $t(v)$ depending on the pattern of v would make the coefficients noncomparable. For example, a tied ranking of $(1, 2, 2, 2, 2)$

would have a coefficient of 1, while $(1,2,3,5,4)$ would have 0.8. The complete ranking is penalized for attempting a more difficult task. Without scaling, the coefficients are 0.4 and 0.8, respectively, which might unfairly penalize the tied ranking. In practice, it depends on how much worse a discordancy is than a tie. In Section 11.7.7, we discuss a model-based compromise. □

11.6 Comparing samples

Suppose we have two independent samples of incomplete rankings. To compare their populations, we could follow Chapter 4 and look at the difference of some statistic. That is, suppose $z^{(ji)}, i = 1, \ldots, n^{(j)}$ is the sample of $z(v^{(i)})$'s for group j, $j = 1, 2$. One could test whether $E(\overline{z}^{(1)}) = E(\overline{z}^{(2)})$, but that equality may not be the hypothesis of interest. For example, suppose two populations of full rankings are exactly the same, then change the second population by tying the first ranks. That is, if the original population is $\{123, 123, 312\}$, then tying yields the population of tied rankings $\{112, 112, 211\}$. The mean, $E(\overline{Y})$, for the complete rankings is $(5/3, 5/3, 8/3)$, and that for tied rankings is $(2, 3/2, 5/2)$. Thus even though the complete rankings underlying the populations are the same, the tying in the second population causes the means to be different. The question arises, "What is meant by the populations being the same?" If the populations have different types of incomplete rankings, then does it make sense to ask the question? It may, but then only if some parameters of the populations are compared, in which case the question becomes one of determining parameters that are comparable for different types of incomplete rankings.

We will consider the following formulation. (We will present just the two-sample case, but the multisample case can be treated similarly.) As in Section 11.3, suppose both populations have support in the union of $\mathcal{P}^{(1)}, \ldots, \mathcal{P}^{(B)}$, and let $p^{(bj)}$ be the proportion of the population j with observations in group $\mathcal{P}^{(b)}$. The null hypothesis H_0 we consider is that the distribution within each $\mathcal{P}^{(b)}$ is the same for both populations. To test H_0, find $\overline{z}^{(bj)}$ for each group $\mathcal{P}^{(b)}$ and population j, then use the statistic

$$W = \sum_{b=1}^{B} c_b(\overline{z}^{(b1)} - \overline{z}^{(b2)}), \tag{11.66}$$

where the c_b's are positive weights. For example, $c_b = (p_{b1}+p_{b2})/2$, or its estimate, is a reasonable choice. The covariance of W is estimated by

$$\widehat{Cov}(W) = \sum_{b=1}^{B} c_b^2 \left(\frac{1}{n^{(b1)}} + \frac{1}{n^{(b2)}}\right)\widehat{\Sigma}^{(b)}, \tag{11.67}$$

where $\widehat{\Sigma}^{(b)}$ is the pooled covariance estimator for $\mathcal{P}^{(b)}$,

$$\widehat{\Sigma}^{(b)} = \frac{1}{n^{(b1)} + n^{(b2)} - 2}\Big(\sum_{i=1}^{n^{(b1)}} (z^{(b1i)} - \overline{z}^{(b1)})(z^{(b2i)} - \overline{z}^{(b1)})' + \sum_{i=1}^{n^{(b1)}} (z^{(b2i)} - \overline{z}^{(b2)})(z^{(b2i)} - \overline{z}^{(b2)})'\Big), \tag{11.68}$$

where $z^{(bji)}$ is the ith observation in group $\mathcal{P}^{(b)}$ and population j. Then under H_0, if all $n^{(bj)} \to \infty$,

$$W'\widehat{Cov}(W)^{-1}W \longrightarrow \chi_t^2 \tag{11.69}$$

if W is $t \times 1$, and the covariance is invertible. Note that this approach will not work unless there are a number of observations from both populations in each $\mathcal{P}^{(b)}$, or at least in many of them.

11.7 Models

We will consider two basic approaches to extending models on full rankings to those on incomplete rankings. The first approach consists of *ad hoc* extensions of the model based on first principles. For example, extending Babington Smith to complete rankings on subsets of objects involves placing a regular Babington Smith model on the subset. The other approach, taken in censored data situations, can be used for any model, but makes some reasonably strong assumptions. We suppose that there is a population of judges, and that each judge has a complete ranking, but for some reason independent of the actual ranking, the ranking is only partially observed. It may be that one has a complete ranking, but is only asked for the top three choices or to make a few paired comparisons; or one does not have the time to perform a detailed ranking;

or ranks are based on some real-valued variables, but roundoff error creates some ties. See Beckett (1993) for an interesting medical example.

The formal censored data model assumes a model $P(\theta)$ on $\mathcal{S}_m$, the space of complete rankings, and a given set $\mathcal{P}^{(1)}, \ldots, \mathcal{P}^{(B)}$ of types of incomplete rankings such that every possible observed v is in one and only one $\mathcal{P}^{(b)}$. We further assume the presence of a mechanism $\mathcal{E}$ that turns a complete ranking into an incomplete one: $\mathcal{E}$ takes values in $\{1, \ldots, B\}$, and the observation is $h(Y, \mathcal{E})$, where the function h is given by

$$\begin{aligned} h(y,\epsilon) : \mathcal{S}_m \times \{1,\ldots,B\} &\longrightarrow \cup_{b=1}^{B} \mathcal{P}^{(b)}, \\ h(y,\epsilon) &= v, \;\; y \in v \in \mathcal{P}^{(\epsilon)}. \end{aligned} \tag{11.70}$$

By our assumptions, there is a unique v for each y, ϵ pair. The sample consists of $(Y^{(1)}, \mathcal{E}^{(1)})$, ..., $(Y^{(n)}, \mathcal{E}^{(n)})$, but we observe only $V^{(1)}, \ldots, V^{(n)}$, where $V^{(i)} = h(Y^{(i)}, \mathcal{E}^{(i)})$. The model makes the further assumptions:

1) $Y^{(1)}, \ldots, Y^{(n)}$, are independent and identically distributed as $P(\theta)$;

2) $\underline{\mathcal{E}} \equiv (\mathcal{E}^{(1)}, \ldots, \mathcal{E}^{(n)})$ and $(Y^{(1)}, \ldots, Y^{(n)})$ are independent;

3) The $\underline{\mathcal{E}}$ is ancillary, that is, its distribution does not depend on θ.

In Examples 11.2, 11.3, and 11.4, the pattern of ties or incomplete rankings is fixed by the experimenter; hence $\underline{\mathcal{E}}$ satisfies 2) and 3) automatically. In Example 11.1, each person decides on what pattern of ties to present; hence Assumption 2) in particular may be suspect. In Example 11.1, in fact, one can prove statistically that Assumption 2) fails.

One observes V, and from that can infer $\underline{\mathcal{E}}$, but is generally interested only in θ. Since $\underline{\mathcal{E}}$ is ancillary, instead of the full likelihood for V, we will look at just the partial likelihood of V conditional on $\underline{\mathcal{E}}$, $g_\theta(v|\epsilon)$. Now for a single observation,

$$\begin{aligned}
\frac{\#v}{m!} g_\theta(v|\epsilon) &= P[V = v \mid \mathcal{E} = \epsilon; \theta] \\
&= P[V = v \;\&\; \mathcal{E} = \epsilon; \theta]/P[\mathcal{E} = \epsilon] \\
&= \sum_{\{y|h(y;\epsilon)=v\}} P[Y = y \;\&\; \mathcal{E} = \epsilon; \theta]/P[\mathcal{E} = \epsilon] \\
&= \sum_{\{y|h(y;\epsilon)=v\}} P[Y = y; \theta] P[\mathcal{E} = \epsilon]/P[\mathcal{E} = \epsilon] \\
&= \sum_{\{y|h(y;\epsilon)=v\}} P[Y = y; \theta].
\end{aligned} \tag{11.71}$$

That is, the conditional likelihood for V is just the sum of the likelihoods for the Y's that could have produced V. The joint conditional likelihood for the sample of $v^{(i)}$'s is

$$\prod_{i=1}^{n} g_\theta(v^{(i)}|\epsilon^{(i)}). \tag{11.72}$$

In the remainder of this section, we present examples of models for incomplete rankings. In some cases we use the censored data approach, and in other cases we use *ad hoc* methods.

11.7.1 Thurstonian models

In general, the density (11.71) is very complicated since each g_θ term is a sum of several multidimensional integrals. However, if the incomplete rankings are based on subsets of objects as in Section 11.1.2, then each g_θ is just the product of the Thurstonian ranking probabilities of the smaller groups of objects. That is, suppose one is to rank within the groups $\{A, B, C, D\}$ and $\{E, F, G, H\}$. Then the probability that the first group is in the order $ADBC$ and the second group is in the order $HFEG$ is $P[Z_A < Z_D < Z_B < Z_C$ and $Z_H < Z_F < Z_E < Z_G]$, which by independence is $P[Z_A < Z_D < Z_B < Z_C] \times P[Z_H < Z_F < Z_E < Z_G]$.

11.7.2 Plackett–Luce and vase models

Section 8.4 illustrated vase models for q-permutations, that is, one gives only the top q rankings. For the Plackett–Luce model, we have

$$P[V = v; \nu] = \frac{\nu_{w_1}}{\nu_{w_1} + \cdots + \nu_{w_m}} \times \frac{\nu_{w_2}}{\nu_{w_2} + \cdots + \nu_{w_m}} \times \cdots \times \frac{\nu_{w_q}}{\nu_{w_q} + \cdots + \nu_{w_m}}, \tag{11.73}$$

where the observed *ordering* of the top q choices is $(w_1, ..., w_q)$. A similar formula works for the backwards Plackett model if one ranks only the bottom q objects. Note that q can be different for different judges. The general vase models and other extensions of Plackett–Luce also easily extend to q-permutations by using only the first q stages.

Example 11.8: Crackers
Critchlow (1985) reports on an experiment in which 22 boys and 16 mothers ranked five types of crackers. Each person was to give their top first, second, and third choices, so that the data consist of q-permutations for $q = 3$. Letting $\theta_i = \log(\nu_i)$ as in Chapter 8, the likelihood for the Plackett–Luce model for one sample is (8.7) where the summation is over all subsets S of $\{1, \ldots, m\}$ with at least $m - q + 1$ elements. Fitting the likelihood equations follows much as for the Plackett–Luce model for complete rankings. Table 11.1 contains the results for the boys and the mothers. The parameters are scaled to sum to 100. The LRS's are 34.94 and 9.67, on 4 degrees of freedom, for the boys and mothers, respectively. The mothers are almost uniform in their choices.

The biggest differences are that the boys like Animal Crackers much better, and Saltines much worse, than do the mothers. They also like Cheese Crackers quite a bit better. □

11.7.3 Babington Smith

There are several ways one might imagine extending Babington Smith models to incomplete rankings. One way is to use the censored data approach. That may be a bit suspect, however. Suppose the observed tied ranking a is $(2, 1, 2, 2, 3)$. The censored data model posits that first the judge makes all 10 paired comparisons,

Table 11.1 *Crackers: Plackett–Luce model. The coefficients are normalized to sum to 100.*

Type of Cracker ↓	Mothers $\hat{\nu}_i$	Mothers se	Boys $\hat{\nu}_i$	Boys se
Animal	**16.93**	**5.24**	**42.45**	**8.16**
Ritz	**14.56**	**4.95**	**5.80**	**2.43**
Saltine	**41.89**	**9.30**	**5.49**	**2.30**
Cheese	**17.93**	**5.90**	**31.60**	**6.98**
Graham	**8.68**	**3.61**	**14.66**	**4.23**

perhaps obtaining a complete ranking, perhaps not. If not, the judge repeats the paired comparisons. When a complete ranking is finally found, the judge then ignores the A-C-D comparisons. Why go to all the bother of making sure all comparisons are consistent when some are ignored?

Another approach is to assume for each pair three possibilities: prefer the first, prefer the second, or prefer them equally. Thus each pair of objects will have a little triple of probabilities attached. The ranking proceeds by making all comparisons until a consistent ranking, possibly with ties, results. Rao and Kupper (1967) and Davidson (1970) suggest particular parametrizations of the paired comparisons when extending the Bradley–Terry parameters. Rao and Kupper add the parameter θ, where $P[Y_i < Y_j] = \nu_i/(\nu_i+\theta\nu_j)$, and Davidson adds c_{ij}'s so that $P[Y_i < Y_j] = \nu_i/(\nu_i + \nu_j + c_{ij})$. He suggests taking $c_{ij} = c\sqrt{\nu_i\nu_j}$ for parameter c.

As mentioned above, when each judge gives a full ranking of a subset of the objects, then the Babington Smith model can be placed on the subset. This idea makes sense since the judge does not make comparisons using the objects not presented. The parameters p_{ij} in (5.7) are the same as for the complete ranking model. If $r \in \mathcal{S}_h$ is a ranking of the objects $\mathcal{O}' = \{i_1, \ldots, i_h\}$, then the density is

$$f_\theta(r;\mathcal{O}') = \exp\Big(\sum_{1\le j}\sum_{<k\le h} \theta_{i_j i_k} I[r_i > r_j] - \psi(\theta;\mathcal{O}')\Big), \qquad (11.74)$$

where $\psi(\theta; \mathcal{O}')$ is the normalizing constant depending on just the θ_{ij}'s for $i, j \in \mathcal{O}'$. The density for a sample is the product of such f's, where the ψ's depend on which objects are ranked. Submodels of Babington Smith, including orthogonal contrast ϕ, BTM, Spearman's distance, and Mallows' ϕ models, can be handled in this manner.

11.7.4 Orthogonal contrast ϕ models: Tied rankings

Assume A is a tied ranking with pattern λ. Suppose the underlying vector Y is distributed according to an orthogonal contrast ϕ model, $Y \sim Phi(\theta; C^{(1)}, \ldots, C^{(T)})$. The function g_θ has a fairly nice form. First, define the "d" of a contrast C in (5.29) at the tied ranking a by

$$d(C(a)) = \sum_{1 \leq k < l \leq K} \sum_{i \in \mathcal{O}_k} \sum_{j \in \mathcal{O}_l} I[a_i > a_j]. \tag{11.75}$$

This definition is actually the same as for a complete ranking y, except that now the indicator function will be 0 if either Object i is preferred to Object j or they are tied. It is not hard to show that $d(C(a)) = \min_{y \in v} d(C(y))$, where $v \in \mathcal{W}(\mathcal{S}_m)$ is the partial ranking associated with a. From Chung and Marden (1991), the density g_θ turns out to be

$$g_\theta(a) = \exp\left(\sum_{t=1}^{T} \theta_t \cdot d(C^{(t)}(a)) - \sum_{t=1}^{T} \psi(\theta_t; C^{(t)}, a)\right), \tag{11.76}$$

where for contrast C with L groups and partition λ^C, and real γ,

$$\begin{aligned}\psi(\gamma; C, a) = \psi(\gamma; \lambda_+) &- \sum_{l=1}^{L} \psi(\gamma, \lambda_l^C) \\ &- \sum_{k=1}^{K} \psi(\gamma; \lambda_k) + \sum_{l=1}^{L} \sum_{k=1}^{K} \psi(\gamma; m_{kl}),\end{aligned} \tag{11.77}$$

where ψ is given in (6.3). Here, $m_{kl} = \#\{i | a_i = k \text{ and } i \in \mathcal{O}_l\}$. Although the ψ looks a bit formidable, it is actually easy to compute since it is just a linear combination of the ϕ's in (5.17). The right-hand side of (11.77) is analogous to inclusion–exclusion formulas.

Mallows ϕ model is a special case of the orthogonal contrast ϕ model, so the corresponding model with ties should be straightforward from (11.76). It is. The density is

$$g_\theta(a) = \exp(\theta \cdot d_{\text{Ken}}(a,\omega) - \psi(\theta;\lambda)). \tag{11.78}$$

This equation follows from (11.77) by setting $C = Brack(\omega)$ of (5.30), so that $\lambda^C = 1_m$, and noting that $\phi(\gamma; 1) = 0$.

Now suppose we have a sample $A^{(1)}, \ldots, A^{(n)}$ of tied rankings, with corresponding partitions $\lambda^{(1)}, \ldots, \lambda^{(n)}$. The loglikehood conditional on the $\mathcal{E}_i$'s is

$$l(\theta) = \theta \sum_{i=1}^{n} d_{\text{Ken}}(a^{(i)},\omega) - \sum_{i=1}^{n} \psi(\theta; \lambda^{(i)}). \tag{11.79}$$

Let $\mu_n(\theta)$ and $\sigma_n^2(\theta)$ be the first and second derivatives of $\frac{1}{n}\sum_{i=1}^{n} \psi(\theta;\ \lambda^{(i)})$, so that the MLE satisfies $\mu_n(\widehat{\theta}) = \sum_{i=1}^{n} d_{\text{Ken}}(a^{(i)},\omega)/n$, and asymptotically $\sqrt{n}\sigma_n(\widehat{\theta})\ (\widehat{\theta} - \theta) \to N(0,1)$.

11.7.5 Orthogonal contrast ϕ models: Subsets

For subsets of rankings, the censored data approach is not as nice. However, using the Babington Smith interpretation, we can extend the orthogonal contrast ϕ model by defining the model on just the observed objects, where the contrasts are redefined by eliminating the unobserved objects. For example, if we apply the contrast $C = (\{A,B\},\{C,D,E\})$ to an observation that ranks only the objects in $\mathcal{O}' = \{A,C,D\}$, then the reduced contrast is $C' = (\{A\},\{C,D\})$. If after reduction there is no comparison, e.g., $\mathcal{O}' = \{C,D\}$, then the contrast is ignored. Given a set of orthogonal contrasts, the reduced set for a given subset of observed objects will also be orthogonal. Each observation has its own reduced orthogonal contrast ϕ model. The models for the different observations share parameters; hence their sufficient statistics can be combined, and likelihood estimation proceeds as for the regular model.

11.7.6 *Exponential family models*

As for complete rankings, one can define an exponential family model by specifying a set of statistics and the dominating density. With complete rankings, the obvious candidate for dominating density is *Uniform*($\mathcal{S}_m$), so obvious that we never even considered any alternatives. With incomplete rankings, there may be problems choosing the statistic and choosing the dominating measure. The latter problem is the same as that when choosing uniformity for test statistics in Section 11.3. If all rankings are of the same type $\mathcal{P}$, then *Uniform*($\mathcal{P}$) is a natural dominating measure, and reasonable statistics can be chosen. It is when there are different types arising that the problems are more difficult. One solution is to condition on the type, but the problem of the statistic still remains.

We will assume all observations are of the same type $\mathcal{P}$. Given a set of sufficient statistics, $t = (t_1, ..., t_p)$ on $\mathcal{P}$, The exponential family density is then

$$g_\theta(v) = \exp(\theta' t(v) - \psi(\theta)) \tag{11.80}$$

with respect to *Uniform*($\mathcal{P}$), where now

$$\psi(\theta) = \log(E(\exp(\theta' t(V))),\ V \sim Uniform(\mathcal{P}). \tag{11.81}$$

The sufficient statistics for a sample are then the sums of those for the individual observations.

Silverberg's (1980) first-order model for q-permutations are described in Section 11.7.2. They have sufficient statistics for a sample being the numbers of observations with Object i ranked j, where $i = 1, \ldots, m$ and $j = 1, \ldots, q$. Second-order models additionally depend on the numbers of times each pair of objects is ranked j_1 and j_2 for $1 \leq j_1, j_2 \leq q$, and similarly for higher-(but-no-higher-than-q-) order models.

An analog of the Bradley–Terry/Mallows model is illustrated next.

Example 11.9. Crackers: Bradley–Terry/Mallows
Recall that the exponential family model with sufficient statistic being y itself is the Bradley–Terry/Mallows model. See Lemma 7.1. The Cracker data of Example 11.8 have the fourth and fifth

Table 11.2 *Crackers: Bradley–Terry/Mallows model.*

	Mothers		Boys	
Type of Cracker ↓	$\hat{\theta}_i$	**se**	$\hat{\theta}_i$	**se**
Animal	**0.0643**	**0.1519**	**−0.5766**	**0.1630**
Ritz	**0.0925**	**0.1531**	**0.4146**	**0.1642**
Saltine	**−0.4488**	**0.1688**	**0.5217**	**0.1791**
Cheese	**−0.0448**	**0.1494**	**−0.3902**	**0.1465**
Graham	**0.3368**	**0.1734**	**0.0251**	**0.1354**

ranks tied, but we can still mimic the BTM model by using the tied ranking a or the midranks $y(v)$ as the sufficient statistics. The only difference is whether "4" or "4.5" is inserted for the objects ranked fourth and fifth. The results for the two choices are very similar, so we will present only the midrank version.

The sufficient statistic is $(y_1(v), \ldots, y_{m-1}(v))$ since the components of $y(v)$ always add to $m(m+1)/2$. The X matrix in the general linear model consists of the $m!/2 = 60$ possible tied rankings. The calculations proceed as with any generalized linear model. Table 11.2 contains the MLE's. Recall that lower values of θ_i indicate greater preference for Object i. The LRS's for the boys and mothers are, respectively, 34.60 and 10.54, on 4 degrees of freedom. Thus the boys are certainly not uniform in their choices, but the mothers may be. Otherwise, the interpretation is much as for the Plackett–Luce model in Example 11.8. □

For distance models, Critchlow (1985) suggests using the Hausdorf metric (11.5) to extend metrics on $\mathcal{S}_m$ to metrics on spaces of tied rankings. See Diaconis (1988), too.

11.7.7 Coefficients and comparing populations

In Section 11.4, we saw some ambiguity in defining analogs of the τ coefficients when there are incomplete rankings. Another solution arises when using a censored data model. We define τ as a parameter of the population of complete rankings. Given the sample of (possibly) incomplete rankings, the goal is to estimate τ. The estimation may be difficult, but the interpretation of the parameter is independent of whether, or in what way, the data are incomplete.

Table 11.3 *Kendall's* τ*'s for tied rankings. The modal ranking is* $\omega = (1, 2, 3, 4, 5)$.

Tied Ranking	t	$t^* \equiv t/(1-\gamma_2)$	$\hat{\tau}$
12345	1.00	1.00	1.00
13425	0.60	0.60	0.60
12333	0.70	1.00	1.00
21312	0.00	0.00	0.00
12132	0.40	0.50	0.46
33312	−0.50	−0.71	−0.67
32211	−0.80	−1.00	−1.00

Consider Kendall's τ in Mallows' ϕ model. For complete rankings, the parameter of interest is

$$\begin{aligned}\tau(\theta) = E_\theta(t_{Ken}(Y)) &= 1 - \frac{4}{m(m-1)} E_\theta(d_{\text{Ken}}(Y, \omega)) \\ &= 1 - \frac{4}{m(m-1)} \psi'_m(\theta),\end{aligned} \tag{11.82}$$

where ω is the fixed modal ranking. Now suppose we have a sample of tied rankings from (11.78). (The rankings need not have the same patterns of ties.) Using the joint density, find $\widehat{\theta}$, the MLE of θ. Then

$$\widehat{\tau} = \tau(\widehat{\theta}) = 1 - \frac{4}{m(m-1)} \psi'_m(\widehat{\theta}), \tag{11.83}$$

i.e., the ψ used in estimating τ need not be that in the density of the tied rankings. From (11.83), using the Δ-method (Exercise 11.16), we have that

$$\sqrt{n} \frac{m(m-1)}{4} \frac{1}{\psi''_m(\widehat{\theta}) \sigma_m(\widehat{\theta})} (\widehat{\tau} - \tau) \longrightarrow N(0, 1). \tag{11.84}$$

Thus $\widehat{\tau}$ is a reasonable estimator of τ. The coefficient $\widehat{\tau}$ is a compromise between the unscaled (11.64) and scaled $(t(v)/(1 - \gamma_2))$ coefficients defined in Remark 11.1. Table 11.3 exhibits some values of the three coefficients for various values of v when $n = 1$.

Some facts are illustrated by the table. If there are no ties, then all three coefficients are the same. If one is 0, they all are. If there are no discordancies, then t^* and $\hat{\tau}$ are 1, but $t < 1$ unless there are no ties. If t^* is neither -1, 0, nor $+1$, and there are ties, then $\hat{\tau}$ is strictly between t and t^*. The basic idea is that for t, ties are broken half as concordancies and half as discordancies, while for $\hat{\tau}$, ties are broken in favor of one or the other depending on $\hat{\theta}$. The higher the $\hat{\theta}$, the more are broken in favor of discordancies.

A similar approach works when comparing populations. In Section 11.6, we had trouble comparing statistics from different populations when the incomplete rankings were of different types. Instead, we can define parameters for the populations of complete rankings, and compare their estimates based on the samples with possible incomplete rankings. Alternatively, the likelihood ratio test for comparing the populations is available.

11.8 Exercises

11.1. Take $v = \{\ (1,2,\ldots, m), (2, 3, \ldots, m, 1), \ldots, (m, 1, \ldots, m-1)\ \}$. Show that $y(v) = c_m$ and $Q(v) = \overline{J}_m$.

11.2. Show that $\hat{\alpha} = \hat{P}'D\hat{P}$ in (11.18). [Write $d(v^{(i)}, v^{(j)}) = \zeta(v^{(i)})'\ D\ \zeta(v^{(j)})$, then average over i and j.]

11.3. Show that $\overline{J}^\lambda$ (11.25) is idempotent, and $\overline{J}^\lambda H_m \overline{J}^\lambda = H^\lambda$ in (11.27).

11.4. Show that H^λ is idempotent, and $\text{tr}(H^\lambda) = K - 1$.

11.5. Verify (11.28)

11.6. Verify (11.36) and (11.37).

11.7. Show that applying (11.40) to Hamming and Spearman distances yields (11.41). [Recall (3.36).]

11.8. Use (11.9) and (11.30) to prove (11.42).

11.9. Show that (11.49) holds. [Use the combinatorical identity obtained from (11.47) that $\sum_{j=i}^{m-h+i} \binom{j-1}{i-1} \binom{m-j}{h-i} = \binom{m}{h}$ for any $i \leq h$.]

11.10. Consider the incomplete ranking v that corresponds to a complete ranking of a subset of objects $\mathcal{O}'$ as in Section 11.1.2. Verify (11.59). [The $\kappa_{ij}(v) = P[Y_i < Y_j \mid Y \in v]$. If i and j are both in $\mathcal{O}'$, then the answer is $I[Y_i < Y_j]$ since Y_i and Y_j are fixed. Equation (11.59) then gives this value in terms of the ranks relative to the objects in $\mathcal{O}'$. If $i, j \notin \mathcal{O}'$, then $I[Y_i < Y_j]$ is independent of the values of Y_k for $k \in \mathcal{O}'$. Suppose $i \in \mathcal{O}'$ and $j \notin \mathcal{O}'$. Then Y_i is fixed at r_s. Now Y_j is equally likely to be in any of the $h+1$ positions relative to $\{Y_k | k \in \mathcal{O}'\}$, so that $I[Y_i < Y_j]$ conditioned on $Y \in v$ is $Bernoulli(1 - r_s/(h+1))$, which yields the value in (11.59). The other case is similar.]

11.11. For the situation in Exercise 11.10, show that $Var(\kappa_{ij}(V))$ equals

$$\begin{cases} 1/4 & \text{if } i, j \in \mathcal{O}' \\ (h-1)/(12(h+1)) & \text{if } i \in \mathcal{O}' \text{ and } j \notin \mathcal{O}' \\ (h-1)/(12(h+1)) & \text{if } i \notin \mathcal{O}' \text{ and } j \in \mathcal{O}' \\ 0 & \text{if } i, j \notin \mathcal{O}'. \end{cases}$$

11.12. Next, consider the covariance when $i = k$ and $j \neq l$. Specifically, show that $Cov(\kappa_{ij}(V), \kappa_{il}(V))$ equals

$$\begin{cases} 1/12 & \text{if } i, j \in \mathcal{O}' \text{ or } i, l \in \mathcal{O}' \\ -1/(12(h+1)) & \text{if } i \notin \mathcal{O}' \text{ and } j, l \in \mathcal{O}' \\ (h-1)/(12(h+1)) & \text{if } i \in \mathcal{O}' \text{ and } j, l \notin \mathcal{O}' \\ 0 & \text{otherwise.} \end{cases}$$

[For the first line, if all objects are in $\mathcal{O}'$, then the covariance is as for complete rankings. If $i, j \in \mathcal{O}'$ but $l \notin \mathcal{O}'$, then from (11.59) the covariance is $-Cov(I[R_s < R_t], R_s)/(h+1)$, where $i = q_r$ and $j = q_s$. By conditioning on R_s, since $R \sim Uniform\ (\mathcal{S}_h)$, $Cov(I[R_s < R_t], R_s)$ can be shown to be $-(h+1)/12$. Similarly for $i, l \in \mathcal{O}'$ and $j \notin \mathcal{O}'$. For the second line, use (11.59) to show that the covariance equals $Cov(R_s, R_t)/(h+1)^2$, where $j = q_r$ and $l = q_s$. The third line equals $Var(1 - R_s/(h+1))$, where $i = q_r$, and the final line follows since at least one of κ_{ij} and κ_{il} is constant.]

11.13. Now suppose i, j, k, l are distinct. Show that if either $i, j \in \mathcal{O}'$ or $i, j \notin \mathcal{O}'$ or $k, l \in \mathcal{O}'$ or $k, l \notin \mathcal{O}'$, then the covariance is 0. Cases left include those in which exactly one of i, j is in $\mathcal{O}'$, and exactly one of k, l is in $\mathcal{O}'$. Show that if $i, k \in \mathcal{O}'$ and $j, l \notin \mathcal{O}'$, or $i, k \notin \mathcal{O}'$ and $j, l \in \mathcal{O}'$, the covariance is $-1/(12(h+1))$, and in the other two cases, it is $1/(12(h+1))$.

11.14. Suppose the conditions on the pairs, triples, and quadruples of objects at the beginning of Section 11.5 hold. Verify (11.62).

[Using Exercise 11.11 for the variance, we have

$$n \cdot Var(\widehat{K}_{ij}) = \frac{1}{4}p(i,j \in \mathcal{O}') + \frac{h-1}{12(h+1)}(p(i \in \mathcal{O}', j \notin \mathcal{O}') \\ + p(i \notin \mathcal{O}', j \in \mathcal{O}')),$$

where $p(\cdot)$ is the proportion of blocks in which the event occurs. In this case, for example, $p(i,j \in \mathcal{O}') = h(h-1)/((m(m-1))$, and $p(i \in \mathcal{O}', j \notin \mathcal{O}') = h(m-h)/((m(m-1))$. Similar considerations yield the covariance terms. Note that when i, j, k, l are distinct, the condition that each quadruple of objects appears in the same number of blocks causes the $\pm 1/(12(h+1))$'s for the final cases in Exercise 11.10 to cancel.]

11.15. Refer to (11.42) to show that the maximum and minimum of $d(e_m, v)$ for tied ranking v are $(1/2)\sum \overline{\lambda}_k$ and $\overline{m}-$ $(1/2) \sum \overline{\lambda}_k$, respectively.

11.16. If $\sqrt{n}(\widehat{\theta} - \theta) \rightarrow N(0, \sigma^2)$, and $h(x)$ is a function with a continuous derivative at $x = \theta$, then the "Δ-method" allows that $\sqrt{n}(h(\widehat{\theta}) - h(\theta)) \rightarrow N(0, (h'(\theta))^2\sigma^2)$. Use this result to prove (11.84).

CHAPTER 12

Appendix

12.1 Some linear algebra

This section contains some basic definitions about matrices. More involved material is in Section 12.2. A complete overview is found in Rao (1973).

The elements of a $p \times q$ matrix A are denoted a_{ij}, where i indicates the row and j the column:

$$A = \begin{pmatrix} a_{11} & a_{12} & \dots & a_{1q} \\ a_{21} & a_{22} & \dots & a_{2q} \\ \vdots & \vdots & \ddots & \vdots \\ a_{p1} & a_{p2} & \dots & a_{pq} \end{pmatrix}. \tag{12.1}$$

If A is a $p \times q$ matrix, then its *transpose* B, denoted $B = A'$, is the $q \times p$ matrix with $b_{ji} = a_{ij}$. Thus the rows of A become the columns of A', and vice versa. The *inverse* of a square $p \times p$ matrix A is the $p \times p$ matrix A^{-1} that satisfies $AA^{-1} = A^{-1}A = I_p$. If it exists it is unique, but it may not exist.

Trace. If A is $p \times p$, then the *diagonals* are the elements a_{ii}, $i = 1, \dots, p$. The *trace*, denoted $\text{tr}(A)$, is the sum of the diagonals, $\text{tr}(A) = a_{11} + \dots + a_{pp}$. If A is $p \times q$ and B is $q \times p$, then

$$\text{tr}(AB) = \sum_{i=1}^{p} \sum_{j=1}^{q} a_{ij} b_{ji} = \text{tr}(BA). \tag{12.2}$$

Definiteness. A $p \times p$ matrix A is *positive definite* if $x'Ax > 0$ for all $p \times 1$ non-zero vectors x. It is called *positive semi-definite* if $x'Ax \geq 0$ for all x. It can be shown that if A is positive semi-definite, then so is BAB' for any $r \times p$ matrix B. Similarly, A

is *negative definite* if $-A$ is positive definite, and *negative semi-definite* if $-A$ is positive semi-definite.

12.1.1 Eigenvalues and eigenvectors

For any $p \times p$ matrix A, the $p \times 1$ vector v with $||v|| = 1$, where $||v|| = (v_1^2 + \cdots + v_p^2)^{1/2}$, is an *eigenvector* of A if there exists a constant λ such that

$$Av = \lambda v. \tag{12.3}$$

The λ is called the *eigenvalue* associated with v. If A is symmetric, then any eigenvalue is real. If A is not symmetric, eigenvalues may have imaginary components. We will be using them only for symmetric matrices, so we do not have to worry about imaginary numbers. There are p eigenvalues for A, although there may be multiplicities (sets of eigenvalues with the same value). The next result is a convenient way to collect the eigenvalues and eigenvectors. An *orthogonal* matrix O is one whose columns have length 1 and are orthogonal, so that $OO' = O'O = I_p$.

Lemma 12.1: Spectral decomposition theorem
If A is a symmetric $p \times p$ matrix, then there exists an orthogonal matrix O and diagonal matrix Λ with diagonal elements $\lambda_1 \geq \lambda_2 \geq \cdots \geq \lambda_p$ such that $A = O\Lambda O'$. □

For the proof, see any linear algebra book, or Rao (1973). It is not hard to show that the columns of O are eigenvectors of A, and the diagonals of Λ are the eigenvalues. Also, (12.2) shows that the trace is the sum of the eigenvalues, that is, $\text{tr}(A) = \lambda_1 + \cdots + \lambda_p$. The matrix Λ is unique, that is, there is only one diagonal matrix with nonincreasing diagonals that satisfies $A = O\Lambda O'$ for some orthogonal O. The O will not be unique unless all the eigenvalues are distinct.

Eigenvalues have a switching property like (12.2) for traces. That is, if λ is an eigenvalue of AB, then there exists a vector v as in (12.3) such that $ABv = \lambda v$. Then $BABv = \lambda Bv$, so that λ is the eigenvector for BA associated with the eigenvector $Bv/||Bv||$.

Idempotent Matrices. A square matrix A is idempotent if $AA = A$. Idempotent matrices arise often in statistics, e.g., as the projection matrices in linear regression. The matrix H_m in (3.4) is such. The next result finds their eigenvalues.

Lemma 12.2
Suppose A is symmetric and idempotent. Then its eigenvalues consist of k 1's and $(p-k)$ 0's, where $k = tr(A)$. □

Proof
Write A as in Lemma 12.1, so that $A = AA$ implies that $O\Lambda O' = O\Lambda^2 O'$, from which we obtain that $\lambda_i = \lambda_i^2$ for each i. Thus each λ_i is either 1 or 0. The number of 1's is tr(A). □

12.2 Means and covariances for vectors and matrices

12.2.1 Definitions

In the following, we are assuming all random variables have finite means and variances. Hogg and Craig (1978) amplify on this material. If X is a $p \times 1$ vector of random variables, then its expected value is the corresponding vector of expected values, i.e.,

$$E(X) = (E(X_1), \ldots, E(X_p))'. \tag{12.4}$$

Similarly, if W is a $p \times q$ matrix of random variables, its expectation is the $p \times q$ matrix of expected values:

$$E(W) = \begin{pmatrix} E(W_{11}) & E(W_{12}) & \ldots & E(W_{1q}) \\ E(W_{21}) & E(W_{22}) & \ldots & E(W_{2q}) \\ \vdots & \vdots & \ddots & \vdots \\ E(W_{p1}) & E(W_{p2}) & \ldots & E(W_{pq}) \end{pmatrix}. \tag{12.5}$$

Variances and covariances are more involved. For vector X, the covariance matrix $Cov(X)$ is a $p \times p$ matrix that contains the variances along the diagonal and the covariances outside the diagonals, so that the ij^{th} element is $Cov(X_i, X_j)$, which equals $Var(X_i)$ if $i = j$. Thus

$$Cov(X) = E((X - E(X))(X - E(X))')$$
$$= \begin{pmatrix} Var(X_1) & Cov(X_1, X_2) & \ldots & Cov(X_1, X_p) \\ Cov(X_2, X_1) & Var(X_2) & \ldots & E(W_{2q}) \\ \vdots & \vdots & \ddots & \vdots \\ Cov(X_p, X_1) & Cov(X_p, X_2) & \ldots & Var(X_p) \end{pmatrix}. \quad (12.6)$$

It is symmetric, $Cov(X)' = Cov(X)$, since $Cov(X_i, X_j) = Cov(X_j, X_i)$. The notation is convenient for linear transformations. Thus if A is $r \times p$ and b is $r \times 1$, then

$$E(AX+b) = AE(X)+b \text{ and } Cov(AX+b) = A\,Cov(X)A'. \quad (12.7)$$

The covariance matrix for a matrix W is obtained as in (12.6) where X is replaced by the vector obtained by stringing out the columns of W into one $pq \times 1$ vector. We will denote this string by underlining W, i.e.,

$$\underline{W} = (W_{11}, W_{21}, \ldots, W_{p1}, W_{12}, W_{22}, \ldots, W_{p2}, \ldots, W_{1q}, W_{2q}, \ldots, W_{pq})'. \quad (12.8)$$

We used this notation in (2.56). Thus W_{ij} appears in the $[i+p(j-1)]^{th}$ slot of $\underline{W}$. Then we define

$$Cov(W) = Cov(\underline{W}). \quad (12.9)$$

12.2.2 Kronecker products

The Kronecker product $A \otimes B$ of matrices A, $r \times s$, and B, $u \times v$, is given in Definition 3.2 to be the $ru \times sv$ matrix formed by multiplying each element in B by A. See (3.14). (Rao, 1973, defines Kronecker product by multiplying the elements of A by B, but all the results carry through.) That is, $a_{ij}b_{kl}$ is the $[r(k-1)+i, s(l-1)+j]^{th}$ element of $A \otimes B$. Equation (3.15) contains some basic properties. The first requires that AB and CD be possible products, and the third assumes that A and B are invertible. Note that the third follows from the first.

When the covariance matrix of a matrix can be written as a Kronecker product, as for Q in (3.16), then certain calculations can be simplified. One convenient fact is that if W is $p \times q$, A is $r \times p$ and B is $s \times q$, then

$$\underline{AWB'} = (A \otimes B)\underline{W}. \tag{12.10}$$

Note that (12.7) and (12.10) immediately imply that

$$E(AWB') = AE(W)B', \tag{12.11}$$

as well as the next lemma.

Lemma 12.3
If $Cov(W) = \Sigma \otimes \Omega$, where Σ is $p \times p$ and Ω is $q \times q$, then with A and B as above, $Cov(AWB') = (A\Sigma A') \otimes (B\Omega B')$. □

Proof
Use Equations (12.7) and (12.10) to show that $Cov(AWB') = (A \otimes B)\ (\Omega \otimes \Sigma)\ (A \otimes B)'$, which with (3.15) implies the result. □

The next lemma helps to find the eigenvalues of Kronecker products.

Lemma 12.4
Suppose $A = \Sigma \otimes \Omega$, where both Σ $(p \times p)$ and Ω $(q \times q)$ are symmetric. If $\lambda_1, \ldots, \lambda_p$ are the eigenvalues of Σ, and $\mu_1, \ldots, \mu_q$ are those of Ω, then the eigenvalues of A are the pq values $\{\lambda_i \mu_j | 1 \leq i \leq p, 1 \leq j \leq q\}$. □

The following is needed in Chapter 3.

Lemma 12.5
If $Cov(U) = \Sigma \otimes \Omega$, where U is $p \times q$, and A is $q \times p$, then $Cov(tr(UA)) = tr(\Sigma A' \Omega A)$. □

Proof
First use spectral decomposition, Lemma 12.1, to write $\Sigma = O\Lambda O'$ and $\Omega = GLG'$. Then with $V = O'UG$, we have that $Cov(V) = \Lambda \otimes L$ by Lemma 12.3, and $\text{tr}(UA) = \text{tr}(VB)$, where $B = G'AO$.

Now since Λ and L are diagonal, the covariances among the elements of V are 0; hence by (12.2) and Lemma 12.3,

$$Cov(\text{tr}(VB)) = \sum_{i=1}^{p}\sum_{j=1}^{q} Var(V_{ij})b_{ji}^2 = \sum_{i=1}^{p}\sum_{j=1}^{q} \lambda_i l_j b_{ji}^2.$$

The final expression can be shown to be $\text{tr}(\Lambda B'LB)$ by writing out the matrices. The definition of B above shows that $\text{tr}(\Lambda B'LB) = \text{tr}(\Sigma A'\Omega A)$, which was to be proved. □

12.3 Normality and chi-squares

This section will briefly summarize results for the normality and chi-square-ness of many of the statistics we use. Hogg and Craig (1978) give a basic introduction. See Rao (1973) or any multivariate analysis book, e.g., Mardia, Kent and Bibby (1979), for further details. We start with Z, a $q \times 1$ vector of independent $Normal(0,1)$'s, so that $E(Z) = 0$ (the $q \times 1$ vector of all 0's) and $Cov(Z) = I_q$. [The $Normal(0,1)$ density is $\phi(z) = \exp(-z^2/2)/\sqrt{2\pi}$ for $-\infty < z < \infty$, which has mean 0 and variance 1.]

Definition 12.1: Multivariate normal
For any $p \times q$ matrix A and $p \times 1$ vector μ, the vector $X = AZ + \mu$ is *multivariate normal* with mean b and covariance $\Sigma \equiv AA'$, written $X \sim N_p(\mu, \Sigma)$. □

The fact that μ and Σ are the mean and covariance follows from (12.7). What is not so obvious is that the distribution of X depends on (A, μ) only through (AA', μ), so that if $AA' = CC'$, $AZ + \mu$ and $CZ + \mu$ have the same distribution. Any vector μ and positive semi-definite symmetric matrix (Section 12.1) Σ will yield a valid multivariate normal distribution since one can always find a matrix A with $AA' = \Sigma$. Also, from the definition it immediately follows that if $X \sim N_p(\mu, \Sigma)$ and $Y = CX + b$, where C is $r \times p$ and b is $r \times 1$, then $Y \sim N_r(C\mu + b, C\Sigma C')$. Definition 12.1 can also be used to show that X_i and X_j are independent if and only if $Cov(X_i, X_j) = 0$.

Definition 12.2: Chi-squared
If $Z \sim N_p(0, I)$ then $W = ||Z||^2$ is "chi-squared on p degrees of freedom," written $W \sim \chi_p^2$. □

It is easy to see that $E(W) = p$, and that the sum of two independent chi-squares is also chi-squared with degrees of freedom being the sum of those of the summands. The density is

$$g_p(w) = \frac{1}{\Gamma(p/2)2^{p/2}} w^{p/2-1} \exp(-w/2), \quad w > 0. \qquad (12.12)$$

Here, $\Gamma(q)$ is the gamma function, $\Gamma(q) = \int_0^\infty z^{q-1} \exp(-z) dz$ for $q > 0$.

The next result is important for proving Lemma 3.1.

Lemma 12.6
If $X \sim N_v(0, \Sigma)$ and A is a symmetric $v \times v$ matrix, then $X'AX \sim \chi^2[A\Sigma]$, where $\chi^2[\cdot]$ is given in Definition 3.1. □

Proof
Since A is symmetric, there exists a matrix B, $q \times p$, such that $A = B'B$. (For example, take $B = L^{1/2}G$ where $A = G'LG$ is the spectral decomposition of A in Lemma 12.1.) Let $Y = BX$, so that $X'AX = Y'Y$ and $Y \sim N_q(0, B\Sigma B')$. Now let $B\Sigma B'$ have spectral decomposition $O\Lambda O'$. Set $V = O'Y$, so that $V \sim N_q(0, \Lambda)$. Since O is orthogonal, $Y'Y = V'V$. Note that Λ is diagonal; hence the V_i's are independent. Thus $X'AX = \sum_{i=1}^q V_i^2 = \sum_{i=1}^q \lambda_i U_i^2 \sim \chi^2[\Lambda]$, where the U_i's are independent $N(0,1)$'s. The last implication is Definition 3.1. The eigenvalues of $A\Sigma$ are the same as those of $B\Sigma B'$ (Section 12.1.1), which proves the result. □

12.4 Some asymptotics

12.4.1 Central limit theorem

Throughout the book, we find central limit theorems, i.e., theorems that show that some sequence of random variables is asymptotically normal. Serfling (1980) gives a nice, concise treatment of such theorems. The main one is next, known as the Lindeberg–Levy Theorem.

Recall the sequence $W_1, W_2, \ldots,$ *converges in distribution* to W if the distribution functions of W_i converge to that of W for all points of continuity of the distribution function of W.

Theorem 12.1: Central limit theorem
Suppose $X_1, X_2, \ldots,$ are independent and identically distributed random $k \times 1$ vectors with finite mean μ and covariance matrix

Σ. Then $\sqrt{n}(\overline{X}_n - \mu)$ converges in distribution to $N_k(0, \Sigma)$ as $n \to \infty$, where $\overline{X}_n = (1/n)\sum_{i=1}^{n} X_i$ is the sample mean of the first n vectors. □

Other central limit theorems we use include Hoeffding's, given as Theorem 3.3, and a corollary of the Lindeberg–Feller Theorem, given as Theorem 3.4. The following mapping theorem is found on page 24 of Serfling (1980).

Lemma 12.7
If $X_n \to X$ in distribution, and g is a continuous function of x, then $g(X_n) \to g(X)$ in distribution. □

Finally, we are ready for the following.

Proof of Lemma 3.1
By Lemma 12.7, $T^2 \equiv W_n' A W_n$ approaches in distribution the random variable $U'AU$ where $U \sim N_k(0, \Sigma_0)$. Then Lemma 12.6 shows $U'AU$ is distributed $\chi^2[A\Sigma_0]$ of Definition 3.1. □

12.4.2 Convergence in probability

A sequence of random variables (vectors) $X_1, X_2, \ldots$ is said to *converge in probability* to a constant (vector) c if for each $\epsilon > 0$, $P[||X_n - c|| > \epsilon] \to 0$. We use this convergence in the following.

Theorem 12.2: Weak law of large numbers
Suppose $U_1, U_2, \ldots$ are independent and identically distributed vectors with finite mean μ. Then as $n \to \infty$, $\overline{U}_n \to \mu$ in probability, where $\overline{U}_n = \sum_{i=1}^{n} U_i/n$. □

See Theorem A on page 27 in Serfling (1980).

The mapping Lemma 12.7 extends to variables converging in probability since X_n converges in probability to c if and only if X_n converges in distribution to the random variable X with $P[X = c] = 1$.

Theorem 12.3: Mapping
Suppose $X_n \to X$ in distribution and $Y_n \to c$ in probability, and $g(x, y)$ is continuous at points (x, c) for x in the support of X. Then $g(X_n, Y_n) \to g(X, c)$ in distribution. □

Bibliography

Abramowitz, M. and Stegun, I. A. (1972) *Handbook of Mathematical Functions.* U.S. Government Printing Office: Washington, D.C.

Agresti, A. (1984) *Analysis of Ordinal Categorical Data.* Wiley: New York.

Agresti, A. (1990) *Categorical Data Analysis.* Wiley: New York.

Alvo, M. and Cabilio, P. (1991) On the balanced incomplete block design for rankings. *Annals of Statistics*, **19**, 1597 – 1613.

Alvo, M. and Cabilio, P. (1995) Rank correlation methods for missing data. To appear in *Canadian Journal of Statistics.*

Alvo, M. and Cabilio, P. (1995a) A note on the balanced incomplete block design for rankings. Technical Report No. 200, Laboratory for Research in Statistics and Probability, University of Ottawa and Carleton University.

Alvo, M. and Cabilio, P. (1995b) Testing ordered alternatives in the presence of incomplete data. *Journal of the American Statistical Association*, **90**. To appear.

Alvo, M., Cabilio, P. and Feigin, P. D. (1982) Asymptotic theory for measures of concordance with special reference to average Kendall tau. *Annals of Statistics*, **10**, 1269 – 1276.

Alwin, D. F. and Jackson, D. J. (1982) Adult values for children: an application of factor analysis to ranked preference data. *Social Structure and Behavior*, 311 – 329. Hauser, R. M., Mechanic, D., Haller, A. O. and Hauser, T. S., eds. Academic Press: New York.

Anderson, R. L. (1959) Use of contingency tables in the analysis of consumer preference studies. *Biometrics*, **15**, 582 – 590.

Arrow, K. J. (1963) *Social Choice and Individual Values.* Wiley: New York.

Baer, R. M. and Brock, P. (1968) Natural sorting over permutation spaces. *Mathematics of Computation*, **22**, 385 – 410.

Barnes, S. H. and Kaase, M. (1979) *Political Action: Mass Participation in Five Western Countries.* Sage: Beverly Hills.

Barton, D. E., David, F. N. and Mallows, C. L. (1958) Non-randomness in a sequence of two alternatives: I. Wilcoxon and allied test statistics. *Biometrika*, **45**, 166 – 177.

Beckett, L. A. (1993) A censored ranking problem. *Probability Models and Statistical Analyses for Ranking Data*, 92 – 107. Fligner, M. A. and Verducci, J. S., eds. Springer–Verlag: New York.

Beran, R. and Walker, M. E. (1990) A scaling approach to social distance. Presented at the Sixty-second Annual Meeting, Midwestern Psychological Association, Chicago. May, 1990.

Bickel, P. J. and Doksum, K. A. (1977) *Mathematical Statistics*. Holden–Day: New York.

Bishop, Y. M. M., Fienberg, S. E. and Holland, P. W. (1975) *Discrete Multivariate Analysis*. MIT Press: Cambridge, Massachusetts.

Block, H. W., Chhetry, D., Fang, Z. and Sampson, A. R. (1990) Partial orders on permutations and dependence orderings on bivariate empirical distributions. *Annals of Statistics*, **18**, 1840 – 1850.

Block, H. W., Chhetry, D., Fang, Z. and Sampson, A. R. (1991) Partial orderings on permutations. *Topics in Statistical Dependence*, 45 – 55. Block, H. W., Sampson, A. R. and Savits, T. H. (eds.), IMS: Hayward, California.

Block, H. W., Chhetry, D., Fang, Z. and Sampson, A. R. (1992) Some new partial orderings on S_n and $S_n \times S_n$. *Journal of Statistical Planning and Inference*, **30**, 351 – 357.

Böckenholt, U. (1992) Thurstonian representation for partial ranking data. *British Journal of Mathematical and Statistical Psychology*, **45**, 31 – 49.

Böckenholt, U. (1993) Applications of Thurstonian models to ranking data. *Probability Models and Statistical Analyses for Ranking Data*, 157 – 172. Fligner, M. A. and Verducci, J. S., eds. Springer–Verlag: New York.

Böckenholt, U. and Gaul, W. (1986) Analysis of choice behavior via probabilistic ideal point and vector models. *Applied Stochastic Models and Data Analysis*, **2**, 209 – 226.

Box, G. E. P. and Tiao, G. C. (1973) *Bayesian Inference in Statistical Analysis*. Addison–Wesley: Massachusetts.

Bradley, R. A. and Terry, M. A. (1952) Rank analysis of incomplete block designs. I. *Biometrika*, **39**, 324 – 345.

Brady, H. (1989) Factor and ideal point analysis for interpersonally incomparable data. *Psychometrika*, **54**, 181 – 202.

Brook, D. and Upton, G. J. G. (1974) Biases in local government elections due to position on the ballot paper. *Applied Statistics*, **23**, 414 – 419.

Burr, E. J. (1960) The distribution of Kendall's score S for a pair of tied rankings. *Biometrika*, **47**, 151 – 171.

Chung, L. (1989) *The Use of Nonnull Models for Rank Data in Nonparametric Statistics*. Ph. D. Thesis, Statistics, University of Illinois at Urbana–Champaign.

Chung, L. and Marden, J. I. (1990) Finding the marginal distribution of ranks in some nonnull ranking models, with applications to goodness-of-fit statistics. Department of Statistics, University of Illinois at Urbana–Champaign.

Chung, L. and Marden, J. I. (1991) Use of nonnull models for rank statistics in bivariate, two-sample, and analysis-of-variance problems. *Journal of the American Statistical Association*, **86**, 188 – 200.

Chung, L. and Marden, J. I. (1993) Extensions of Mallows' ϕ model. *Probability Models and Statistical Analyses for Ranking Data*, 108 – 139. Fligner, M. A. and Verducci, J. S., eds. Springer–Verlag: New York.

Cleveland, W. S. (1985) *The Elements of Graphing Data*. Wadsworth: Monterey.

Cochran, W. G. and Cox, G. M. (1957) *Experimental Designs*. Wiley: New York.

Cohen, A. (1982) Analysis of large sets of ranking data. *Communications in Statistics — Theory and Methods*, **11**, 235 – 256.

Cohen, A. and Mallows, C. L. (1980) Analysis of ranking data. Technical Report, Bell Laboratories, Murray Hill, New Jersey.

Cohen, A. and Mallows, C. L. (1983) Assessing goodness of fit of ranking models to data. *The Statistician*, **32**, 361 – 373.

Coombs, C. H. (1964) *A Theory of Data*. Wiley: New York.

Critchlow, D. E. (1985) *Metric Methods for Analyzing Partially Ranked Data*. Springer–Verlag: New York.

Critchlow, D. E. (1986) A unified approach to constructing nonparametric rank tests. Technical Report #86-15, Department of Statistics, Purdue University.

Critchlow, D. E. and Verducci, J. S. (1992) Detecting a trend in paired rankings. *Applied Statistics*, **41**, 17 – 29.

Critchlow, D. E., Fligner, M. A. and Verducci, J. S. (1991) Probability models on rankings. *Journal of Mathematical Psychology*, **32**, 294 – 373.

Critchlow, D. E. and Fligner, M. A. (1993) Ranking models with item covariates. *Probability Models and Statistical Analyses for Ranking Data*, 1 – 19. Fligner, M. A. and Verducci, J. S., eds. Springer–Verlag: New York.

Croon, M. A. (1989a) The analysis of partial rankings by log-linear and latent-class models. *Multiway Data Analysis*, 497 – 506. North–Holland: Amsterdam.

Croon, M. A. (1989b) Latent class models for the analysis of rankings. *New Developments in Psychological Choice Modeling*, 99 – 121. Feger, Klauer and de Soete, eds. North–Holland: Amsterdam.

Croon, M. A. and Luijkx, R. (1993) Latent structure models for ranking data. *Probability Models and Statistical Analyses for Ranking Data*, 53 – 74. Fligner, M. A. and Verducci, J. S., eds. Springer–Verlag: New York.

Daniels, H. E. and Kendall, M. G. (1947) The significance of rank correlations where parental correlation exists. *Biometrika*, **34**, 197 – 208.

Daniels, H. E. (1950a) Rank correlation and population methods. *Biometrika*, **33**, 129 – 135.

Daniels, H. E. (1950b) Rank correlation and population models. *Journal of the Royal Statistical Society B*, **12**, 171 – 181.

Darroch, J. N. and Ratcliff, D. (1972) Generalized iterative scaling for log-linear models. *Annals of Mathematical Statistics*, **43**, 1470 – 1480.

David, H. A. (1988) *The Method of Paired Comparisons.* Oxford University Press: New York.

Davidson, R. R. (1970) On extending the Bradley–Terry model to accommodate ties in paired comparison experiments. *Journal of the American Statistical Association*, **65**, 317 – 328.

DeSoete, G., Carroll, J. D. and DeSarbo, W. S. (1986) The wandering ideal point model: a probabilistic multidimensional unfolding model for paired comparison data. *Journal of Mathematical Psychology*, **30**, 28 – 41.

Dempster, A. P., Laird, N. M. and Rubin, D. B. (1977) Maximum likelihood from incomplete data via the EM algorithm (with discussion). *Journal of the Royal Statistical Society B*, **39**, 1 – 38.

Diaconis, P. (1988) *Group Representations in Probability and Statistics.* Institute of Mathematical Statistics: Hayward, California.

Diaconis, P. (1989) A generalization of spectral analysis with application to ranked data. *Annals of Statistics*, **17**, 949 – 979.

Diaconis, P. and Graham, R. L. (1977) Spearman's footrule as a measure of disarray. *Journal of the Royal Statistical Society B*, **39**, 262 – 268.

Duncan, O. D. and Brody, C. (1982) Analyzing n rankings of three items. *Social Structure and Behavior*, 269 – 310. Hauser, R. M., Mechanic, D., Haller, A. O. and Hauser, T. S., eds. Academic Press: New York.

Durbin, J. (1951) Incomplete blocks in ranking experiments. *British Journal of Psychology*, **4**, 85 – 90.

Durbin, J. and Stuart, A. (1951) Inversions and rank correlations. *Journal of the Royal Statistical Society B*, **13**, 303 – 309.

Efron, B. (1982) *The Jackknife, the Bootstrap and Other Resampling Plans.* SIAM: Philadelphia.

Ehrenberg, A. C. S. (1952) On sampling from a population of rankers. *Biometrika*, **39**, 82 – 87.

Everitt, B. S. and Hand, D. J. (1981) *Finite Mixture Distributions.* Chapman and Hall: London.

Feigin, P. and Cohen, A. (1978) On a model of concordance between judges. *Journal of the Royal Statistical Association B*, **40**, 203 – 213.

Feigin, P. D. (1993) Modelling and analysing paired ranking data. *Probability Models and Statistical Analyses for Ranking Data*, 75 – 91. Fligner, M. A. and Verducci, J. S., eds. Springer–Verlag: New York.

Feigin, P. D. and Alvo, M. (1986) Intergroup diversity and concordance for ranking data: an approach via metrics for permutations. *Annals of Statistics*, **14**, 691 – 707.

Feller, W. (1968) *An Introduction to Probability Theory and Its Applications*. Wiley: New York.

Fienberg, S. (1971) Randomization and social affairs: the 1970 draft lottery. *Science*, **171**, 255 – 261.

Fishburn, P. C. and Brams, S. J. (1983) Paradoxes of preferential voting. *Mathematics Magazine*, **56**, 207 – 214.

Fligner, M. A. and Verducci, J. S. (1990) Posterior probabilities for a consensus ordering. *Psychometrika*, **55**, 53 – 63.

Fligner, M. A., and Verducci, J. S. (1986) Distance based ranking models. *Journal of the Royal Statistical Society B*, **48**, 359 – 369.

Fligner, M. A., and Verducci, J. S. (1988) Multistage ranking models. *Journal of the American Statistical Association*, **83**, 892 – 901.

Fligner, M. A., and Verducci, J. S., eds. (1993) *Probability Models and Statistical Analyses for Ranking Data*. Springer–Verlag: New York.

Friedman, M. (1937) The use of ranks to avoid the assumption of normality implicit in the analysis of variance. *Journal of the American Statistical Association*, **32**, 675 – 701.

Gabriel, K. R. (1982) Biplot. *Encyclopedia of Statistical Science*, **1**, 263 – 271.

Gibbons, J. D. and Chakraborti, S. (1992) *Nonparametric Statistical Inference*. Dekker, New York.

Goldberg, A. I. (1976) The relevance of cosmopolitan/local orientations to professional values and behavior. *Sociology of Work and Occupation*, **3**, 331 – 356.

Goodman, L. A. (1969) How to ransack social mobility tables and other kinds of cross-classification tables. *The American Journal of Sociology*, **75**, 1 – 39.

Gordon, A. D. (1979) A measure of the agreement between rankings. *Biometrika*, **66**, 7 – 15.

Gurari, E. M. (1994) *TeX and LATeX: Drawing and Literate Programming*. McGraw Hill: New York.

Hartigan, J. A. (1975) *Clustering Algorithms*. Wiley: New York.

Hays, W. L. (1960) A note on average tau as a measure of concordance. *Journal of the American Statistical Association*, **55**, 331 – 341.

Henery, R. J. (1981) Permutation probabilities as models for horse races. *Journal of the Royal Statistical Society B*, **43**, 86 – 91.

Henery, R. J. (1983) Permutation probabilities for gamma random variables. *Applied Probability*, **20**, 822 – 834.

Herstein, I. N. (1964) *Topics in Algebra*. Blaisdell: Massachusetts.

Hoeffding, W. (1951) A combinatorial central limit theorem. *Annals of Mathematical Statistics*, **22**, 558 – 566.

Hogg, R. V. and Craig, A. T. (1978) *Introduction to Mathematical Statistics*. MacMillan: New York.

Hollander, M. and Sethuraman, J. (1978) Testing for agreement between groups of judges. *Biometrika*, **65**, 403 – 411.

Huber, P. J. (1985) Projection pursuit. *Annals of Statistics*, **13**, 435 – 475.

Joe, H. and Verducci, J. S. (1993) Babington Smith models. *Probability Models and Statistical Analyses for Ranking Data*, 37 – 52. Fligner, M. A. and Verducci, J. S., eds. Springer–Verlag: New York.

Kendall, M. G. and Smith, B. Babington (1940) On the method of paired comparisons. *Biometrika*, **31**, 324 – 345.

Kendall, M. G., and Gibbons, J. D. (1990) *Rank Correlation Methods*. Edward Arnold: London.

Kerov, S. and Vershick, A. (1985) The characters of the infinite symmetric group and probability properties of the Robinson–Schensted–Knuth algorithm. *SIAM Journal on Algebraic and Discrete Methods*, **7**, 116 – 124.

Kruskal, W. H. (1958) Ordinal measures of association. *Journal of the American Statistical Association*, **53**, 814 – 861.

Lantermann, E. D. and Feger, H., eds. (1980) *Similarity and Choice: Papers in Honour of Clyde Coombs*. Huber: Bern.

Lehmann, E. L. (1975) *Statistical Methods Based on Ranks*. Holden–Day: San Francisco.

Lehmann, E. L. (1983) *Theory of Point Estimation*. Wiley: New York.

Li, L. and Schucany, W. R. (1975) Some properties of a test for concordance of two groups of rankings. *Biometrika*, **62**, 417 – 423.

Lindsay, B. G. (1983a) The geometry of mixture likelihoods: a general theory. *Annals of Statistics*, **11**, 86 – 94.

Lindsay, B. G. (1983b) The geometry of mixture likelihoods, Part II: the exponential family. *Annals of Statistics*, **11**, 783 – 792.

Luce, R. D. (1959) *Individual Choice Behavior*. Wiley: New York.

Luce, R. D. (1977) Thurstone's discriminal processes fifty years later. *Psychometrika*, **42**, 461 – 489.

Mallows, C. L. (1957) Non-null ranking models: I. *Biometrika*, **44**, 114 – 130.

Mann, H. B. (1945) Non-parametric tests against trends. *Econometrica*, **13**, 245 – 259.

Marden, J. I. (1992) Use of nested orthogonal contrasts in analyzing rank data. *Journal of the American Statistical Association*, **87**, 307 – 318.

Marden, J. I. (1994) Comparing distances for rank vectors. Department of Statistics, University of Illinois at Urbana–Champaign.

Mardia, K. V., Kent, J. T. and Bibby, J. M. (1979) *Multivariate Analysis*. Academic Press: London.

Marley, A. A. J. (1993) Aggregation theorems and the combination of probabilistic rank orders. *Probability Models and Statistical Analyses for Ranking Data*, 216 – 240. Fligner, M. A. and Verducci, J. S., eds. Springer–Verlag: New York.

Marshall, A. W. and Olkin, I. (1979) *Inequalities: Theory of Majorization and its Applications.* Academic Press: New York.

McCullagh, P. (1993a) Permutations and regression models. *Probability Models and Statistical Analyses for Ranking Data*, 196 – 215. Fligner, M. A. and Verducci, J. S., eds. Springer–Verlag: New York.

McCullagh, P. (1993b) Models on spheres and models for permutations. *Probability Models and Statistical Analyses for Ranking Data*, 278 – 283. Fligner, M. A. and Verducci, J. S., eds. Springer–Verlag: New York.

McCullagh, P. and Nelder, J. A. (1989) *Generalized Linear Models.* Chapman and Hall: London.

McLachlan, G. J. and Basford, K. E. (1988) *Mixture Models.* Marcel Dekker: New York.

Moran, P. A. P. (1950) Recent developments in ranking theory. *Journal of the Royal Statistical Society B*, **12**, 153 – 162.

Nombekela, S. W., Murphy, M. R., Gonyou, H. W. and Marden, J. I. (1993) Dietary preferences in early lactation cows as affected by primary tastes and some common feed flavors. *Journal of Dairy Science*, **77**, 2393 – 2399.

Pendergrass, R. N. and Bradley, R. A. (1960) Ranking in triple comparisons. *Contributions to Probability and Statistics*, 331 – 351. Olkin, O., Ghurye, S. G., Hoeffding, W., Madow, W. G. and Mann, H. B., eds. Stanford University Pres.

Plackett, R. L. (1975) The analysis of permutations. *Applied Statistics*, **24**, 193 – 202.

Press, W. H., Teukolsky, S. A., Vetterling, W. T. and Flannery, B. P. (1992) *Numerical Recipes in C.* Cambridge University Press.

Rao, C. R. (1973) *Linear Statistical Inference and its Applications.* Wiley: New York.

Rao, C. R. (1982) Diversity and dissimilarity coefficients: a unified approach. *Journal of Theoretical Population Biology*, **21**, 24 – 43.

Rao, P. V. and Kupper, L. L. (1967) Ties in paired-comparison experiments: a generalization of the Bradley–Terry model. *Journal of the American Statistical Association*, **62**, 194 – 204.

Rodrigues, O. (1839) *Louiville Journal of Mathematics*, **4**, 236.

Rogers, T. (1991) Students as literary critics: the interpretive experiences, beliefs, and processes of ninth-grade students. *Journal of Reading Behavior*, **23**, 391 – 423.

Santner, T. J. and Duffy, D. E. (1989) *The Statistical Analysis of Discrete Data.* Springer–Verlag: New York.

Schucany, W. R. and Frawley, W. H. (1973) A rank test for two group concordance. *Psychometrika*, **38**, 249 – 258.

Schulman, R. R. (1979) Ordinal Data: an alternative distribution. *Psychometrika*, **44**, 3 – 20.

Serfling, R. J. (1980) *Approximation Theorems of Mathematical Statistics.* Wiley: New York.

Silverberg, A. R. (1980) *Statistical Models for q-permutations.* Ph. D. Thesis, Department of Statistics, Princeton University.

Silverberg, A. R. (1984) Statistical models for q-permutations. *Proceedings of the Biopharmaceutical Section, American Statistical Association.*, 107 – 112.

Smith, B. Babington (1950) Discussion of Professor Ross's paper. *Journal of the Royal Statistical Society B*, **12**, 53 – 56.

Stern, H. (1990) Models for distributions on permutations. *Journal of the American Statistical Association*, **85**, 558 – 564.

Stern, H. (1993) Probability models on rankings and the electoral process. *Probability Models and Statistical Analyses for Ranking Data*, 173 – 195. Fligner, M. A. and Verducci, J. S., eds. Springer–Verlag: New York.

Tallis, G. M. and Dansie, B. R. (1983) An alternative approach to the analysis of permutations. *Applied Statistics*, **32**, 110 – 114.

Tanner, M. A. (1993) *Tools for Statistical Inference.* Springer–Verlag: New York.

Terpstra, T. J. (1952) The asymptotic normality and consistency of Kendall's test against trend, when ties are present in one ranking. *Indagationes Mathematicae*, **14**, 327 – 333.

Thompson, G. L. (1993a) Generalized permutation polytopes and exploratory graphical methods for ranked data. *Annals of Statistics*, **21**, 1401 – 1430.

Thompson, G. L. (1993b) Graphical techniques for ranked data. *Probability Models and Statistical Analyses for Ranking Data*, 294 – 298. Fligner, M. A. and Verducci, J. S., eds. Springer–Verlag: New York.

Thurstone, L. L. (1927) A law of comparative judgement. *Psychological Reviews*, **34**, 273 – 286.

Tubbs, J. D. (1993) Assessing changes in repeated ranked lists. Department of Mathematical Sciences, University of Arkansas.

Upton, G. J. G. and Brook, D. (1975) The determination of the optimum position on a ballot paper. *Applied Statistics*, **24**, 279 – 287.

Ury, H. and Kleinecke, D. (1979) Tables of the distribution of Spearman's footrule. *Applied Statistics*, **28**, 271 – 275.

Valz, P. D. and Thompson, M. E. (1993) Exact inference for Kendall's S and Spearman's ρ with extension to Fisher's exact test in $r \times c$ contingency tables. Technical Report STAT-93-05, University of Waterloo.

Van Blokland-Vogelesang, A. W. (1989) Unfolding and consensus ranking: a prestige ladder for technical occupations. *New Developments in*

Psychological Choice Modeling, 237 – 258. De Soete, G., Feger, H. and Klauer, K. C., eds. North–Holland: Amsterdam.

Van Blokland-Vogelesang, A. W. (1993) A nonparametric distance model for unidimensional unfolding. *Probability Models and Statistical Analyses for Ranking Data*, 241 – 276. Fligner, M. A. and Verducci, J. S., eds. Springer–Verlag: New York.

Van Blokland-Vogelesang, A. W. and Van Blokland, P. J. (1990) UNFOLD: unidimensional unfolding of preference data. User's Manual. I. E. C. ProGAMMA: Groningen, The Netherlands.

Verducci, J. S. (1982) *Discriminating between two probabilities on the basis of ranked preferences.* Ph.D. Thesis, Statistics, Stanford University.

Verducci, J. S. (1989) Minimum majorization decomposition. *Contributions to Probability and Statistics*, 160 – 173. Gleser, L. J., Perlman, M. D., Press, S. J., and Sampson, A. R., eds. Springer–Verlag: New York.

Walker, M. E. (1993) Social distance: analysis of race preferences. Unpublished manuscript, University of Illinois at Urbana–Champaign.

Weisberg, S. (1985) *Applied Linear Regression.* Wiley: New York.

Whitfield, J. W. (1950) Uses of the ranking method in psychology. *Journal of the Royal Statistical Society*, **12**, 163 – 170.

Wu, C. F. J. (1983) On the convergence properties of the EM algorithm. *Annals of Statistics*, **11**, 95 – 103.

Yao, K. G. (1994) Bayesian estimation of Thurstonian ranking models: applications via the Gibbs sampling and the importance sampling algorithms. Department of Psychology, University of Illinois at Urbana–Champaign.

Ye, J. and McCullagh, P. (1993) Matched pairs and ranked data. *Probability Models and Statistical Analyses for Ranking Data*, 299 – 306. Fligner, M. A. and Verducci, J. S., eds. Springer–Verlag: New York.

Yellott, J. (1977) The relationship between Luce's choice axiom, Thurstone's theory of comparative judgment, and the double exponential distribution. *Journal of Mathematical Psychology*, **15**, 109 – 144.

Yemelichev, V. A., Kovalev, M. M. and Kravtsov, M. K. (1984) *Polytopes, Graphs and Optimisation.* Cambridge University Press.

Author Index

Subject Index